# Complex Variables

*Pure and Applied*
UNDERGRADUATE TEXTS · 16

# Complex Variables

Joseph L. Taylor

American Mathematical Society
Providence, Rhode Island

2010 *Mathematics Subject Classification.* Primary 30–01, 30Axx, 30Bxx, 30Dxx, 30Exx.

For additional information and updates on this book, visit
**www.ams.org/bookpages/amstext-16**

**Library of Congress Cataloging-in-Publication Data**

Taylor, Joseph L., 1941–
Complex variables / Joseph L. Taylor.
p. cm. — (Pure and applied undergraduate texts ; v. 16)
Includes bibliographical references and index.
ISBN 978-0-8218-6901-7 (alk. paper)
1. Functions of complex variables. I. Title.

QA331.7.T389 2011
515′.9—dc23

2011019541



∞ The paper used in this book is acid-free and falls within the guidelines
established to ensure permanence and durability.
Visit the AMS home page at http://www.ams.org/

10 9 8 7 6 5 4 3 2 1 16 15 14 13 12 11

# Contents

# Preface

Basic complex variables is a very popular subject among mathematics faculty and students. There is one big theorem (the Cauchy Integral Theorem) with a somewhat difficult proof, but this is followed by a host of easy to prove consequences that are both surprising and powerful. Furthermore, for undergraduates, the subject is refreshingly new and different from the analysis courses which precede it.

Undergraduate complex variables is our favorite course to teach, and we teach it as often as we can. Over a period of years we developed notes for use in such a course. The course is a one-semester undergraduate course on complex variables at the University of Utah. It is designed for junior level students who have completed three semesters of calculus and have also had some linear algebra and at least one semester of foundations of analysis.

Over the past several years, these notes have been expanded and modified to serve other audiences as well. We have used the expanded notes to teach a course in applied complex variables for engineering students as well a course at the first year graduate level for mathematics graduate students. The topics covered, the emphasis given to topics, and the choice of exercises were different for each of these audiences.

When taught as a one-semester junior level course for mathematics students, the course is a transitional course between freshman and sophomore level calculus, linear algebra, and differential equations and the much more sophisticated senior level mathematics courses taught at Utah. The students are expected to understand definitions and proofs, and the exercises assigned will include proofs as well as computations. The course moves at a leisurely pace, and the material covered includes only Chapters 1, 2, and 3 and selected sections from Chapters 4, 5, and 6. A full year course could easily cover the entire text.

When we teach a one-semester undergraduate course for engineers using these notes, we cover essentially the same material as in the course for mathematics majors, but not all the proofs are done in detail, and there is more emphasis on computational examples and exercises. The applications in Chapter 6 are given special emphasis.

When taught as a one-semester graduate course, we assume students have some knowledge of complex numbers and so Chapter 1 is given just a brief review. Chapters 2, 3, and 4 are covered completely, parts of Chapters 5 and 6 are covered, along with Chapter 7. If time allows, Chapters 8 and 9 are summarized at the end of the course. In the homework, the emphasis is on the theoretical exercises.

We have tried to present this material in a fashion which is both rigorous and concise, with simple, straightforward explanations. We feel that the modern tendency to expand textbooks with ever more material, excessively verbose explanations, and more and more bells and whistles, simply gets in the way of the student's understanding of the material.

The exercises differ widely in level of abstraction and level of difficulty. They vary from the simple to the quite difficult and from the computational to the theoretical. There are exercises that ask students to prove something or to construct an example with certain properties. There are exercises that ask students to apply theoretical material to help do a computation or to solve a practical problem. Each section contains a number of examples designed to illustrate the material of the section and to teach students how to approach the exercises for that section.

This text, in its various incarnations, has been used by the author and his colleagues for several years at the University of Utah. Each use has led to improvements, additions, and corrections.

The text begins, in Chapter 1, with a discussion of the fact that the real number system is insufficient for some purposes (solving polynomial equations). The system of complex numbers is developed in an attempt to remedy this problem. We then study basic arithmetic of complex numbers, convergence of sequences and series of complex numbers, power series, the exponential function, polar form for complex numbers and the complex logarithm.

The core material of a complex variables course is the material covered here in Chapters 2 and 3. Analytic functions are introduced in Chapter 2 as functions which have a complex derivative. This leads to a discussion of the Cauchy-Riemann equations and harmonic functions. We then introduce contour integrals and the index function (winding number) for closed paths. We prove the Cauchy-integral theorem for triangles and then for convex sets. We believe that this approach leads to the simplest and quickest rigorous proof of a form of Cauchy's theorem that can be used to prove the existence of power series expansions for analytic functions.

The proof that analytic functions have power series expansions occurs early in Chapter 3. This is followed by a wide range of powerful applications with simple proofs – Morera's Theorem, Liouville's Theorem, the Fundamental Theorem of Algebra, the characterization of zeroes and singularities of analytic functions, the Maximum Modulus Principle, and Schwarz's Lemma.

Chapter 4 begins with a proof of the general form of Cauchy's Theorem and Cauchy's Formula. These theorems involve functions which are analytic on a general open set. The integration takes place around a cycle (a generalization of a closed path) which is required to have index zero about any point not in the set. We go on to study Laurent series, the Residue Theorem, Rouché's Theorem, inverse functions and the Open Mapping Theorem. The chapter ends with a discussion of homotopy and its relationship to integrals around closed paths. We normally do not cover this last section in our undergraduate course, but it would certainly be appropriate for a graduate course using this text.

Residue theory is covered in Chapter 5. We discuss techniques for computing residues as well as a wide variety of applications of residue theory to problems involving the calculation of integrals. We normally do not cover all of this in our undergraduate course. We typically skip the last two sections of this chapter in favor of covering most of Chapter 6.

Chapter 6 deals with conformal maps. We prove the Riemann Mapping Theorem and show how it can be used to transform problems involving analytic or harmonic functions on a simply connected subset of the plane to the analogous problems on the unit disc. We use this technique to study the Dirichlet problem on open, proper, simply connected subsets of $\mathbb{C}$. We discuss applications to heat flow, electrostatics, and hydrodynamics. This material is of particular interest to students in a course on complex variables for engineering students.

The goal of Chapter 7 on analytic continuation is to prove the Picard theorems concerning meromorphic functions at essential singularities. Along the way, we prove the Schwarz Reflection Principle and the Monodromy Theorem, and discuss lifting analytic functions through a covering map.

Chapters 8 and 9 are normally not covered in our one-semester undergraduate complex variables course. Some of the topics in these chapters are, however, often covered in the graduate course. In Chapter 8 we use Weierstrass products to construct an analytic function with a given discrete set of zeroes, with prescribed multiplicities. This is the Weierstrass Theorem. It leads directly to the Weierstrass Factorization Theorem for entire functions. We also prove the Mittag-Leffler Theorem – which gives the existence of a meromorphic function with a prescribed set of poles and principle parts. The final result of the chapter is the proof of Hadamard's Theorem characterizing entire functions of finite order. This is a key ingredient in the proof of the Prime Number Theorem in Chapter 9.

In Chapter 9 we introduce the gamma and zeta functions, develop their basic properties, discuss the Riemann Hypothesis, and prove the Prime Number Theorem. Course notes by our colleague Dragan Miličić provided the original basis for this material. It then went through several years of expansion and refinement before reaching its present form. There is a great deal of technical calculation in this material and we do not cover it in our undergraduate course. However, a few lectures summarizing this material have proved to be a popular way to end the graduate course.

Several standard texts in complex variables and related topics were useful guides in preparing this text. These may also be of interest to the student who wishes to learn more about the subject. They are listed in a short bibliography at the end of the text.

Chapter 1

# The Complex Numbers

## 1.1. Definition and Simple Properties

The number system is a tool devised by humans to aid in the description of quantities of the various things humans have to deal with. It has evolved as human culture has evolved, beginning with something very primitive like: *1, 2, 3, many,* moving on to the natural numbers, then the integers, the rational numbers, the real numbers and then the complex numbers.

At each stage of development, the number system was expanded in response to the need to describe quantities that the old number system could not. For example, the negative numbers were introduced in order to be able to describe a loss as opposed to a gain, or moving backward rather than forward. The rational numbers were introduced because we do not always deal with whole numbers of things (we have 2/3 of a pie left). The real number system evolved from the rational number system out of a need to be able to describe such things as the length of the hypotenuse of a right triangle (this involves square roots) and the area or circumference of a circle (this involves $\pi$).

In this course, we will assume students are familiar with the real number system and its properties. We will define the complex number system as a needed extension of the real number system and develop its properties. We will then go on to study functions of a complex variable.

**The Real Numbers are Insufficient.** The complex number system was developed in response to the need for solutions to polynomial equations. The simplest polynomial equation that does not have a solution in the real number system is the equation

$$x^2 + 1 = 0,$$

which has no real solution because $-1$ has no real square root. More generally, a quadratic equation

$$ax^2 + bx + c = 0, \tag{1.1.1}$$

where $a, b$ and $c$ are real numbers, formally has two solutions given by the quadratic formula

$$x = \frac{-b \pm \sqrt{b^2 - 4ac}}{2a}, \tag{1.1.2}$$

but these will not be real numbers if $b^2 - 4ac$ is negative. If we could take square roots of negative numbers, then the quadratic formula would give us solutions to (1.1.1) for all choices of real coefficients $a, b, c$. To make this possible, we expand the real number system in the following way, thus creating the complex number system $\mathbb{C}$.

**Constructing $\mathbb{C}$.** We begin by adjoining a single new number to our old number system $\mathbb{R}$. We will denote it by $i$ and declare it to be a square root of $-1$. Thus,

$$i^2 = -1.$$

Our new number system is to contain both $\mathbb{R}$ and the new number $i$ and it should be closed under addition and multiplication. If it is to be closed under multiplication, we need a number $iy$ for every real number $y$. Likewise, if it is to be closed under addition, there should be a number $x + iy$ in our new number system for each pair of real numbers $(x, y)$. It turns out that this is enough. If we define the set of complex numbers $\mathbb{C}$ to be the set of all symbols of the form $x + iy$ where $(x, y)$ is a pair of real numbers, and if we define addition and multiplication appropriately, then the resulting number system is a field in which every polynomial equation has a root. We will be a long time proving the latter half of this statement, but it is not hard to prove the first part.

To define the operations of addition and multiplication in $\mathbb{C}$, we begin by noting that, as a set, $\mathbb{C}$ may be identified with $\mathbb{R}^2$ – the set of all pairs $(x, y)$ of real numbers. Obviously, each pair $(x, y)$ determines a symbol $x + iy$ and vice versa. This identification makes $\mathbb{C}$ into a vector space over $\mathbb{R}$ and gives us operations of addition and scalar multiplication by reals which satisfy the usual associative and distributive rules. The resulting operation of addition is

$$(x_1 + iy_1) + (x_2 + iy_2) = (x_1 + x_2) + i(y_1 + y_2).$$

It remains to define a product on $\mathbb{C}$.

We have already declared that $i^2 = -1$. If we also require that the associative and distributive laws of multiplication should hold and that the multiplication of real numbers should remain as before, then the product of two complex numbers $x_1 + iy_1$ and $x_2 + iy_2$ must be

$$(x_1+iy_1)(x_2+iy_2) = x_1x_2+ix_1y_2+iy_1x_2+i^2y_1y_2 = (x_1x_2-y_1y_2)+i(x_1y_2+y_1x_2).$$

We formalize this conclusion in the following definition.

**Definition 1.1.1.** We define the system $\mathbb{C}$ of complex numbers to be the set of all symbols of the form $x + iy$ with $(x, y) \in \mathbb{R}^2$, with addition and multiplication defined by

$$(x_1 + iy_1) + (x_2 + iy_2) = (x_1 + x_2) + i(y_1 + y_2)$$

and

$$(x_1 + iy_1)(x_2 + iy_2) = (x_1x_2 - y_1y_2) + i(x_1y_2 + y_1x_2).$$

A complex number of the form $x + i0$, with $x \in \mathbb{R}$ will be denoted simply as $x$. This identifies $\mathbb{R}$ as a subset of $\mathbb{C}$. Similarly, a complex number of the form $0 + iy$ with $y$ real will be denoted simply as $iy$. The numbers of this form are traditionally called the *imaginary* numbers.

Note that, from the above definition, if $x, y \in \mathbb{R}$, then

$$yi = (y + i0)(0 + i) = 0 + iy = iy$$

and so $x + iy$ and $x + yi$ are the same complex number. Which form is used to describe this number is usually dictated by which looks best typographically. When specific numbers replace $x$ and $y$, the latter seems to look best. Thus, we usually write $2 + 3i$ rather than $2 + i3$.

**Example 1.1.2.** If $z_1 = 5 + 2i$ and $z_2 = 3 - 4i$, find $z_1 + z_2$ and $z_1 z_2$.

**Solution:**

$$\begin{aligned} z_1 + z_2 &= 5 + 3 + (2 - 4)i = 8 - 2i, \\ z_1 z_2 &= (5 \cdot 3 - 2 \cdot (-4)) + (5 \cdot (-4) + 3 \cdot 2)i = 23 - 14i. \end{aligned}$$

**Example 1.1.3.** Show that the quadratic equation (1.1.1) has solutions which are complex numbers.

**Solution:** If $b^2 - 4ac \geq 0$, the quadratic formula (1.1.2) tells us the solutions are

$$\frac{-b + \sqrt{b^2 - 4ac}}{2a} \quad \text{and} \quad \frac{-b - \sqrt{b^2 - 4ac}}{2a}.$$

On the other hand, if $b^2 - 4ac < 0$, then $4ac - b^2$ is positive and has real square roots. By squaring both sides, and using $i^2 = -1$, it is easy to see that

$$\pm\sqrt{b^2 - 4ac} = \pm i\sqrt{4ac - b^2}.$$

This suggests that the solutions to the quadratic equation in this case are the two complex numbers

$$-\frac{b}{2a} + i\frac{\sqrt{4ac - b^2}}{2a} \quad \text{and} \quad -\frac{b}{2a} - i\frac{\sqrt{4ac - b^2}}{2a}.$$

That these two numbers are, indeed, solutions to the quadratic equation may be verified by directly substituting them in for $x$ in (1.1.1). We leave this as an exercise (Exercise 1.1.7).

**Field Properties.** In our definition of the product of two complex numbers, we were guided by the desire to have the usual rules of arithmetic hold – that is, the commutative and associative laws for addition and for multiplication and the distributive law. Did we succeed? These are some of the properties of a *field*. Do these laws actually hold in $\mathbb{C}$ with the operations as defined above? The following theorem says they do.

**Theorem 1.1.4.** *If $z_1, z_2, z_3$ are complex numbers, then*

(a) $z_1 + z_2 = z_2 + z_1$ – *commutative law of addition;*

(b) $(z_1 + z_2) + z_3 = z_1 + (z_2 + z_3)$ – *associative law of addition;*

(c) $z_1 z_2 = z_2 z_1$ – *commutative law of multiplication;*

(d) $(z_1 z_2) z_3 = z_1(z_2 z_3)$ – *associative law of multiplication;*

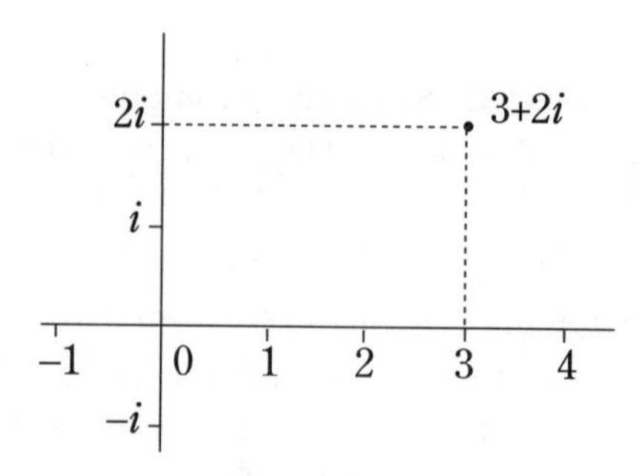


**Figure 1.1.1.** Plot of the Complex Number $3 + 2i$.

(e) $z_1(z_2 + z_3) = z_1 z_2 + z_1 z_3$ – *distributive law.*

**Proof.** As far as the operation of addition is concerned, $\mathbb{C}$ is just $\mathbb{R}^2$, which is a vector space over $\mathbb{R}$. Parts (a) and (b) of the theorem follow directly from this. Part (c) is obvious from the definition of multiplication. We will prove part (d) and leave part (e) as an exercise (Exercise 1.1.8).

If $z_j = x_j + iy_j$ for j= 1, 2, 3, then

$$\begin{aligned}(z_1z_2)z_3 &= ((x_1 + iy_1)(x_2 + iy_2))(x_3 + iy_3)\\ &= (x_1x_2 - y_1y_2 + i(x_1y_2 + y_1x_2))(x_3 + iy_3)\\ &= x_1x_2x_3 - y_1y_2x_3 - x_1y_2y_3 - y_1x_2y_3\\ &\qquad + i(x_1x_2y_3 - y_1y_2y_3 + x_1y_2x_3 + y_1x_2x_3),\end{aligned}$$

while

$$\begin{aligned}z_1(z_2z_3) &= (x_1 + iy_1)((x_2 + iy_2)(x_3 + iy_3))\\ &= (x_1 + iy_1)(x_2x_3 - y_2y_3 + i(x_2y_3 + y_2x_3))\\ &= x_1x_2x_3 - x_1y_2y_3 - y_1x_2y_3 - y_1y_2x_3\\ &\qquad + i(x_1x_2y_3 + x_1y_2x_3 + y_1x_2x_3 - y_1y_2y_3).\end{aligned}$$

Since the results are the same, the proof of (d) is complete. □

The properties described in the above theorem are some of the properties that must hold in a field. A field must also have additive and multiplicative identities – that is, elements 0 and 1 which satisfy

(1.1.3) $$z + 0 = z$$

and

(1.1.4) $$1 \cdot z = z$$

for every element $z$ in the field. That this holds for $\mathbb{C}$ follows immediately from the fact that $\mathbb{C}$ is a vector space over $\mathbb{R}$.

A field must also have the properties that every element $z$ has an additive inverse, that is, an element $-z$ such that

(1.1.5) $$z + (-z) = 0,$$

and every non-zero element $z$ has a mutiplicative inverse, that is, an element $z^{-1}$ such that

$$z \cdot z^{-1} = 1. \tag{1.1.6}$$

The first of these follows immediately from the fact that $\mathbb{C}$ is a vector space over $\mathbb{R}$. The second is nearly as easy. If $z = x + iy \neq 0$, then a direct calculation shows that

$$z^{-1} = \frac{x - iy}{x^2 + y^2} = \frac{x}{x^2 + y^2} - \frac{y}{x^2 + y^2} i$$

satisfies (1.1.6). We conclude:

**Theorem 1.1.5.** *With addition and multiplication defined as in Definition* 1.1.1, *the complex numbers form a field.*

### Complex Conjugation and Modulus.

**Definition 1.1.6.** If $z = x + iy$ is a complex number, then its complex conjugate, denoted $\overline{z}$, is defined by

$$\overline{z} = x - iy,$$

while its modulus, denoted $|z|$, is defined by

$$|z| = \sqrt{x^2 + y^2}.$$

Note that the modulus, as defined above, is just the usual Euclidean norm in the vector space $\mathbb{R}^2$. Thus, if $z_1, z_2 \in \mathbb{C}$, then $|z_1 - z_2|$ is the Euclidean distance from $z_1$ to $z_2$. The term *modulus* is traditional, but the terms *norm* and *absolute value* are also commonly used to mean the same thing. We will use all three.

Note also that the two solutions of a quadratic equation with real coefficients given in Example 1.1.3 are complex conjugates of each other. Thus, the solutions to a quadratic equation with real coefficients occur in conjugate pairs. Quadratic equations with complex coefficients also have roots and they are also given by the quadratic formula. However, we cannot prove this until we prove that every complex number has a square root. In fact, in Section 1.4 we will prove that every complex number has roots of all orders.

For a complex number $z = x + iy$, the real number $x$ is called the *real part* of $z$ and is denoted $\mathrm{Re}(z)$, while the number $y$ is called the *imaginary part* of $z$ and is denoted $\mathrm{Im}(z)$. In graphing complex numbers using a rectilinear coordinate system, $x$ determines the coordinate on the horizontal axis, while $y$ determines the coordinate on the vertical axis.

Note that a complex number $z$ is real if and only if $\overline{z} = z$, and it is purely imaginary if and only if $\overline{z} = -z$. Note also, that if $z = x + iy$, then

$$\mathrm{Re}(z) = x = \frac{z + \overline{z}}{2} \quad \text{and} \quad \mathrm{Im}(z) = y = \frac{z - \overline{z}}{2i}.$$

The elementary properties of conjugation and modulus are gathered together in the next theorem.

**Theorem 1.1.7.** *If $z$ and $w$ are complex numbers, then*

(a) $\overline{\overline{z}} = z$;

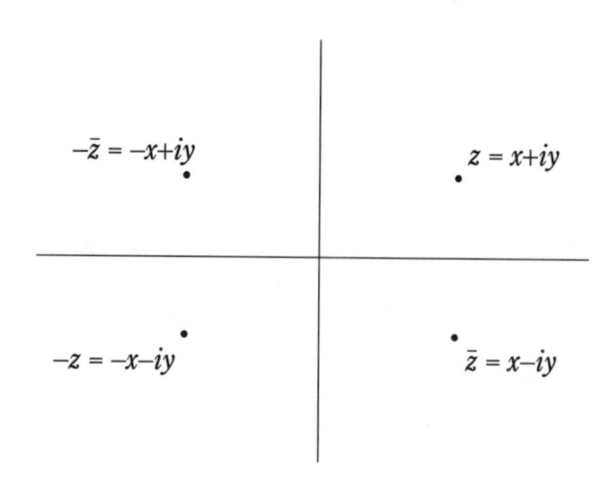


**Figure 1.1.2.** Plot of the Complex Numbers $z$, $\overline{z}$, $-z$, and $-\overline{z}$.

(b) $z\overline{z} = |z|^2$;

(c) $\overline{z+w} = \overline{z} + \overline{w}$;

(d) $\overline{zw} = \overline{z}\,\overline{w}$;

(e) $|zw| = |z||w|$ *and* $|\overline{z}| = |z|$;

(f) $|z|$ *is a non-negative real number and is* 0 *if and only if* $z = 0$;

(g) $|\operatorname{Re}(z\overline{w})| \le |z||w|$;

(h) $|z+w| \le |z| + |w|$.

**Proof.** We will prove (g) and (h). The other parts are elementary computations or observations and will be left as exercises.

Parts (g) and (h) are the *Cauchy-Schwarz inequality* and the *triangle inequality* for the vector space $\mathbb{R}^2$. Versions of these inequlities hold in general Euclidian space $\mathbb{R}^n$. The proofs we give here are specializations to $\mathbb{C}$ of the standard proofs of these inequalities in $\mathbb{R}^n$.

To prove (g), we begin with the observation that (a) and (d) imply that

$$z\overline{w} + \overline{z}w = 2\operatorname{Re}(z\overline{w}).$$

We then let $t$ be an arbitrary real number and note that, by Parts (c), (d), and (f),

$$0 \le |zt+w|^2 = (zt+w)(\overline{z}t+\overline{w}) = |z|^2t^2 + 2\operatorname{Re}(z\overline{w})t + |w|^2$$

for all values of $t$. This implies that the quadratic polynomial in $t$ given by

$$|z|^2t^2 + 2\operatorname{Re}(z\overline{w})t + |w|^2$$

is never negative and, therefore, has at most one real root. This is only possible if the expression under the radical in the quadratic formula is negative or zero. Thus,

$$4(\operatorname{Re}(z\overline{w}))^2 - 4|z|^2|w|^2 \le 0.$$

Part (g) follows immediately from this.

Part (h) follows directly from Part (g) and the other parts of the theorem. In fact,

$$\begin{aligned}|z+w|^2 &= (z+w)(\overline{z}+\overline{w}) \\ &= |z|^2 + 2\,\mathrm{Re}(z\overline{w}) + |w|^2 \\ &\leq |z|^2 + 2|z||w| + |w|^2 = (|z|+|w|)^2.\end{aligned}$$

On taking square roots, we conclude $|z+w| \leq |z|+|w|$, which is Part (h). □

The inequalities in the following theorem are used extensively – particularly in the next section. The proofs are very simple and are left as an exercise (Exercise 1.1.12).

**Theorem 1.1.8.** *If* $z = x + iy$, *then* $\max\{|x|,|y|\} \leq |z| \leq |x| + |y|$.

**Inversion and Division.** Recall that the inverse of a non-zero complex number $z = x + iy$ is

$$z^{-1} = \frac{x-iy}{x^2+y^2} = \frac{\overline{z}}{|z|^2} = \frac{\overline{z}}{z\overline{z}}.$$

Stating this in the last form makes the identity $zz^{-1} = 1$ obvious.

This also suggests the right way to do complex division problems in general: to express $w/z$ as a complex number in standard form (as a real number plus $i$ times a real number), simply multiply both numerator and denominator by $\overline{z}$. That is,

$$\frac{w}{z} = \frac{w\overline{z}}{z\overline{z}} = \frac{w\overline{z}}{|z|^2}.$$

The number $w\overline{z}$ is then easily put in standard form and the problem is finished by dividing by the real number $|z|^2$.

**Example 1.1.9.** Express $\dfrac{1}{2+3i}$ in the standard form $x + yi$.

**Solution:**

$$\frac{1}{2+3i} = \frac{2-3i}{(2+3i)(2-3i)} = \frac{2-3i}{|2+3i|^2} = \frac{2}{13} - \frac{3}{13}i.$$

**Example 1.1.10.** Express $\dfrac{3+4i}{3-4i}$ and $\dfrac{3-4i}{3+4i}$ in standard form.

**Solution:**

$$\begin{aligned}\frac{3+4i}{3-4i} &= \frac{(3+4i)^2}{|3+4i|^2} = -\frac{7}{25} + \frac{24}{25}i, \\ \frac{3-4i}{3+4i} &= \frac{(3-4i)^2}{|3-4i|^2} = -\frac{7}{25} - \frac{24}{25}i.\end{aligned}$$

### Exercise Set 1.1

1. Express $(3+i)+(2-7i)$ and $(3+i)(2-7i)$ in the standard form $x+yi$.
2. Express $\dfrac{1}{3+5i}$ in the standard form $x+yi$.
3. Express $(1+2i)^2$ and $(1+2i)^{-2}$ in the standard form $x+yi$.
4. Express $\dfrac{2-3i}{3+2i}$ in the standard form $x+yi$.
5. Find a square root for $i$.
6. If $z=x+iy$, express $z^3$ in standard form.
7. By direct substitution, prove that the two solutions to the quadratic equation given in Example 1.1.3 really do satisfy equation (1.1.1).
8. Prove Part (e) of Theorem 1.1.4.
9. Prove Parts (a), (b) and (c) of Theorem 1.1.7.
10. Prove Parts (d), (e) and (f) of Theorem 1.1.7.
11. For $z \in \mathbb{C}$ and $a \in \mathbb{R}$ prove the following:
    (a) $\mathrm{Re}(az) = a\,\mathrm{Re}(z)$ and $\mathrm{Im}(az) = a\,\mathrm{Im}(z)$;
    (b) $\mathrm{Re}(iz) = -\mathrm{Im}(z)$ and $\mathrm{Im}(iz) = \mathrm{Re}(z)$.
12. Prove Theorem 1.1.8 – that is, prove that if $z=x+iy$, then
$$\max\{|x|,|y|\} \le |z| \le |x|+|y|.$$
13. Graph the set of points $z \in \mathbb{C}$ which satisfy the equation $|z-i|=1$.
14. Graph the set of points $z \in \mathbb{C}$ which satisfy the equation $z^2+\overline{z}^2=2$. Hint: If $z=x+iy$, rewrite this equation as an equation in $x$ and $y$.
15. Prove that if $z$ is a non-zero complex number, then $\overline{1/z}=1/\overline{z}$ and $|1/z|=1/|z|$.
16. If $z$ is any non-zero complex number, prove that $z/\overline{z}$ has modulus one.
17. Prove that every complex number of modulus 1 has the form $\cos\theta + i\sin\theta$ for some angle $\theta$.
18. Prove that every line or circle in $\mathbb{C}$ is the solution set of an equation of the form
$$a|z|^2+\overline{w}z+w\overline{z}+b=0,$$
where $a$ and $b$ are real numbers and $w$ is a complex number. Conversely, show that every equation of this form has a line, circle, point, or the empty set as its solution set.

## 1.2. Convergence in $\mathbb{C}$

We assume the reader is familiar with the basics concerning convergent sequences and series of real numbers – particularly those results which follow from the completeness of the real number system, such as the fact that bounded monotone sequences converge and the various convergence tests for series. We also assume a familiarity with the basics of power series in a real variable. The purpose of this section is to extend these ideas and results to sequences and series of complex numbers and power series in a complex variable. This is an introductory section. A deeper study of complex power series will come later in the text.

There is no natural order relation on the complex numbers. The statement $z < w$ makes no sense for complex numbers $z$ and $w$. However, if these numbers happen to be real, then the inequality does make sense, because $\mathbb{R}$ is an ordered field. We make heavy use of inequalities in this section, but note they are always inequalities between real numbers. Thus, if an inequality of the form $a < b$ occurs in this text the numbers $a$ and $b$ are assumed to be real numbers, even if there is no explicit statement to that effect.

**Convergence of Sequences.** A sequence of complex numbers converges if and only if it converges as a sequence of vectors in $\mathbb{R}^2$. The formal definition is the familiar one:

**Definition 1.2.1.** A sequence $\{z_n\}$ of complex numbers is said to converge to the number $w$ if, for every $\epsilon > 0$, there exists an integer $N$ such that

$$|z_n - w| < \epsilon \quad \text{whenever} \quad n \geq N.$$

In this case, we say that $w$ is the *limit* of the sequence $z_n$ and write $\lim_{n\to\infty} z_n = w$, or $\lim z_n = w$, or simply $z_n \to w$.

**Remark 1.2.2.** There are a couple of simple observations about convergence of sequences that will prove to be very useful.

(1) If $\{z_n\}$ is a sequence of complex numbers, then $\lim z_n = w$ if and only if $\lim |z_n - w| = 0$.

(2) If $\{a_n\}$ and $\{b_n\}$ are sequences of real numbers with $0 \leq a_n \leq b_n$ and if $b_n \to 0$, then also $a_n \to 0$.

These both follow immediately from the definition of limit of a sequence. The second is one form of what is sometimes called the *squeeze principle.*

The first of these observations reduces the problem of showing that a sequence of complex numbers converges to a given complex number to showing that a certain sequence of non-negative real numbers converges to 0. This is useful because we have many tools at our disposal to show that a sequence of non-negative numbers converges to 0. One of the most useful of these tools is the second observation above. The next example and the proof of the following theorem are excellent examples of how this works.

**Example 1.2.3.** Prove that $\lim z^n = 0$ if $|z| < 1$.

**Solution:** Note that $|z^n| = |z|^n$ follows from Part (e) of Theorem 1.1.7. If $|z| < 1$ then $\lim |z|^n = 0$. That $\lim z^n = 0$, as well, follows from (1) of Remark 1.2.2.

**Theorem 1.2.4.** *A sequence of complex numbers $\{z_n\}$ converges to a complex number $w$ if and only if $\{\operatorname{Re}(z_n)\}$ converges to $\operatorname{Re}(w)$ and $\{\operatorname{Im}(z_n)\}$ converges to $\operatorname{Im}(w)$.*

**Proof.** By Theorem 1.1.8 we know that

$$|\operatorname{Re}(z_n) - \operatorname{Re}(w)| \le |z_n - w|,$$
$$|\operatorname{Im}(z_n) - \operatorname{Im}(w)| \le |z_n - w| \quad \text{and}$$
$$|z_n - w| \le |\operatorname{Re}(z_n) - \operatorname{Re}(w)| + |\operatorname{Im}(z_n) - \operatorname{Im}(w)|$$

for each $n$. The first two of these inequalities, together with Remark 1.2.2, imply that if $z_n \to w$, then $\operatorname{Re}(z_n) \to \operatorname{Re}(w)$ and $\operatorname{Im}(z_n) \to \operatorname{Im}(w)$. The third inequality, together with the fact that the sum of two sequences converging to zero also converges to zero, implies the converse – that is, if $\operatorname{Re}(z_n) \to \operatorname{Re}(w)$ and $\operatorname{Im}(z_n) \to \operatorname{Im}(w)$, then $z_n \to w$. □

**Example 1.2.5.** Show that the sequence $\{2^{-n} + in/(n+1)\}$ converges to $i$.

**Solution:** This follows from the previous theorem and the fact that $2^{-n} \to 0$ and $n/(n+1) \to 1$.

**Example 1.2.6.** Show that if $\{z_n\}$ and $\{w_n\}$ are two convergent sequences of complex numbers with $z_n \to z$ and $w_n \to w$, then $z_n + w_n \to z + w$ and $z_n w_n \to zw$.

**Solution:** If $z_n = x_n + iy_n$, $z = x + iy$, $w_n = u_n + iv_n$ and $w = u + iv$, then

$$z_n + w_n = x_n + u_n + i(y_n + v_n),$$
$$z + w = x + u + i(y + v),$$
$$z_n w_n = x_n u_n - y_n v_n + (x_n v_n + y_n u_n)i$$

and

$$zw = xu - yv + (xv + yu)i.$$

We know that $x_n \to x$, $y_n \to y$, $u_n \to u$ and $v_n \to v$. We also know the rules about limits of products, sums and differences of sequences of real numbers. These rules imply $x_n + u_n \to x + u$, $y_n + v_n \to y + v$, $x_n u_n - y_n v_n \to xu - yv$ and $x_n v_n + y_n u_n \to xv + yu$. Since the real and imaginary parts of $z_n + w_n$ and $z_n w_n$ converge to the real and imaginary parts, respectively, of $z+w$ and $zw$, the previous theorem implies that $z_n + w_n \to z + w$ and $z_n w_n \to zw$.

**Series of Complex Numbers.** A series of complex numbers is a formal sum of the form

$$\sum_{k=0}^{\infty} z_k = z_0 + z_1 + z_2 + \cdots + z_k + \cdots,$$

with $z_k \in \mathbb{C}$. The series *converges* if its sequence of partial sums $\{s_n\}$ converges, where

$$s_n = \sum_{k=0}^{n} z_k. \tag{1.2.1}$$

In this case, if $s = \lim_{n\to\infty} s_n$, then we say $s$ is the sum of the series and write

$$s = \sum_{k=0}^{\infty} z_k.$$

Just as with real series, if a series $\sum_{k=0}^{\infty} z_k$ converges, then its terms must tend to zero. Thus, if $\{z_k\}$ fails to have limit 0, then the series diverges. This test for divergence is called the *term test*. Its proof for complex series is the same as the proof for real series. We leave it as an exercise (Exercise 1.2.9).

**Example 1.2.7.** For what values of $z$ does the complex geometric series

$$\sum_{k=0}^{\infty} z^k$$

converge and what does it converge to?

**Solution:** We first note that if $|z| \geq 1$, then $|z|^k \geq 1$ for all $k$ and the series diverges by the term test. For $|z| < 1$ we use the same trick that is used to study the real geometric series. The $n$th partial sum of the series is

$$s_n = \sum_{k=0}^{n} z^k.$$

If we multiply this by $(1 - z)$, a vast cancellation of terms occurs and we obtain

$$(1 - z)s_n = 1 - z^{n+1},$$

so that

$$s_n = \frac{1 - z^{n+1}}{1 - z}.$$

This sequence converges to $\dfrac{1}{1-z}$ if $|z| < 1$, since

$$\left| s_n - \frac{1}{1-z} \right| = \frac{|z|^n}{|1-z|} \to 0$$

in this case.

A series $\sum_{k=0}^{\infty} z_k$ is said to *converge absolutely* if the series of positive terms $\sum_{k=0}^{\infty} |z_k|$ converges. From calculus, we know a great deal about convergence of positive termed series (comparison test, ratio test, root test, etc.), and so the following theorem will obviously play an important role.

**Theorem 1.2.8.** *If a series* $\sum_{k=0}^{\infty} z_k$ *of complex numbers converges absolutely, then it converges. Furthermore,*

$$\left| \sum_{k=0}^{\infty} z_k \right| \leq \sum_{k=0}^{\infty} |z_k|.$$

**Proof.** By hypothesis, the series $\sum_{k=0}^{\infty} |z_k|$ converges. If $z_k = x_k + iy_k$, then $|x_k| < |z_k|$ and $|y_k| < |z_k|$. By the comparison test, the two series

$$\sum_{k=0}^{\infty} |x_k| \quad \text{and} \quad \sum_{k=0}^{\infty} |y_k|$$

both converge. This, in turn, implies that

$$\sum_{k=0}^{\infty} x_k \quad \text{and} \quad \sum_{k=0}^{\infty} y_k$$

converge, since an absolutely convergent series of real numbers converges. This tells us that the real and imaginary parts of the sequence $\{s_n\}$ of partial sums of

$\sum_{k=0}^{\infty} z_k$ are convergent and, by Theorem 1.2.4, so is the sequence $\{s_n\}$ itself. Thus, the series converges.

It follows from the triangle inequality that

$$|s_n| = \left|\sum_{k=0}^{n} z_k\right| \leq \sum_{k=0}^{n} |z_k|.$$

The inequality of the theorem follows when we pass to the limit as $n \to \infty$ in this inequality and use the result of Exercise 1.2.7. □

**Example 1.2.9.** Prove that the complex series $\sum_{k=1}^{\infty}(k + k^2 i)^{-1}$ converges absolutely.

**Solution:** By the second form of the triangle inequality (see Exercise 1.2.2) we have

$$|k + k^2 i| \geq k^2 - k \geq \frac{k^2}{2} \quad \text{if} \quad k \geq 2.$$

Thus,

$$|k + k^2 i|^{-1} \leq 2k^{-2} \quad \text{if} \quad k \geq 2.$$

Since the $p$-series $\sum_{k=1}^{\infty} k^{-2}$ converges, so does $\sum_{k=1}^{\infty} 2k^{-2}$ and, by comparison, $\sum_{k=1}^{\infty} |k + k^2 i|^{-1}$. Thus, $\sum_{k=1}^{\infty}(k + k^2 i)^{-1}$ converges absolutely.

**Power Series.** A complex power series is a series of the form

$$\sum_{n=0}^{\infty} a_n (z - z_0)^n, \tag{1.2.2}$$

where the coefficients $a_n$ are complex numbers, $z_0$ is a complex number and $z$ is a complex variable. A power series of this form is said to be *centered* at $z_0$. It defines a complex function of the complex variable $z$, with domain the set of those $z \in \mathbb{C}$ for which the series converges.

**Remark 1.2.10.** We know from calculus that the set on which a real power series

$$\sum_{n=0}^{\infty} a_n (x - x_0)^n$$

converges is an interval – *the interval of convergence* – consisting of an open interval centered at $x_0$ and possibly one or both of its endpoints. The power series converges absolutely at each point of the open interval. The radius of this interval is called the *radius of convergence* of the power series and is computed using the root test or the ratio test. As we shall see in Chapter 3, a similar result is true for complex power series, but the interval of convergence is replaced by a disc of convergence. If for $r > 0$ we set

$$D_r(z_0) = \{z \in \mathbb{C} : |z| < r\} \quad \text{and} \quad \overline{D}_r(z_0) = \{z \in \mathbb{C} : |z| \leq r\},$$

then $D_r(z_0)$ is called the open disc of radius $r$, centered at $z_0$, while $\overline{D}_r(z_0)$ is called the closed disc of radius $r$, centered at $z_0$. Given a power series, centered at $z_0$, there is a number $R \geq 0$, called the radius of convergence of the power series, such that the series converges absolutely for each $z \in D_R(z_0)$ and diverges for each $z \notin \overline{D}_R(z_0)$. When we study power series in detail, we will prove this result and tell how to calculate $R$ in general. However, for most of the series we will be studying, there are elementary ways to show that this result holds and to find $R$. One simply uses the standard convergence tests from calculus – particularly the ratio test.

**Example 1.2.11.** Find the radius of convergence of the series

$$\sum_{n=1}^{\infty} \frac{z^n}{n}. \tag{1.2.3}$$

**Solution:** If we apply the ratio test to the series $\sum_{n=1}^{\infty} |z|^n/n$, we conclude that this series converges for $|z| < 1$ and diverges for $|z| > 1$. Hence, by Theorem 1.2.8, the series (1.2.3) also converges for $|z| < 1$. If $|z| > 1$, then the sequence of terms of this series fails to converge to zero (since $|z|^n/n \to +\infty$) and so the series diverges by the term test. Thus our series converges on $D_1(0)$ and diverges outside of $\overline{D}_1(0)$. We conclude that it has radius of convergence 1.

**Example 1.2.12.** Show that the radius of convergence of the series

$$\sum_{n=0}^{\infty} \frac{z^n}{n!} \tag{1.2.4}$$

is $+\infty$ – that is, the series converges for all $z$.

**Solution:** We apply the ratio test to the series $\sum_{n=0}^{\infty} |z|^n/n!$. We have

$$\left(\frac{|z|^{n+1}}{(n+1)!}\right)\left(\frac{|z|^n}{n!}\right)^{-1} = \frac{|z|}{n+1}$$

which converges to 0 for all $z$. Hence, by the ratio test, the power series (1.2.4) converges for all $z \in \mathbb{C}$. The radius of convergence is, thus, $+\infty$.

**Example 1.2.13.** Find the radius of convergence of the power series

$$\sum_{n=0}^{\infty} a_n z^n$$

where $a_n = 2^n$ if $n$ is prime and $a_n = 0$ if $n$ is not prime.

**Solution:** We cannot use the ratio test on this one because most of the terms of the series are 0. However, if we compare the series $\sum_{n=0}^{\infty} a_n |z|^n$ with the series $\sum_{n=0}^{\infty} 2^n |z|^n$ – a series to which we can apply the ratio test – we conclude that the series converges for $|z| < 1/2$. If $|z| > 1/2$, then the terms of our series do not tend to 0 and so the series diverges by the term test. We conclude that the radius of convergence is $1/2$.

## Exercise Set 1.2

1. Show that the sequence $(2 + ni)^{-1}$ converges to 0.
2. Prove the second form of the triangle inequality for complex numbers: $\big||z| - |w|\big| \le |z - w|$.
3. Show that $\left|\dfrac{1}{z+5}\right| \le \dfrac{1}{|z| - 5}$.
4. Does the sequence $\left(1/\sqrt{2} + i/\sqrt{2}\right)^n$ converge?

5. For which values of $z$ does the sequence $\{z^n\}$ converge?
6. Prove that if a sequence $\{z_n\}$ converges, then it is bounded – that is, there is a positive number $M$ such that $|z_n| \leq M$ for all $n$. Hint: Show that there is an $N$ such that $|z_n| \leq |z| + 1$ for all $n \geq N$. This gives an upper bound for $|z_n|$ for all but finitely many of the $z_n$.
7. Prove that if $\{z_n\}$ is a sequence with $\lim z_n = w$, then $\lim \overline{z_n} = \overline{w}$ and $\lim |z_n| = |w|$.
8. For the sequence of the previous exercise, prove that $\lim 1/z_n = 1/w$ provided $w \neq 0$.
9. Prove the term test for divergence of a series. That is, if the sequence of terms $\{z_k\}$ does not have limit 0, then the series $\sum_{k=0}^{\infty} z_k$ diverges.
10. Does the series $\sum_{n=0}^{\infty} n/(3 + 2ni)$ converge?
11. Does the series $\sum_{n=1}^{\infty} n/(n^3 + 2i)$ converge?
12. For which values of $z$ does the series $\sum_{n=0}^{\infty} 1/(n^2 + z^2)$ converge?
13. Find the radius of convergence of the power series $\sum_{n=0}^{\infty} nz^n$.
14. Find the radius of convergence of the power series $\sum_{n=0}^{\infty} z^n/3^n$.
15. Find the radius of convergence of the power series $\sum_{n=0}^{\infty} z^n/(1 + 2^n)$.
16. Find the radius of convergence of the power series $\sum_{n=0}^{\infty}(n!/n^n)z^n$.

## 1.3. The Exponential Function

There are many real-valued functions of a real variable that have natural extensions to complex-valued functions of a complex variable. In fact, this is true of all real functions that have convergent power series expansions. If $f(x) = \sum_{n=0}^{\infty} a_n(x-x_0)^n$ is a real power series which converges on an interval of radius $R$ about $x_0$, then the complex power series $f(z) = \sum_{n=0}^{\infty} a_n(z - x_0)^n$ converges in the open disc of radius $R$ about $x_0$ (Exercise 1.3.10) and serves to extend $f$ to a function $f(z)$ defined for $z$ in the disc $D_R(x_0)$.

One of the most important examples of the use of this technique is provided by the exponential function. We know from calculus that

$$e^x = \exp x = \sum_{n=0}^{\infty} \frac{x^n}{n!},$$

where the series converges to $e^x$ on the entire real line. We saw in Example 1.2.12 that this same series, with $x$ replaced by the complex variable $z$, converges absolutely on the entire complex plane. Thus, we get an extension of $e^x$ to a function $e^z$ defined on $\mathbb{C}$ as follows:

**Definition 1.3.1.** For each $z \in \mathbb{C}$, we define $\exp(z) = e^z$ by

$$e^z = \sum_{n=0}^{\infty} \frac{z^n}{n!}.$$

This is the complex exponential function. It has many important properties, some expected and some surprising, as we shall see below.

**The Law of Exponents.** A property that we should expect of an exponential function is the following.

**Theorem 1.3.2.** *If $z, w \in \mathbb{C}$, then $\mathrm{e}^{z+w} = \mathrm{e}^z\, \mathrm{e}^w$.*

**Proof.** The proof uses the complex form of the binomial formula:

$$(z+w)^n = \sum_{j=0}^{n} \frac{n!}{j!(n-j)!} z^j w^{n-j}. \tag{1.3.1}$$

This is proved using induction on $n$. We leave it as an exercise (Exercise 1.3.11).

Proceeding with the proof of the theorem, and using (1.3.1), we have

$$\begin{aligned} \mathrm{e}^{z+w} &= \sum_{n=0}^{\infty} \frac{(z+w)^n}{n!} \\ &= \sum_{n=0}^{\infty}\sum_{j=0}^{n} \frac{1}{j!(n-j)!} z^j w^{n-j}. \end{aligned} \tag{1.3.2}$$

If we make a change of variables by setting $k = n - j$ in the inside summation, then (1.3.2) becomes

$$\sum_{n=0}^{\infty} \left( \sum_{j+k=n} \frac{z^j w^k}{j!k!} \right).$$

This is precisely what we get if we expand the product

$$\mathrm{e}^z\, \mathrm{e}^w = \left( \sum_{j=0}^{\infty} \frac{z^j}{j!} \right) \left( \sum_{k=0}^{\infty} \frac{w^k}{k!} \right).$$

and collect terms of degree $n$. Provided this operation is valid, we conclude that $\mathrm{e}^{z+w} = \mathrm{e}^z\, \mathrm{e}^w$. It turns out that it is valid to expand the product of two infinite series in this fashion if they are both absolutely convergent. We prove this in the following lemma, which will complete the proof of the theorem. □

**Lemma 1.3.3.** *Let $\sum_{j=0}^{\infty} a_j$ and $\sum_{k=0}^{\infty} b_k$ be two absolutely convergent series of complex numbers. Then*

$$\left( \sum_{j=0}^{\infty} a_j \right) \left( \sum_{k=0}^{\infty} b_k \right) = \sum_{n=0}^{\infty} \left( \sum_{j+k=n} a_j b_k \right). \tag{1.3.3}$$

**Proof.** The partial sums of the series involved here are

$$s_J = \sum_{j=0}^{J} a_j, \quad t_K = \sum_{k=0}^{K} b_k, \quad u_N = \sum_{n=0}^{N} \left( \sum_{j+k=n} a_j b_k \right).$$

The left side of (1.3.3) is, by definition, $(\lim s_J)(\lim t_K)$, while the right side is $\lim u_N$. We know $\lim s_J$ and $\lim t_K$ both exist since the series defining them converge absolutely. We must prove that $\lim u_N$ exists and equals $(\lim s_J)(\lim t_K)$.

For a given pair $J, K$, we let $N = J+K$. Then $u_N$ is the sum of all terms $a_j b_k$ for which $j+k \le N$, while $s_J t_K$ is the sum of those terms $a_j b_k$ for which $j \le J$ and $k \le K$. Thus,

$$(1.3.4) \qquad \begin{aligned} |u_N - s_J t_K| &\le \sum \{|a_j b_k| : j+k \le N, \text{ and either } j > J \text{ or } k > K\} \\ &\le \sum_{j=J+1}^{\infty} |a_j| \sum_{k=0}^{\infty} |b_k| + \sum_{j=0}^{\infty} |a_j| \sum_{k=K+1}^{\infty} |b_k|. \end{aligned}$$

The sum $\sum_{j=J+1}^{\infty} |a_j|$ converges to zero as $J \to \infty$ because the series $\sum_{j=0}^{\infty} a_j$ converges absolutely, while $\sum_{k=K+1}^{\infty} |b_k|$ converges to 0 as $K \to \infty$ because $\sum_{k=0}^{\infty} b_k$ converges absolutely.

To complete the proof, we choose for each non-negative integer $N$ a $J$ and $K$ with $J+K = N$, and we do this in such a way that as $N \to \infty$ the corresponding $J$ and $K$ both also tend to $\infty$. For example, we could choose $J = K = N/2$ if $N$ is even, and $J = K+1 = N/2 + 1/2$ if $N$ is odd. Then, from (1.3.4) it is clear that $\lim |u_N - s_J s_K| = 0$, which implies $\lim u_N = (\lim s_J)(\lim s_K)$. □

**The Exponential of an Imaginary Number.** If $z = x + iy$, then the law of exponents (Theorem 1.3.2) implies that

$$e^z = e^x \, e^{iy}.$$

We understand the behavior of $e^x$ for real $x$ from calculus. It is 1 at 0, is everywhere positive, increasing and concave upward; it rapidly approaches $+\infty$ as $x \to \infty$ and rapidly approaches 0 as $x \to -\infty$. What about $e^{iy}$ – the exponential of a purely imaginary number? Here there are some surprises.

By the law of exponents, $e^{iy} e^{-iy} = e^0 = 1$. Thus, $1/e^{iy} = e^{-iy}$. But also, for any $z \in \mathbb{C}$, $\overline{e^z} = e^{\bar{z}}$ (Exercise 1.3.1) and, in particular, $\overline{e^{iy}} = e^{-iy} = 1/e^{iy}$. This implies that $|e^{iy}|^2 = 1$ and, consequently, $|e^{iy}| = 1$. Thus, $e^{iy}$ is always a point on the circle of radius 1 centered at 0 (we call this the unit circle).

A more explicit description of the number $e^{iy}$ comes from examination of its power series definition. If we group together the even numbered terms (terms with $n$ of the form $n = 2k$) and the odd numbered terms (terms with $n$ of the form $n = 2k+1$) in this power series, we derive the identity

$$\begin{aligned} e^{iy} = \sum_{n=0}^{\infty} \frac{(iy)^n}{n!} &= \sum_{k=0}^{\infty} (-1)^k \frac{y^{2k}}{(2k)!} + i \sum_{k=0}^{\infty} (-1)^k \frac{y^{2k+1}}{(2k+1)!} \\ &= \cos y + i \sin y. \end{aligned}$$

Thus, we have proved the following theorem, which is known as *Euler's Identity.*

**Theorem 1.3.4.** *The identity* $e^{iy} = \cos y + i \sin y$ *holds for all* $y \in \mathbb{R}$.

This shows that, not only is $e^{iy}$ a point on the unit circle, it is the point which is reached by rotating through an angle $y$ (measured in radians) from the initial point $(1, 0)$.

**Example 1.3.5.** Express the complex numbers $e^{2\pi i}$, $e^{\pi i}$ and $e^{\pi i/2}$ in standard form.

**Solution:** By Euler's identity, we have

$$e^{2\pi i} = \cos 2\pi + i \sin 2\pi = 1,$$
$$e^{\pi i} = \cos \pi + i \sin \pi = -1,$$

while

$$e^{\pi i/2} = \cos \pi/2 + i \sin \pi/2 = i.$$

**Properties of the Exponential Function.** The law of exponents and Euler's identity immediately imply:

**Theorem 1.3.6.** *If $z = x + iy$ is a complex number, then $e^z = e^x(\cos y + i \sin y)$.*

There are a number of properties of the exponential function which follow easily from this characterization of $e^z$. We collect them together in the following theorem whose proof is left to the exercises.

**Theorem 1.3.7.** *The exponential function has the following properties:*

(a) $e^z$ *is never* $0$;

(b) $|e^z| = e^{\mathrm{Re}(z)}$;

(c) $|e^z| \le e^{|z|}$;

(d) $e^z$ *is periodic of period* $2\pi i$, *meaning* $e^{z+2\pi i} = e^z$ *for every* $z \in \mathbb{C}$;

(e) $e^z = 1$ *if and only if* $z = 2\pi n i$ *for some integer* $n$.

**Complex Trigonometric Functions.** If we write out Euler's identity with $y$ replaced by $\theta$ and then by $-\theta$, we obtain a pair of identities

$$e^{i\theta} = \cos\theta + i\sin\theta,$$
$$e^{-i\theta} = \cos\theta - i\sin\theta.$$

If we solve this system of equations for $\cos\theta$ and $\sin\theta$, we obtain

$$\cos\theta = \frac{e^{i\theta} + e^{-i\theta}}{2}, \quad \sin\theta = \frac{e^{i\theta} - e^{-i\theta}}{2i}. \tag{1.3.5}$$

This suggests defining $\sin z$ and $\cos z$ for a complex variable $z$ in the following way:

**Definition 1.3.8.** For each $z \in \mathbb{C}$, we set

$$\cos z = \frac{e^{iz} + e^{-iz}}{2},$$
$$\sin z = \frac{e^{iz} - e^{-iz}}{2i}.$$

These are the same functions that one would get by replacing $x$ by $z$ in the power series expansions of $\sin x$ and $\cos x$.

With $\sin z$ and $\cos z$ defined, it is easy to define complex versions of the other trigonometric functions. For example,

$$\tan z = \frac{\sin z}{\cos z} = -i\frac{e^{iz} - e^{-iz}}{e^{iz} + e^{-iz}}. \tag{1.3.6}$$

**Example 1.3.9.** How would you define a complex version of the arctan function?

**Solution:** The power series which converges to $\arctan x$ on $(-1,1)$ is

$$\arctan x = \sum_{n=0}^{\infty} (-1)^n \frac{x^{2n+1}}{2n+1}.$$

If we replace $x$ by the complex variable $z$, we obtain a series which converges on the disc $D_1(0)$. We then use it to define $\arctan z$ on this disc:

$$\arctan z = \sum_{n=0}^{\infty} (-1)^n \frac{z^{2n+1}}{2n+1}.$$

## Exercise Set 1.3

1. Using the power series for $e^z$, prove that $\overline{e^z} = e^{\bar{z}}$ for each $z \in \mathbb{C}$.
2. Express the complex numbers $e^{\pi i/4}$, $e^{3\pi i/2}$, and $e^{13\pi i/6}$ in standard form.
3. Express $1/\sqrt{2} + i/\sqrt{2}$ in the form $e^z$ for some complex number $z$.
4. Find all values of $z$ for which $e^z = 1 + \sqrt{3}\, i$.
5. Prove that $e^z$ is never 0 (Part (a) of Theorem 1.3.7).
6. Prove that $|e^z| = e^{\mathrm{Re}(z)} \le e^{|z|}$ for each $z \in \mathbb{C}$ (Parts (b) and (c) of Theorem 1.3.7).
7. Verify that $e^{z+2\pi i} = e^z$ for every $z \in \mathbb{C}$ (Part (d) of Theorem 1.3.7).
8. Verify that $e^z = 1$ if and only if $z = 2\pi n i$ for some interger $n$ (Part (e) of Theorem 1.3.7).
9. For which values of $z$ is $e^z$ a real number? For which values is it an imaginary number?
10. Show that if the real power series $\sum_{n=0}^{\infty} a_n x^n$ has radius of convergence $R$, then the corresponding complex power series $\sum_{n=0}^{\infty} a_n z^n$ also has radius of convergence $R$ (you may assume radius of convergence for a complex power series makes sense and has the properties described in Remark 1.2.10).
11. Use induction on $n$ to prove the complex binomial formula (1.3.1).
12. Derive the trigonometric identities

$$\sin(z+w) = \sin z \cos w + \cos z \sin w,$$
$$\cos(z+w) = \cos z \cos w - \sin z \sin w$$

from the law of exponents (Theorem 1.3.2) and Euler's identity (Theorem 1.3.4).

13. Show that $\cos ix = \cosh x$ and $\sin ix = i \sinh x$, where cosh and sinh are the hyperbolic cosine and hyperbolic sine from calculus.
14. Show that

$$\sin(x+iy) = \sin x \cosh y + i \cos x \sinh y \quad \text{and}$$
$$\cos(x+iy) = \cos x \cosh y - i \sin x \sinh y.$$

15. How would you define a complex version of the function $\log(1+x)$?
16. Verify that the power series defining arctan in Example 1.3.9 converges on the disc $D_1(0)$.

## 1.4. Polar Form for Complex Numbers

If $u$ is a complex number of modulus 1, then $u = \cos\theta + i\sin\theta = e^{i\theta}$, where $\theta$ is an angle from the positive $x$ axis to the ray from the origin through $u$. This angle is measured in radians and is the signed length of an arc on the unit circle joining $(1,0)$ to $u$. It is positive if the arc is traversed in the counterclockwise direction and negative if it is traversed in the clockwise direction.

If $z$ is any non-zero complex number, then $z/|z|$ is a complex number of modulus 1 and so it has the form $e^{i\theta}$ for some $\theta$. If we set $r = |z|$, then

$$z = r\,e^{i\theta}.$$

This is the *polar form* of the complex number $z$. The angle $\theta$ is called the *argument* of $z$ and is denoted $\arg z$.

Comments:

(1) By Euler's identity (Theorem 1.3.4), $r\,e^{i\theta} = r(\cos\theta + i\sin\theta)$.

(2) The number $z$ must be non-zero for the argument to be defined. The polar form of $z = 0$ is just $z = 0$.

(3) For $z \neq 0$, there are infinitely many angles $\theta$ for which $z = r\,e^{i\theta}$. Given one such $\theta$ the others are all of the form $\theta + 2\pi n$, where $n$ is an integer. Thus, $\arg(z)$ is not a single number, but a collection of numbers that differ from one another by multiples of $2\pi$. We can specify a particular one of these by insisting it lie in a particular half-open interval of length $2\pi$ – such as $[0, 2\pi)$ or $(-\pi, \pi]$.

(4) The numbers $(r, \theta)$, where $r = |z|$ and $\theta$ is a value of $\arg(z)$, are polar coordinates for $z = x + iy$ considered as a point $(x, y)$ in $\mathbb{R}^2$.

(5) We now have two useful ways of expressing a complex number $z$, the *standard form* $x + iy$ and the *polar form* $r\,e^{i\theta}$. It is important to be able to easily change from one to the other.

**Example 1.4.1.** Find the polar form for the complex number $z = 2 + 2i$.

**Solution:** We have

$$r = \sqrt{2^2 + 2^2} = 2\sqrt{2} \quad \text{and} \quad \theta = \arctan(1) = \pi/4.$$

Thus, $z = 2\sqrt{2}\,e^{\pi i/4}$ is the polar form of $z$.

**Example 1.4.2.** Put the complex number $z = 2\,e^{\pi i/6}$ in standard form $x + iy$.

**Solution:** We have

$$x = 2\cos\pi/6 = \sqrt{3} \quad \text{and} \quad y = 2\sin\pi/6 = 1$$

and so $z = \sqrt{3} + i$.

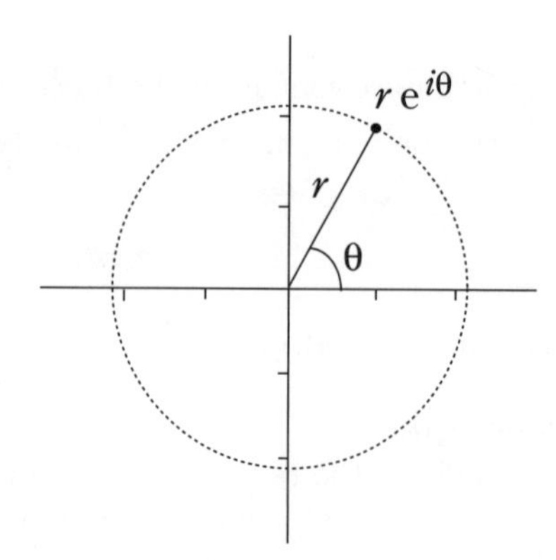


**Figure 1.4.1.** The Polar Form $r\,e^{i\theta}$ of a Complex Number.

**Products, Powers, and Roots.** Polar form is particularly useful in dealing with products of complex numbers, since the product has such a simple expression if the numbers are given in polar form. If $z_1 = r_1\,e^{i\theta_1}$ and $z_2 = r_2\,e^{i\theta_2}$, then the law of exponents implies that

(1.4.1) $$z_1 z_2 = r_1 r_2\,e^{i(\theta_1+\theta_2)}\,.$$

Thus, $z_1z_2$ is the complex number whose norm (distance from the origin) $|z_1z_2|$ is the product of $|z_1|$ and $|z_2|$ and whose argument is the sum of the arguments of $z_1$ and $z_2$.

Similarly, the quotient of $z_1$ and $z_2$ is given by

(1.4.2) $$z_1/z_2 = (r_1/r_2)\,e^{i(\theta_1-\theta_2)}\,.$$

**Example 1.4.3.** Find $z_1z_2$ and $z_1/z_2$ if $z_1 = 2\,e^{\pi i/3}$ and $z_2 = 3\,e^{2\pi i/3}$.

**Solution:** By (1.4.1) and (1.4.2), we have

$$z_1z_2 = 6\,e^{\pi i} = -6 \quad\text{and}\quad z_1/z_2 = (2/3)\,e^{-\pi i/3} = 1/3 - i\sqrt{3}/3.$$

It is evident from (1.4.1) that the $n$th power of a complex number $z = r\,e^{i\theta}$ is

(1.4.3) $$z^n = r^n\,e^{in\theta}\,.$$

From this we conclude that if $z = r\,e^{i\theta}$, and we choose $w = r^{1/n}\,e^{i\theta/n}$, then $w^n = z$. Thus, $w$ is an $n$th root of $z$. It is not the only one, however. Since $z$ can also be written as $z = e^{i(\theta+2\pi k)}$ for any integer $k$, each of the numbers

$$w_k = r^{1/n}\,e^{i(\theta/n+2\pi k/n)},$$

where $k$ is an integer, is also an $n$th root of $z$. Of course, these numbers are not all different. Those whose arguments differ by an integral multiple of $2\pi$ are the same. The numbers $w_0, w_1, w_2, \cdots, w_{n-1}$ are all distinct, but every other $w_k$ is equal to one of these. This proves the following theorem.

**Theorem 1.4.4.** *If $z = r\,e^{i\theta}$ is a non-zero complex number, then $z$ has exactly $n$ $n$th roots. They are the numbers*

$$r^{1/n}\,e^{i(\theta/n+2\pi k/n)} \quad \textit{for} \quad k = 0, 1, 2, \cdots, n-1.$$

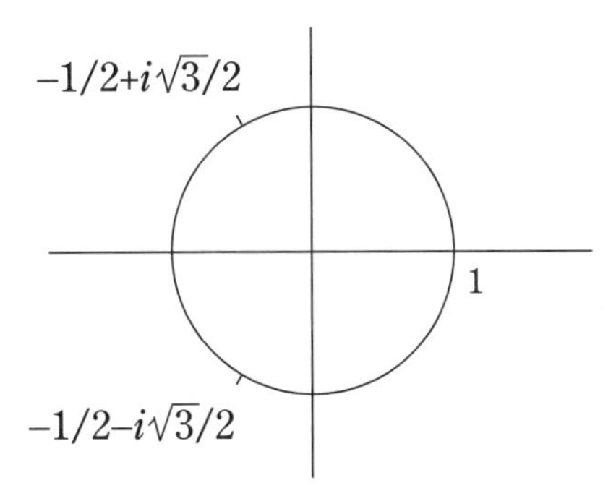


**Figure 1.4.2.** The Cube Roots of Unity.

Note that the $n$ numbers $e^{i(\theta/n+2\pi k/n)}$ that appear in this result are evenly spaced around the unit circle, with each successive pair separated by an angle of $2\pi/n$.

The $n$th roots of the number 1 play a special role. They are called *the nth roots of unity*. If we apply the above theorem in the special case where $r = 1$ and $\theta = 0$, it tells us that the $n$th roots of unity are the numbers

$$e^{2\pi ki/n} \quad \text{for} \quad k = 0, 1, 2, \cdots, n-1. \tag{1.4.4}$$

**Example 1.4.5.** Find the cube roots of unity.

**Solution:** In the particular case where $n = 3$, (1.4.4) tells us that the roots of unity are

$$\begin{aligned} e^0 &= 1, \\ e^{2\pi i/3} &= -1/2 + (\sqrt{3}/2)i, \quad \text{and} \\ e^{4\pi i/3} &= -1/2 - (\sqrt{3}/2)i. \end{aligned}$$

**Example 1.4.6.** Find the cube roots of $2i$.

**Solution:** Since $2i = 2\,e^{\pi i/2}$, Theorem 1.4.4 implies that the cube roots of $2i$ are

$$\begin{aligned} 2^{1/3}\,e^{\pi i/6} &= 2^{1/3}(\sqrt{3}/2 + i/2), \\ 2^{1/3}\,e^{i(\pi/6+2\pi/3)} &= 2^{1/3}\,e^{5\pi i/6} = 2^{1/3}(-\sqrt{3}/2 + i/2), \quad \text{and} \\ 2^{1/3}\,e^{i(\pi/6+4\pi/3)} &= 2^{1/3}\,e^{3\pi i/2} = -2^{1/3}i. \end{aligned}$$

**The Logarithm.** If $z = r\,e^{i\theta}$, and $\log r$ is the natural logarithm of the positive number $r$, then the law of exponents implies that

$$z = e^{\log r + i\theta}.$$

Thus, it would make sense to define $\log z$ to be $\log r + i\theta = \log|z| + i\arg z$. There is a problem with this, however. There are infinitely many possible choices for $\arg z$, and so $\log z$ is not well defined, just as $\arg z$ is not well defined.

The solution to this problem is to restrict $\theta = \arg z$ to lie in a specific half-open interval of length $2\pi$.

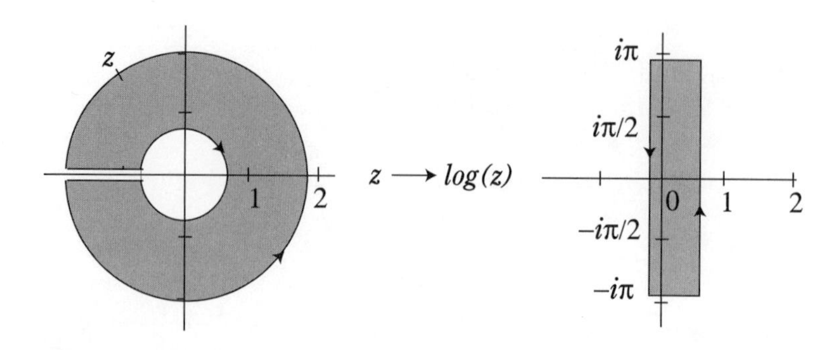


**Figure 1.4.3.** Principal Branch of the Log Function on an Annulus.

**Definition 1.4.7.** Given a half-open interval $I$ of length $2\pi$ on the line $\mathbb{R}$, let $\arg_I z$, for $z \neq 0$, be the value of $\arg z$ that lies in the interval $I$. Then the function defined for $z \neq 0$ by

$$\log z = \log |z| + i \arg_I z$$

will be called the *branch* of the log function defined by $I$. In the special case where $I = (-\pi, \pi]$, this function will be called the *principal branch* of the log function.

The following properties of the various branches of the log function follow easily from the definition and the properties of the exponential function. The proofs are left to the exercises.

**Theorem 1.4.8.** *If* log *is the branch of the the log function determined by an interval $I$, then*

(a) *if* $z \neq 0$, *then* $e^{\log z} = z$;

(b) *if* $z \in \mathbb{C}$, *then* $\log e^z = z + 2\pi k i$ *for some integer* $k$;

(c) *if* $z, w \in \mathbb{C}$, *then* $\log zw = \log z + \log w + 2\pi k i$ *for some integer* $k$;

(d) $\log 1 = 2\pi k i$ *for some integer* $k$;

(e) log *agrees with the ordinary natural log function on the positive real numbers if and only if the interval $I$ contains* 0.

Suppose the interval $I$ defining a branch of the log function has endpoints $a$ and $b$ with $a < b$. Then, since $b - a = 2\pi$, the polar coordinate equations $\theta = a$ and $\theta = b$ define the same ray. This ray is called the *cut line* for this branch of the logarithm. Observe that if $z$ and $w$ are two complex numbers with $|z| = |w| > 0$ which are close to each other, but on opposite sides of the cut line – say with $\arg_I z$ near $a$ and $\arg_I w$ near $b$ – then $\log w - \log z$ is nearly $2\pi i$. In other words, as we cross the cut line moving in the clockwise direction, the value of log jumps by $2\pi i$. Thus, log is not continuous at points on the cut line. Later we will show that it is continuous everywhere else.

**Other Functions.** If we fix a branch of the complex log function, we can define a number of other complex functions which are extensions of familiar real functions. We mention some of these briefly.

An $n$th root function can be defined by setting $0^{1/n} = 0$ and

$$z^{1/n} = e^{(1/n)\log z} \quad \text{if} \quad z \neq 0. \tag{1.4.5}$$

A special case of this is the square root function defined by $\sqrt{0} = 0$ and

$$\sqrt{z} = e^{(1/2)\log z} \quad \text{if} \quad z \neq 0. \tag{1.4.6}$$

Note that the $n$th root function, as defined by (1.4.5), is giving only one of the $n$th roots of $z$. Which one is determined by the branch of the log function that is used. The other $n$th roots are obtained by multiplying this one by the $n$th roots of unity (Exercise 1.4.9). For example, given one square root of $z$, the other is obtained by multipling it by $-1$.

**Example 1.4.9.** If $\sqrt{z}$ is defined using the principal branch of the log function, analyze the behavior of $\sqrt{z}$ near the cut line for this branch.

**Solution:** For the principal branch of the log function the interval $I$ is $(-\pi, \pi]$ and so the cut line is the line defined by $\theta = \pi$ in polar coordinates – that is, it is the negative real axis. If $z = r\,e^{i\theta}$ is just above the negative real axis, then $\log z$ is nearly $\log r + i\pi$ and

$$\sqrt{z} = e^{(1/2)\log z}$$

is nearly $i\sqrt{r}$. On the other hand, if $z$ is just below the negative axis, then $\log z$ is nearly $\log r - i\pi$, $(1/2)\log z$ is close to $(\log r - i\pi)/2$ and $\sqrt{z}$ is close to $-i\sqrt{r}$. In other words, as $z$ crosses the cut line, $\sqrt{z}$ jumps from one square root of $z$ to its negative, which is the other square root of $z$.

**Example 1.4.10.** Analyze the function $\sqrt{z^2+1}$, where the square root function is defined, as above, using the principal branch of the log function.

**Solution:** The cut line for $\sqrt{z}$ is the same as for log – the half-line of negative reals. The function $\sqrt{z}$ jumps from values on the positive imaginary axis to their negatives as $z$ crosses this line in the counterclockwise direction. The number $z^2+1$ crosses the negative real half-line in the counterclockwise direction as $z$ crosses either $(i, i\infty)$ or $(-i\infty, -i)$ in the counterclockwise direction. Thus, these two half-lines on the imaginary axes are where discontinuities of $\sqrt{z^2+1}$ occur. As either of these half-lines is crossed in the counterclockwise direction, a typical value of this function jumps from a number $it$ with $t > 0$ to $-it$ (see Figure 1.4.4).

Raising a complex number to a complex power is another function that can be defined using a branch of the log function. If $z \neq 0$ and $a$ is any complex number, then we set

$$z^a = e^{a \log z}. \tag{1.4.7}$$

Here, we are thinking of $a$ as being fixed and $z$ is the independent variable of the function. If we want the exponent to be the variable, we would write

$$a^z = e^{z \log a}. \tag{1.4.8}$$

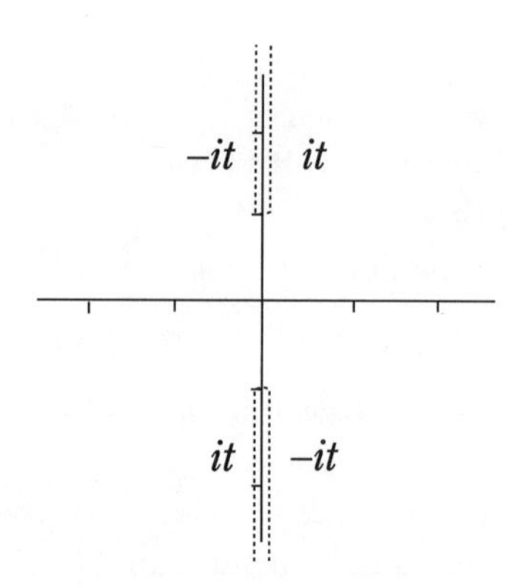


**Figure 1.4.4.** The Cut Line Discontinuities of $\sqrt{z^2+1}$.

Note that for any function defined in terms of a branch $\log z$ of the log function, we can expect trouble along the cut line for log. Since, log has a jump or discontinuity as we cross the cut line, we can expect the same for functions defined in terms of it.

Also, we emphasize that functions defined this way depend on the choice of a branch of the log function. If one wants such a function to agree with the standard one on the positive real axis, then one must choose a branch of the log function which is the ordinary natural logarithm on the positive real numbers. By Theorem 1.4.8, Part(e), this happens if and only if the interval $I$, in which $\arg(z)$ is required to lie, contains 0. In particular, the principal branch has this property.

## Exercise Set 1.4

1. Put each of the complex numbers $-1$, $i$, $-i, 1+\sqrt{3}\,i$, and $5-5i$ in polar form.
2. Put each of the complex numbers $e^{4\pi i}$, $3\,e^{2\pi i/3}$, $5\,e^{5\pi i/2}$, and $2\,e^{-3\pi i/4}$ in standard form.
3. Find all powers of $e^{\pi i/8}$. How many distinct powers of this number are there?
4. Show that $(1-i)^7 = 8(1+i)$ by converting to polar form, taking the seventh power, and then converting back to standard form.
5. Using a calculator, calculate $(1.2\,e^{.5\,i})^n$ for $n = 1, \cdots, 6$ and graph the resulting points. Do the same for $(.8\,e^{.5\,i})^n$.
6. Prove that if $z$ is a number on the unit circle, then $z$ has finitely many distinct powers $z^n$ if and only if the argument of $z$ is a rational multiple of $2\pi$.
7. What are the 4th roots of unity?
8. What are the cube roots of $-9$?
9. Show that, given one $n$th root of $z$, the others are obtained by multiplying it by the $n$th roots of unity.
10. Find $\arg_I z$ if (a) $z = -i$ and $I = (-\pi, \pi]$, (b) $z = -i$ and $I = [0, 2\pi)$, (c) $z = 1$ and $I = [3\pi/2, 7\pi/2)$.

11. For the principal branch of the log function, find $\log(1-i)$.
12. Find $\log(1-i)$ for the branch of the log function determined by the interval $[0, 2\pi)$.
13. Prove (a) and (b) of Theorem 1.4.8.
14. Prove (c), (d), and (e) of Theorem 1.4.8.
15. Analyze the function $z^i$ defined by (1.4.7) using the principal branch of the log function. What kind of a jump, if any, does it have as $z$ crosses the negative real axis?
16. Analyze the function $\sqrt{1-z^2}$, where the square root function is defined by the principal branch of the log function. Where does it have discontinuities (jumps)?
17. Let the square root function be defined by the principal branch of the log function. Compare the functions $\sqrt{z^2-1}$ and $\sqrt{z+1}\sqrt{z-1}$. Where are the discontinuities of each function?
18. The identity $1 + z + z^2 + \cdots + z^n = \dfrac{1-z^{n+1}}{1-z}$ was derived in Example 1.2.7. Use this to derive Lagrange's trigonometric identity:
$$1 + \cos\theta + \cos 2\theta + \cdots + \cos n\theta = \frac{1}{2} + \frac{\sin{(n+1/2)\theta}}{2\sin\theta/2}.$$
Hint: Take the real parts of both sides in the first identity.

Chapter 2

# Analytic Functions

Complex Variables is the study of functions of a complex variable – specifically, analytic functions of a complex variable. These are functions which have a complex derivative in a sense that we shall define shortly. Analytic functions have amazing properties, and deriving these properties will be the main focus of the text. The amazing properties of analytic functions stem from the fact that the existence of a complex derivative is a much more powerful condition than the existence of the real derivative for a function of a real variable.

We have already seen several examples of analytic functions. The exponential function and the sine and cosine functions are analytic functions on the entire plane. The log and $n$th root functions are analytic except on the cut line for the branch of the log function that is used in their definitions.

We must prove one difficult theorem about analytic functions (the Cauchy Integral Theorem), and then a wealth of amazing theorems with simple proofs will follow.

We begin this chapter with a review of the basic facts about continuous functions in the plane. We then define analytic functions and prove some elementary theorems concerning them. Next we introduce contour integration and prepare for our assault on Cauchy's Theorem. Finally, in Sections 2.5 and 2.6 we prove two limited versions of this theorem. This provides us enough of the power of Cauchy's Theorem to allow us to move on to Chapter 3, where we develop a wide range of applications. However, we put off proving the most general version of Cauchy's Theorem until Chapter 4.

## 2.1. Continuous Functions

Since, geometrically, $\mathbb{C}$ is just $\mathbb{R}^2$, notions of distance, convergence of sequences, and continuity of functions are the same for $\mathbb{C}$ as for $\mathbb{R}^2$. In particular, a complex-valued function of a complex variable may be regarded as a function of two real variables with values in $\mathbb{R}^2$. The student who has had a course in multivariable calculus already knows what it means for such a function to be continuous. Still, we will review the basics of continuity in this context, partly in order to set terminology and notation that is peculiar to the complex variables context.

**Open and Closed Sets.** Recall that the open disc $D_r(z_0)$ and closed disc $\overline{D}_r(z_0)$, centered at $z_0$, with radius $r > 0$, are defined by

$$D_r(z_0) = \{z \in \mathbb{C} : |z - z_0| < r\} \quad \text{and} \quad \overline{D}_r(z_0) = \{z \in \mathbb{C} : |z - z_0| \le r\}.$$

Open intervals and closed intervals on the real line play an important part in calculus in one variable. Open and closed discs are the direct analogues in $\mathbb{C}$ of open and closed intervals on the line. However, the geometry of the plane is much more complicated than that of the line. We will need the concepts of open and closed for sets that are far more complicated than discs. This leads to the following definition.

**Definition 2.1.1.** If $W$ is a subset of $\mathbb{C}$, we will say that $W$ is *open* if, for each point $w \in W$, there is an open disc centered at $w$ which is contained in $W$. We will say that a subset of $\mathbb{C}$ is *closed* if its complement is open. A *neighborhood* of a point $z \in \mathbb{C}$ is any open set which contains $z$.

It might seem obvious that open discs are open sets and closed discs are closed sets. However, that is only because we have chosen to call them *open* discs and *closed* discs. We actually have to prove that they satisfy the conditions of the preceding definition. We do this in the next theorem.

**Theorem 2.1.2.** *We have:*

(a) *the empty set $\emptyset$ is both open and closed;*

(b) *the whole space $\mathbb{C}$ is both open and closed;*

(c) *each open disc is open;*

(d) *each closed disc is closed.*

**Proof.** The empty set $\emptyset$ is open because it has no points, and so the condition that a set be open, stated in Definition 2.1.1, is vacuously satisfied. The set $\mathbb{C}$ is open because it contains any open disc centered at any of its points. Thus, $\emptyset$ and $\mathbb{C}$ are both open. Since they are complements of one another, they are also both closed.

To prove (c), we suppose $D_r(z_0)$ is an open disc and $w$ is one of its points. Then $|w - z_0| < r$ and so, if we set $s = r - |w - z_0|$, then $s > 0$. Also, if $z \in D_s(w)$, then $|z - w| < s$ and so

$$|z - z_0| \le |z - w| + |w - z_0| < s + |w - z_0| = r,$$

and so $z \in D_r(z_0)$ (see Figure 2.1.1). Thus, we have shown that, for each $w \in D_r(z_0)$, there is an open disc, $D_s(w)$, centered at $w$, which is contained in $D_r(z_0)$. By definition, this means that $D_r(z_0)$ is open. This completes the proof of (c).

To prove (d), we consider a closed disc $\overline{D}_r(z_0)$. To prove that it is a closed set, we must show its complement is open. Suppose $w$ is a point in its complement. This means $w \in \mathbb{C}$ but $w \notin \overline{D}_r(z_0)$, and so $|w - z_0| > r$. This time we set $s = |w - z_0| - r$ and we claim that the open disc $D_s(w)$ is contained in the complement of $\overline{D}_r(z_0)$.

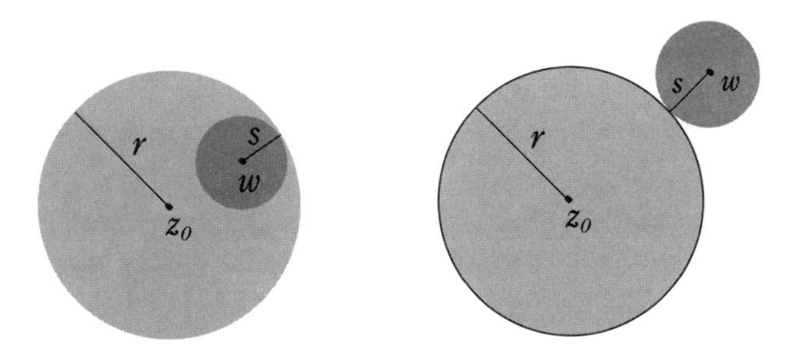


**Figure 2.1.1.** Proving Theorem 2.1.2 (c) and (d).

In fact, if $z \in D_s(w)$, then $|z - w| < s$ and so, by the second form of the triangle inequality (Exercise 1.2.2),

$$|z - z_0| \geq |w - z_0| - |z - w| > |w - z_0| - s = r,$$

which means $z$ is in the complement of $\overline{D}_r(z_0)$. Thus, we have proved that each point of the complement of $D_r(z_0)$ is the center of an open disc contained in the complement of $\overline{D}_r(z_0)$. This proves that this complement is open, hence, that $\overline{D}_r(z_0)$ is closed. □

The next theorem shows that the collection of all open subsets of $\mathbb{C}$ forms what is called a *topology* for $\mathbb{C}$. It says that the collection of open subsets is closed under arbitrary unions and finite intersections.

**Theorem 2.1.3.** *The union of an arbitrary collection of open sets is open, while the intersection of any finite collection of open sets is open. On the other hand, the intersection of an arbitrary collection of closed sets is closed, while the union of any finite collection of closed sets is closed.*

**Proof.** If $z_0 \in \mathbb{C}$, $\mathcal{V}$ is an arbitrary collection of open sets, and $U = \bigcup \mathcal{V}$ is its union, then $z_0$ is in $U$ if and only if it is in at least one of the sets in $\mathcal{V}$. Suppose, it is in $V \in \mathcal{V}$. Then, since $V$ is open, there is a disc $D_r(z_0)$, centered at $z_0$, which is contained in $V$. Then this disc is also contained in $U$. This proves that $U$ is open.

Now suppose $\{V_1, V_2, \cdots, V_k\}$ is a finite collection of open sets and

$$z_0 \in U = V_1 \cap V_2 \cap \cdots \cap V_n.$$

Then, since each $V_k$ is open, there exists for each $k$ a radius $r_k$ such that $D_{r_k}(z_0) \subset V_k$. If $r = \min\{r_1, r_2, \cdots, r_n\}$, then $D_r(z_0) \subset V_k$ for every $k$, which implies that $D_r(z_0) \subset U$. It follows that $U$ is open.

The proofs of the statements for closed sets follow from those for open sets by taking complements. We leave the details to Exercise 2.1.1. □

A consequence of the above theorem is: if $U$ is open and $K$ is closed, then the set-theoretic difference $U \setminus K$ is open. Similarly, $K \setminus U$ is closed (Exercise 2.1.2).

**Example 2.1.4.** If $0 < r < 1$, prove that the annulus $A = \{z \in \mathbb{C} : r < |z| < R\}$ is open.

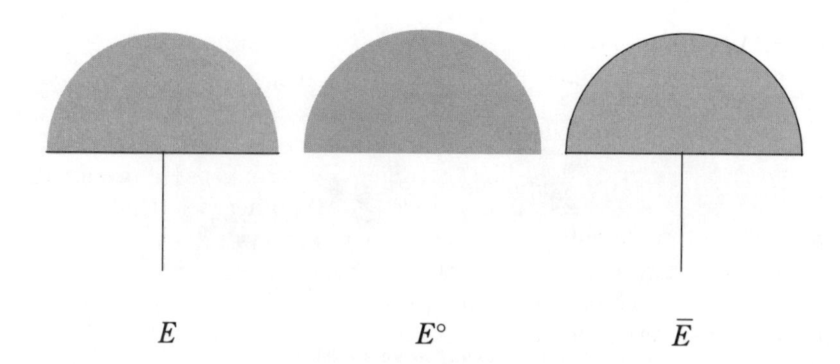


**Figure 2.1.2.** The Set $E$ of Example 2.1.7, its Interior $E^\circ$, and Closure $\overline{E}$.

**Solution:** The disc $D_R(0)$ is open, the disc $\overline{D}_r(0)$ is closed, and $A$ is the set-theoretic difference $D_R(0) \setminus \overline{D}_r(0)$. Thus, by the previous remark, $A$ is open.

**Interior, Closure, and Boundary.** If $E$ is a subset of $\mathbb{C}$, then $E$ contains a largest open subset, meaning an open subset of $E$ that contains all other open subsets of $E$. In fact, the union of all open subsets of $E$ is open, by Theorem 2.1.3, and is a subset of $E$ which contains all open subsets of $E$. Similarly, the intersection of all closed sets containing $E$ is the smallest closed set containing $E$. Thus, the following definition makes sense.

**Definition 2.1.5.** Let $E$ be a subset of $\mathbb{C}$. Then:

(a) the largest open subset of $E$ is called the *interior* of $E$ and is denoted $E^\circ$;

(b) the smallest closed set containing $E$ is called the *closure* of $E$ and is denoted $\overline{E}$;

(c) the set $\overline{E} \setminus E^\circ$ is called the boundary of $E$ and is denoted $\partial E$.

Recall that a neighborhood of a point $z_0 \in \mathbb{C}$ is any open set containing $z_0$. The proof of the following theorem is elementary and is left to the exercises.

**Theorem 2.1.6.** *Let $E$ be a subset of $\mathbb{C}$ and $z$ an element of $\mathbb{C}$. Then:*

(a) *$z \in E^\circ$ if and only if there is a neighborhood of $z$ that is contained in $E$;*

(b) *$z \in \overline{E}$ if and only if every neighborhood of $z$ contains a point of $E$;*

(c) *$z \in \partial E$ if and only if every neighborhood of $z$ contains points of $E$ and points of the complement of $E$.*

**Example 2.1.7.** Find the interior, closure and boundary for the set

$$E = \{z \in \mathbb{C} : |z| < 1,\ \mathrm{Im}(z) \geq 0\} \cup \{-iy : y \in [0,1]\}.$$

**Solution:** It is immediate from the previous theorem that

$$\begin{aligned} E^\circ &= \{z \in \mathbb{C} : |z| < 1,\ \mathrm{Im}(z) > 0\}, \\ \overline{E} &= \{z \in \mathbb{C} : |z| \leq 1,\ \mathrm{Im}(z) \geq 0\} \cup \{-iy : y \in [0,1]\}, \\ \partial E &= \{z \in \mathbb{C} : |z| = 1,\ \mathrm{Im}(z) \geq 0\} \cup [-1,1] \cup \{-iy : y \in [0,1]\}. \end{aligned}$$

**Limits.** We will be primarily concerned with complex-valued functions defined on open subsets of $\mathbb{C}$. However, we will also need to deal with functions whose domain is some other kind of subset of $\mathbb{C}$. For example, in the next section we will be dealing with curves in $\mathbb{C}$. A curve in $\mathbb{C}$ is a continuous complex-valued function whose domain is an interval on the real line. We will also deal eventually with functions defined on a closed disc or a circle or some other closed subset of $\mathbb{C}$. The definitions of limit and continuity we adopt here are general enough to handle all these situations. However, as a result, they are very much domain dependent. That is, the limit of a function at a point and whether or not the function is continuous at a point depend very much on which set is considered to be the domain of the function.

The definition of limit is the familiar one from calculus except for extra care about the domain of the function and a condition concerning isolated points, as defined below.

If $E$ is a subset of $\mathbb{C}$, then an *isolated point* of $E$ is an element $a \in E$ such that there is a neighborhood of $a$ which contains no other points of $E$.

**Definition 2.1.8.** If $f$ is a complex-valued function with domain $E$, and $a \in \overline{E}$ but is not an isolated point of $\overline{E}$, then we say $\lim_{z \to a} f(z) = L$ if, for every $\epsilon > 0$, there is a $\delta > 0$ such that $|f(z) - L| < \epsilon$ whenever $z \in E$ and $0 < |z - a| < \delta$. In this case, we say that $f$ has *limit* $L$ at $a$.

The condition "$z \in E$" in this definition means that whether or not the limit exists may depend on $E$. That is, given a function $f$ defined on a set $E$ and a subset $D$ of $E$, we can always consider a new function $f|_D$ which is $f$ restricted to the new domain $D$. It may be that the limit as $z \to a$ does not exist for $f$ as defined on its original domain $E$, but the limit of $f|_D$ does exist.

**Example 2.1.9.** Show that the function $f(z) = z/|z|$ with domain $D_1(0) \setminus \{0\}$ does not have a limit as $z \to 0$, but if this function is restricted to the domain $(0, 1) \subset \mathbb{R}$, then its limit as $z \to 0$ is 1.

**Solution:** If $f$ is considered as a function with domain $D_1(0) \setminus \{0\}$, then for each $z = r\,e^{i\theta}$ in this domain, $f(z) = e^{i\theta}$. Thus $f$ takes on every value of modulus one in every open disc $D_\delta(0)$ with $\delta < 1$. So there is certainly no one number that $f(z)$ is approaching as $z \to 0$. On the other hand, suppose $f$ is restricted to $(0, 1) \subset \mathbb{R}$. On this interval, $f$ is identically 1 and so its limit as $z \to 0$ is also 1.

A *deleted neighborhood* of $a \in \mathbb{C}$ is a set of the form $V \setminus \{a\}$ where $V$ is neighborhood of $a$. Using neighborhoods and deleted neighborhoods, the limit concept may be rephrased as follows.

**Theorem 2.1.10.** *If $f$ is a complex-valued function with domain $E$, and if $a \in \overline{E}$, then $\lim_{z \to a} f(z) = L$ if and only if for every neighborhood $W$ of $L$ there is a deleted neighborhood $V$ of $a$ such that*

$$f(E \cap V) \subset W. \tag{2.1.1}$$

**Proof.** If $\lim_{z \to a} f(z) = L$ and $W$ is a neighborhood of $L$, then, since $W$ is open and contains $L$, there is an $\epsilon > 0$ such that $D_\epsilon(L) \subset W$. By the definition of limit, there is a $\delta > 0$ such that $|f(z) - L| < \epsilon$ whenever $z \in E$ and $0 < |z - a| < \delta$. This is equivalent to the statement

$$f(E \cap D_\delta(a) \setminus \{a\}) \subset D_\epsilon(L) \subset W.$$

Thus, if we choose $V = D_\delta(a) \setminus \{a\}$, then $V$ is a deleted neighborhood of $a$ satisfying (2.1.1). This proves the "only if" part of the theorem.

To prove the "if" part, we assume that for every neighborhood $W$ of $L$ there is a deleted neighborhood $V$ of $a$ such that $f(E \cap V) \subset W$. Then, given $\epsilon > 0$, let $W = D_\epsilon(L)$ and let $V$ be a deleted neighborhood of $a$ satisfying (2.1.1). Since $V \cup \{a\}$ is open, there exists a $\delta > 0$ such that $D_\delta(a) \subset V \cup \{a\}$. Then

$$f(E \cap D_\delta(a) \setminus \{a\}) \subset f(E \cap V) \subset W = D_\epsilon(L).$$

This is equivalent to the statement that $|f(z) - w| < \epsilon$ whenever $z \in E$ and $0 < |z - a| < \delta$. Thus, by definition, $\lim_{z \to a} f(z) = L$. □

**Continuity.** The definition of a continuous function should also be familiar from calculus.

**Definition 2.1.11.** A function $f$ with domain $E$ is said to be continuous at $a \in E$ if

$$\lim_{z \to a} f(z) = f(a).$$

If $f$ is continuous at every point of $E$, then we say that $f$ is continous on $E$. The set of all functions with domain $E$, which are continuous on $E$, will be denoted $\mathcal{C}(E)$.

**Example 2.1.12.** Show that $|z|$ is a continuous function of $z$ on all of $\mathbb{C}$.

**Solution:** Let $z$ and $a$ be elements of $\mathbb{C}$. By the second form of the triangle inequality, we have

$$||z| - |a|| \leq |z - a|.$$

It follows from this that $\lim_{z \to a} |z| = |a|$ and, hence, that $|z|$ is continuous at $z$. In fact, given $\epsilon > 0$, it suffices to choose $\delta = \epsilon$ in the definition of limit, since $|z - a| < \epsilon$ implies $||z| - |a|| < \epsilon$.

The definition of continuous function is also very domain dependent. For example, if the function $f(z) = z/|z|$ of Example 2.1.9 is given the value 1 at 0, then it is continuous at 0 as a function with domain $[0, 1)$, but if we consider its domain to be $D_1(0)$, then it is not continuous.

The next result characterizes functions which are defined and continuous on an open set $U$ as those functions $f$ on $U$ for which $f^{-1}$ preserves open sets.

**Theorem 2.1.13.** *If $f$ is a complex-valued function defined on an open set $U \subset \mathbb{C}$, then $f$ is continuous on $U$ if and only if $f^{-1}(W)$ is open for every open set $W \subset \mathbb{C}$.*

**Proof.** Suppose $f$ is continous on $U$ and $W$ is an open subset of $\mathbb{C}$. If $a \in f^{-1}(W)$, then $\lim_{z \to a} f(z) = f(a) \in W$. Since $W$ is a neighborhood of $f(a)$, it follows from Theorem 2.1.10 that there is a neighborhood $V$ of $a$ such that $f(U \cap V) \subset W$ (actually, Theorem 2.1.10 guarantees a *deleted* neighborhood $V$ with this property, but since $f(a) \in W$, it does not change things to put $a$ back in the deleted neighborhood and claim there is an actual neighborhood $V$ as above). Thus, $U \cap V$ is an open subset of $f^{-1}(W)$ containing $a$. We have now proved that every element of $f^{-1}(W)$ is contained in an open set contained in $f^{-1}(W)$. The union of all such open sets is open and is equal to $f^{-1}(W)$. Thus, $f^{-1}(W)$ is open.

To prove the converse, we suppose that $f^{-1}(W)$ is open for every open set $W$. It follows that for each $a \in U$ and each neighborhood $W$ of $f(a)$, we have $f^{-1}(W)$ is a neighborhood of $a$. By Theorem 2.1.10, $\lim_{z \to a} f(z) = f(a)$ and so $f$ is continuous at $a$. Since $a$ was any point of $U$, $f$ is continuous on $U$. □

If $f$ and $g$ are functions with domain a set $E$ and they are both continuous at $a \in E$, then $f + g$ and $fg$ are continuous at $a$, and $f/g$ is continuous at $a$ provided $g(a) \neq 0$. The proofs of these facts for complex-valued functions of a complex variable are no different than the proofs, given in calculus, of the corresponding facts for functions of a real variable. Thus, we will accept them without further comment.

Obviously, constants are continuous everywhere, as is the function $z$ itself. It follows that polynomials in $z$ are also continuous everywhere.

If $g$ with domain $E$ is continuous at $a$ and $f$ with domain $D$ is continuous at $b = f(a)$, and if $g(E) \subset D$, then the composite function $f \circ g$, with domain $E$, is continuous at $a$. This is another fact whose proof for functions of a complex variable is no different than the proof from calculus of the analogous result for functions of a real variable.

**Example 2.1.14.** Prove that if $g$ is a continuous non-vanishing function on an open subset $U \subset \mathbb{C}$, then $\log |g|$ is continous on $U$.

**Solution:** The function $|g(z)|$ is continuous on $U$ because it is the composition of the function $g$, which is continuous on $U$, and the function $|\cdot|$, which is continous everywhere. Also, $|g(z)|$ has its values in the positive reals, since $g$ is non-vanishing on $U$. The function log is continuous on the positive reals and so the composition $\log |g|$ is continuous on $U$.

## Exercise Set 2.1

1. Prove the second statement of Theorem 2.1.3. That is, prove that the intersection of an arbitrary collection of closed sets is closed, while the union of any finite collection of closed sets is closed.
2. Prove that if $U$ is open and $K$ is closed, then the set-theoretic difference $U \setminus K$ is open, while $K \setminus U$ is closed.
3. Prove Theorem 2.1.6.
4. Show that the set $A = \{z \in \mathbb{C} : \text{Re}(z) > 0\}$ is open. Hint: You must show that each point $w \in A$ is the center of an open disc which is entirely contained in $A$.
5. Tell which of the following sets are open subsets of $\mathbb{C}$, which are closed, and which are neither (no proof required):
   (a) $\{z \in \mathbb{C} : 1 < |z| < 2\}$;
   (b) $\{z \in \mathbb{C} : \text{Im}(z) = 0, 0 < \text{Re}(z) < 1\}$;
   (c) $\{z \in \mathbb{C} : -1 \le \text{Re}(z) \le 1\}$, $-1 \le \text{Im}(z) \le 1\}$.

6. Find the interior, closure, and boundary for the set $\{z \in \mathbb{C} : 1 \leq |z| < 2\}$ (no proof required).
7. Prove that $w \in \mathbb{C}$ is in the closure of a set $E \subset \mathbb{C}$ if and only if there is a sequence $\{z_n\} \subset E$ such that $\lim z_n = w$. Thus, a set $E$ is closed if and only if it contains all limits of convergent sequences of points in $E$.
8. Does $\lim_{z \to 0} f(z)$ exist if $f(z) = \dfrac{|z - \overline{z}|}{|z|}$ with domain $\mathbb{C} \setminus \{0\}$? How about if the domain is restricted to be just $\mathbb{R} \setminus \{0\}$?
9. Prove that $\mathrm{Re}(z)$, $\mathrm{Im}(z)$, and $\overline{z}$ are continuous functions of $z$.
10. At which points of $\mathbb{C}$ is the function $(1 - z^4)^{-1}$ continuous.
11. Prove that $\arg_I$ is continuous except on its cut line.
12. Use the result of the preceding exercise to prove that a branch of the log function is continuous except on its cut line.
13. Use Theorem 2.1.13 to prove that if $f$ and $g$ are continuous functions with open domains $U_f$ and $U_g$ and if $g(U_g) \subset U_f$, then $f \circ g$ is continuous on $U_g$.
14. Prove that if $f$ is a continuous function defined on an open subset $U$ of $\mathbb{C}$, then sets of the form $\{z \in U : |f(z)| < r\}$ and $\{z \in U : \mathrm{Re}(f(z)) < r\}$ are open.
15. Use the result of the preceding exercise to come up with an open subset of $\mathbb{C}$ that has not been previously described in this text.
16. Prove that a function $f$ with open domain $U$ is continuous at a point $a \in U$ if and only if whenever $\{z_n\} \subset U$ is a sequence converging to $a$, the sequence $\{f(z_n)\}$ converges to $f(a)$.

## 2.2. The Complex Derivative

There is nothing surprising about the definition of the derivative of a function of a complex variable – it looks just like the definition of the derivative of a function of a real variable. What is surprising are the consequences of a function having a derivative in this sense.

**Definition 2.2.1.** Let $f$ be a function defined on a neighborhood of $z \in \mathbb{C}$. If

$$\lim_{w \to z} \frac{f(w) - f(z)}{w - z}$$

exists, then we denote it by $f'(z)$ and we say $f$ is differentiable at $z$ with complex derivative $f'(z)$. If $f$ is defined and differentiable at every point of an open set $U$, then we say that $f$ is *analytic* on $U$.

**Remark 2.2.2.** When convenient, we will make the change of variables $\lambda = w - z$ and write the derivative in the form

$$f'(z) = \lim_{\lambda \to 0} \frac{f(z + \lambda) - f(z)}{\lambda}. \tag{2.2.1}$$

Clearly constant functions are differentiable and have complex derivative 0, since the difference quotient in Definition 2.2.1 is identically 0 for such a function.

The first hint that there is something fundamentally different about this notion of derivative is in the following example.

**Example 2.2.3.** Show that the function $f(z) = z$ is differentiable everywhere on $\mathbb{C}$ with derivative 1 and, hence, is analytic on $\mathbb{C}$, but the function $f(z) = \overline{z}$ is differentiable nowhere.

**Solution:** For $f(z) = z$, the difference quotient in (2.2.1) is

$$\frac{\lambda}{\lambda} = 1,$$

which clearly has limit 1 as $\lambda \to 0$ for every $z$. On the other hand, if $f(z) = \overline{z}$, then the difference quotient is

$$\frac{\overline{\lambda}}{\lambda} = \mathrm{e}^{-2i\theta},$$

if $\lambda = r\,\mathrm{e}^{i\theta}$ in polar form. The limit of this function as $\lambda \to 0$ clearly does not exist, since it has a different fixed value along each ray emanating from 0. This is true no matter what $z$ is, and so $\overline{z}$ is nowhere differentiable.

What makes this example so surprising, at first, is that, as a function of the two real variables $x$ and $y$, $\overline{z} = x - iy$ is of class $\mathcal{C}^\infty$ – meaning that its partial derivatives of all orders exist and are continuous – and yet, its complex derivative does not exist. Thus, existence of the complex derivative involves more than just smoothness of the function.

We will soon prove that a function which has a power series expansion that converges on an open disc is analytic on that disc. This would imply that the exponential function, for example, is analytic on all of $\mathbb{C}$. We do not have to wait, however, to prove this fact. There is an elementary proof that $\mathrm{e}^z$ is analytic on $\mathbb{C}$.

**Example 2.2.4.** Prove that $\mathrm{e}^z$ is an analytic function of $z$ on the entire complex plane and show that it is its own derivative.

**Solution:** Given an arbitrary point $z \in \mathbb{C}$, we will show that $\mathrm{e}^z$ has derivative $\mathrm{e}^z$ at $z$. By the law of exponents

$$\frac{\mathrm{e}^{z+\lambda} - \mathrm{e}^z}{\lambda} = \mathrm{e}^z \, \frac{\mathrm{e}^\lambda - 1}{\lambda}.$$

Thus, to show that the derivative of $\mathrm{e}^z$ is $\mathrm{e}^z$ we need only show that

$$\lim_{\lambda \to 0} \frac{\mathrm{e}^\lambda - 1}{\lambda} = 1. \tag{2.2.2}$$

However, if $t = |\lambda|$, inspection of the power series for $\mathrm{e}^\lambda$ and $\mathrm{e}^t$ shows that

$$\left| \frac{\mathrm{e}^\lambda - 1}{\lambda} - 1 \right| = \left| \frac{\mathrm{e}^\lambda - 1 - \lambda}{\lambda} \right| \le \frac{\mathrm{e}^t - 1 - t}{t}. \tag{2.2.3}$$

Now to show that the expression on the left has limit zero and, thus, verify (2.2.2), we simply apply L'Hôpital's rule to the expression on the right.

**Elementary Properties of the Derivative.** A simple result about derivatives of functions of a real variable that also holds in the context of complex derivatives is the following. The proof is elementary and is left to the exercises.

**Theorem 2.2.5.** *If the complex derivative $f'$ of $f$ exists at $a \in \mathbb{C}$, then $f$ is continuous at $a$.*

The complex derivative has all of the familiar properties in relation to sums, products, and quotients of functions. The proofs of these are in no way different from the proofs of the corresponding results for functions of a real variable. In the following theorem, Part (a) is trivial and we leave Parts (b) and (c) to the exercises.

**Theorem 2.2.6.** *If $f$ and $g$ are functions of a complex variable which are differentiable at $z \in \mathbb{C}$, then*

(a) *$f+g$ is differentiable at $z$ and $(f+g)'(z) = f'(z) + g'(z)$;*

(b) *$fg$ is differentiable at $z$ and $(fg)'(z) = f'(z)g(z) + f(z)g'(z)$;*

(c) *if $g(z) \neq 0$, $1/g$ is differentiable at $z$ and $(1/g)'(z) = -g'(z)/g^2(z)$.*

Parts (a) and (b) of this theorem and the fact that constant functions and the function $z$ are analytic on $\mathbb{C}$ imply that every polynomial in $z$ is analytic on $\mathbb{C}$. Of course, since $\overline{z}$ is not analytic, we cannot expect mixed polynomials that contain powers of both $z$ and $\overline{z}$ to be analytic.

Parts (b) and (c) of the theorem imply that $f/g$ is differentiable at $z$ if $f$ and $g$ are and if $g(z) \neq 0$. They also imply the quotient rule

$$\left(\frac{f}{g}\right)'(z) = \frac{f'(z)g(z) - g'(z)f(z)}{g^2(z)}.$$

The chain rule also holds for the complex derivative.

**Theorem 2.2.7.** *If $g$ is differentiable at $a$ and $f$ is differentiable at $b = g(a)$, then $f \circ g$ is differentiable at $a$ and*

$$(f \circ g)'(a) = f'(g(a))g'(a).$$

**Proof.** Let $U$ be a neighborhood of $b$ on which $f$ is defined. We define a function $h(w)$ on $U$ in the following way

$$h(w) = \begin{cases} \dfrac{f(w) - f(b)}{w - b}, & \text{if } w \neq b; \\ f'(b), & \text{if } w = b. \end{cases}$$

Then $h$ is continuous at $b$, since

$$f'(b) = \lim_{w \to b} \frac{f(w) - f(b)}{w - b}.$$

Also,

$$\frac{f \circ g(z) - f \circ g(a)}{z - a} = h(g(z))\frac{g(z) - g(a)}{z - a} \tag{2.2.4}$$

for all $z$ in the deleted neighborhood $V = g^{-1}(U) \setminus \{a\}$ of $a$. If we take the limit of both sides of (2.2.4) and use the fact that $f$ and $h$ are continuous at $b$ and $g$ is continuous at $a$, we conclude that $(f \circ g)'(a) = f'(g(a))g'(a)$, as required. □

**Example 2.2.8.** Suppose $p(z)$ is a polynomial in $z$. Where is the function $e^{p(z)}$ analytic and what is its derivative?

**Solution:** Since $e^z$ and $p(z)$ both are differentiable everywhere, so is the composition $e^{p(z)}$, by Theorem 2.2.7, and the derivative is

$$\left(e^{p(z)}\right)' = p'(z)\, e^{p(z)}\,.$$

**The Cauchy-Riemann Equations.** Since a function $f$ of a complex variable may be regarded as a complex-valued function on a subset of $\mathbb{R}^2$, we can write it in the form

$$f(x+iy) = u(x,y) + iv(x,y), \tag{2.2.5}$$

where $u$ and $v$ are the real and imaginary parts of $f$, regarded as functions defined on a subset of $\mathbb{R}^2$. It is natural to ask what the existence of a complex derivative for $f$ implies about the functions $u$ and $v$ as functions of the two real variables $x$ and $y$. It is easy to see that it implies the existence of the partial derivatives $u_x$, $u_y$, $v_x$ and $v_y$. In fact, it implies much more as the following discussion will show.

Recall that a function $g$ of two real variables is said to be *differentiable* at $(x,y)$ if there are numbers $A$ and $B$ such that

$$g(x+h, y+k) - g(x,y) = Ah + Bk + \epsilon(h,k),$$

where $\epsilon(h,k)/|(h,k)| \to 0$ as $(h,k) \to (0,0)$. If $g$ is differentiable at $(x,y)$, then the numbers $A$ and $B$ are the partial derivatives $g_x$ and $g_y$ at $(x,y)$.

Suppose $f$ is a complex-valued function defined in a neighborhood of $z \in \mathbb{C}$. If $M = f'(z)$ exists, then we may write

$$f(z+\lambda) - f(z) = M\lambda + \epsilon(\lambda), \tag{2.2.6}$$

where $\epsilon(\lambda)/\lambda \to 0$ as $\lambda \to 0$. In fact, $\epsilon(\lambda)$ is given by

$$\epsilon(\lambda) = f(z+\lambda) - f(z) - M\lambda,$$

and so, the fact that $\epsilon(\lambda)/\lambda \to 0$ as $\lambda \to 0$ is equivalent to the statement that $f'(z)$ exists and is equal to $M$.

If we write $f, M, z, \lambda$, and $\epsilon$ in terms of their real and imaginary parts: $f = u + iv, M = C + iD, z = x + iy, \lambda = h + ik$, and $\epsilon = \rho + i\omega$, then (2.2.6) becomes

$$\begin{aligned} u(x+h, y+k) + iv(x+h, y+k) - u(x,y) - iv(x,y) \\ = (C+iD)(h+ik) + \rho(h,k) + i\omega(h,k). \end{aligned} \tag{2.2.7}$$

On equating real and imaginary parts, this leads to the two equations

$$\begin{aligned} u(x+h, y+k) - u(x,y) &= Ch - Dk + \rho(h,k), \\ v(x+h, y+k) - v(x,y) &= Dh + Ck + \omega(h,k). \end{aligned} \tag{2.2.8}$$

The condition that $\epsilon(\lambda)/\lambda \to 0$ as $\lambda \to 0$ implies that $\rho(h,k)/|(h,k)| \to 0$ and $\omega(h,k)/|(h,k)| \to 0$ (note that $|(h,k)| = \sqrt{h^2+k^2} = |\lambda|$). Thus, we can draw two conclusions from the existence of $f'(z)$: (1) $u$ and $v$ are differentiable at $(x,y)$, and (2) the partial derivatives of $u$ and $v$ at $(x,y)$ are given by

$$\begin{aligned} u_x(x,y) &= C, \quad u_y(x,y) = -D, \\ v_x(x,y) &= D, \quad v_y(x,y) = C. \end{aligned} \tag{2.2.9}$$

A surprising consequence of this is that if $f'$ exists at $z = x + iy$, then

$$\begin{aligned} u_x &= v_y, \\ u_y &= -v_x \end{aligned} \tag{2.2.10}$$

at $(x, y)$. Equations (2.2.10) are the *Cauchy-Riemann equations*. Equations (2.2.9) also show that if $f'$ exists at $z$, then $f'(z) = C + iD = u_x + iv_x = -i(u_y + iv_y)$. If we set $f_x = u_x + iv_x$ and $f_y = u_y + iv_y$, then this can be written as $f' = f_x = -if_y$ wherever $f'$ exists.

The above discussion shows that, at any point where $f$ has a complex derivative, its real and imaginary parts are differentiable functions and satisfy the Cauchy-Riemann equations. The converse is also true: If the real and imaginary parts of $f$ are differentiable and satisfy the Cauchy-Riemann equations at a point $z = x + iy$, then $f'(z)$ exists. The proof of this is a matter of working backwards through the above discussion, beginning with the assumption that $u$ and $v$ are differentiable at $(x, y)$, with partial derivatives that satisfy $u_x = v_y = C$ and $u_y = -v_x = -D$. This leads to (2.2.8), which eventually leads back to the conclusion that $C + iD$ is the derivative of $f$ at $z = x + iy$. We leave the details to the exercises. The result is the following theorem.

**Theorem 2.2.9.** *If $f = u + iv$ is a complex-valued function defined in a neighborhood of $z \in \mathbb{C}$, with real and imaginary parts $u$ and $v$, then $f$ has a complex derivative at $z$ if and only if $u$ and $v$ are differentiable and satisfy the Cauchy-Riemann equations* (2.2.10) *at $z = x + iy$. In this case,*

$$f' = f_x = -if_y.$$

**Example 2.2.10.** We already know that $e^z$ is analytic everywhere. However, give a different proof of this by showing $e^z$ satisfies the Cauchy-Riemann equations.

**Solution:** With $z = x + iy$, we write $e^z = e^x(\cos y + i \sin y)$. The real and imaginary parts of $e^z$ are $u(x, y) = e^x \cos y$ and $v(x, y) = e^x \sin y$. Thus,

$$\begin{aligned} u_x(x, y) &= e^x \cos y = v_y, \quad \text{and} \\ u_y(x, y) &= -e^x \sin y = -v_x. \end{aligned}$$

**Example 2.2.11.** Use the Cauchy-Riemann equations to prove that, for each branch of the log function, $\log(z)$ is analytic everywhere except on its cut line and has derivative $1/z$.

**Solution:** We first prove that the principal branch of the log function is analytic on the right half-plane $H = \{z \in \mathbb{C} : \mathrm{Re}(z) > 0\}$. For $z \in H$ we have $z = x + iy = r\,e^{i\theta}$ where

$$r = \sqrt{x^2 + y^2} \quad \text{and} \quad \theta = \tan^{-1}(y/x).$$

Thus, the principal branch of log on $H$ is

$$\log(x + iy) = (1/2)\ln(x^2 + y^2) + i \tan^{-1}(y/x).$$

Taking partial derivatives yields

$$
\begin{aligned}
\frac{\partial}{\partial x}(1/2)\ln(x^2+y^2) &= \frac{x}{x^2+y^2},\\
\frac{\partial}{\partial x}\tan^{-1}(y/x) &= \frac{-y/x^2}{1+(y/x)^2} = \frac{-y}{x^2+y^2},\\
\frac{\partial}{\partial y}(1/2)\ln(x^2+y^2) &= \frac{y}{x^2+y^2},\\
\frac{\partial}{\partial y}\tan^{-1}(y/x) &= \frac{1/x}{1+(y/x)^2} = \frac{x}{x^2+y^2}.
\end{aligned} \tag{2.2.11}
$$

Thus, the Cauchy-Riemann equations are satisfied by the principal branch of the log function on $H$. Furthermore

$$(\log z)' = \frac{\partial}{\partial x}\log(x+iy) = \frac{x-iy}{x^2+y^2} = \frac{1}{z}.$$

Now if $z$ is any point not on the negative real axis and not in $H$, then we simply rotate $z$ into $H$. That is, we choose $\alpha = \pm\pi/2$ such that $e^{i\alpha}z \in H$. Then

$$\log z = \log(e^{i\alpha} z) - i\alpha.$$

Since log has derivative $1/w$ at $w = e^{i\alpha} z$, it follows from the chain rule that log has derivative $e^{i\alpha}/(e^{i\alpha} z) = 1/z$ at $z$. Thus, the principal branch of the log function is analytic with derivative $1/z$ at any point $z$ not on its cut line.

The analogous statement for other branches of the log function also follows from a rotation argument, as above. That is, each such function is just the principal branch of the log function composed with a rotation.

**Harmonic Functions.** In the next chapter, we will prove that analytic functions are $\mathcal{C}^\infty$– that is, they have continuous complex derivatives of all orders. This, in particular, implies that analytic functions have continuous partial derivatives of all orders with respect to $x$ and $y$. Assuming this result for the moment, we have

**Theorem 2.2.12.** *The real and imaginary parts of an analytic function on $U$ are harmonic functions on $U$, meaning they satisfy Laplace's equation*

$$u_{xx} + u_{yy} = 0.$$

**Proof.** If $f = u + iv$ is an analytic function, then $u$ and $v$ satisfy the Cauchy-Riemann equations and so

$$u_{xx} = (u_x)_x = (v_y)_x = (v_x)_y = (-u_y)_y = -u_{yy}.$$

This shows that the real part of $f$ satisfies Laplace's equation. Since $v$ is the real part of the analytic function $-if$, it follows that $v$ is also harmonic. Thus, both real and imaginary parts of an analytic function are harmonic. □

If $u$ and $v$ are harmonic functions such that the function $f = u+iv$ is analytic, then we say $u$ and $v$ are *harmonic conjugates* of one another.

**Example 2.2.13.** Prove that $u(x,y) = e^x \cos y$ is a harmonic function on all of $\mathbb{R}^2$ and find a harmonic conjugate for it.

**Solution:** The function $u$ is the real part of $f(z) = e^z$ and is, therefore, harmonic by the previous theorem. The imaginary part of $f$ is $v(x, y) = e^x \sin y$, and so this function $v$ is a harmonic conjugate of $u$.

## Exercise Set 2.2

1. Fill in the details in Example 2.2.4 by verifying the inequality (2.2.3) and showing that the limit of the expression on the right is 0.
2. Prove Theorem 2.2.5.
3. Prove Part (b) of Theorem 2.2.6.
4. Prove Part (c) of Theorem 2.2.6.
5. Use induction and Theorem 2.2.6 to show that $(z^n)' = nz^{n-1}$ if $n$ is a non-negative integer.
6. Find the derivative of $z^7 + 5z^4 - 2z^3 + z^2 - 1$. Which results from this section are used in this calculation?
7. Find the derivative of $e^{z^3}$.
8. If we use the principal branch of the log function, at which points of $\mathbb{C}$ does $\dfrac{\log z}{z}$ have a complex derivative? What is its derivative at these points?
9. Finish the proof of Theorem 2.2.9 by showing that if $f = u + iv$, $u$ and $v$ are differentiable at $z$, and $u$ and $v$ satisfy the Cauchy-Riemann equations at $z$, then $f'(z)$ exists.
10. Use the Cauchy-Riemann equations to verify that the function $f(z) = z^2$ is analytic everywhere.
11. Describe all real-valued functions which are analytic on $\mathbb{C}$.
12. Derive the Cauchy-Riemann equations in polar coordinates:
$$u_r = r^{-1} v_\theta,$$
$$u_\theta = -r v_r$$
by using the change of variable formulas $x = r\cos\theta$, $y = r\sin\theta$ and the chain rule.
13. We showed in Example 2.2.11 that each branch of the log function is analytic on the complex plane with its cut line removed. Use the Cauchy-Riemann equations in polar form (previous problem) to give another proof of this fact.
14. Assuming each branch of the log function is analytic, use the chain rule to give another prove that each such function has derivative $1/z$.
15. Use the Cauchy-Riemann equations to prove that if $f$ is analytic on an open set $U$, then the function $g$ defined by $g(z) = \overline{f(\overline{z})}$ is analytic on the set $\{\overline{z} : z \in U\}$.
16. Verify that the function $\log|z|$ is harmonic on $\mathbb{C} \setminus \{0\}$ and find a harmonic conjugate for it on the set consisting of $\mathbb{C}$ with the non-positive real axis removed.

## 2.3. Contour Integrals

Integration plays a key role in this subject – specifically, integration along curves in $\mathbb{C}$. A *curve* or *contour* in the plane $\mathbb{C}$ is a continuous function $\gamma$ from an interval on the line into $\mathbb{C}$. Such an object is sometimes called a *parameterized curve* and the interval $I$ is called the parameter interval. We will be interested in a particular kind of curve, one whose parameter interval is a closed bounded interval which can be subdivided into finitely many subintervals, on each of which $\gamma$ is continuously differentiable.

**Smooth Curves.** Let $I = [a, b]$ be a closed interval on the real line and let $\gamma : I \to \mathbb{C}$ be a complex-valued function on $I$. If $c \in I$, then the derivative $\gamma'(c)$ of $\gamma$ at $c$ is defined in the usual way:

$$\gamma'(c) = \lim_{t \to c} \frac{\gamma(t) - \gamma(c)}{t - c}. \tag{2.3.1}$$

Of course, $\gamma$ is complex-valued and so this limit should be interpreted as the type of limit discussed in Section 2.1. It can be calculated by expressing $\gamma$ in terms of its real and imaginary parts, that is, by writing $\gamma(t) = x(t) + iy(t)$, where $x(t)$ and $y(t)$ are real-valued functions on $I$. Then $\gamma'(t) = x'(t) + iy'(t)$ (Exercise 2.3.6).

What about the endpoints $a$ and $b$ of the interval $I$? Should we either not talk about the derivative at the endpoints or, perhaps, use one-sided derivatives defined in terms of one-sided limits (limit from the right at $a$ and limit from the left at $b$)? Actually, there is no need to do anything special at $a$ and $b$ or to exclude them. If the domain of $\gamma$ is $[a, b]$, then our domain dependent definition of limit takes care of the problem. If $c = a$, the limit as $t \to a$ in (2.3.1) only involves values of $t$ to the right of $a$, since only those are in the domain of the difference quotient that appears in this limit. Similarly, if $c = b$, the limit as $t \to b$ involves only points to the left of $b$. Thus, the derivatives at $a$ and $b$ that our definition leads to are what in calculus would be called the right derivative at $a$ and the left derivative at $b$.

The curve $\gamma$ is *differentiable* at $c$ if the limit defining $\gamma'(c)$ exists. It is *continuously differentiable* or *smooth* on $I$ if it is differentiable at every point of $I$ and if the derivative is a continuous function on $I$. In this case we will write $\gamma \in \mathcal{C}^1(I)$.

**Definition 2.3.1.** A curve $\gamma : [a, b] \to \mathbb{C}$ in $\mathbb{C}$ is called *piecewise smooth* if there is a partition $a = a_0 < a_1 < \cdots < a_n = b$ of $[a, b]$ such that the restriction of $\gamma$ to $[a_{j-1}, a_j]$ is smooth for each $j = 1, \cdots, n$. A curve which is piecewise smooth will be called a *path*.

With appropriate choices of parameterization, familiar geometric objects in $\mathbb{C}$ can be described as the image of a path.

**Example 2.3.2.** Find a path $\gamma$ that traces once around the circle of radius $r$, centered at 0, in the counterclockwise direction. Describe $\gamma'$.

**Solution:** The smooth path $\gamma(t) = r\,e^{it}$, $t \in [0, 2\pi]$ does the job. Its derivative may be obtained by writing it as $r(\cos t + i \sin t)$ and differentiating the real and imaginary parts. The result is $\gamma'(t) = r(-\sin t + i \cos t) = ir\,e^{it}$.

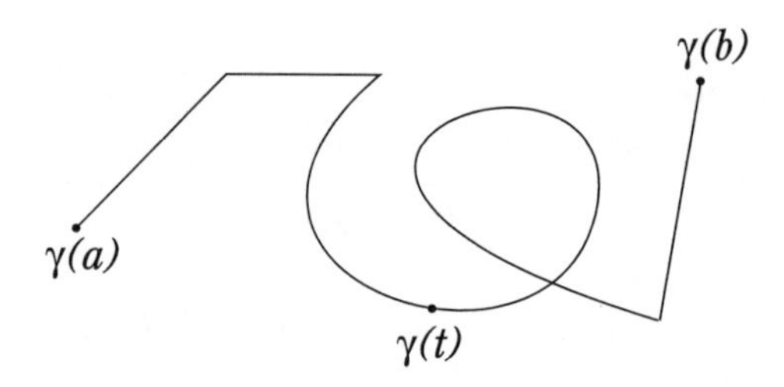


**Figure 2.3.1.** A Path in the Plane.

**Example 2.3.3.** Let $z$ and $w$ be two points in $\mathbb{C}$. Find a path which traces the straight line from $z$ to $w$ and find its derivative.

**Solution:** The path $\gamma$, with parameter interval $[0,1]$, defined by

$$\gamma(t) = (1-t)z + tw = z + t(w-z),$$

satisfies $\gamma(0) = z$ and $\gamma(1) = w$. It is a parametric form of a straight line in the plane, and its derivative is $\gamma'(t) = w - z$.

**Example 2.3.4.** Find a path that traces once around the square with vertices $0, 1, 1+i, i$ in the counterclockwise direction. Find $\gamma'(t)$ on the subintervals where $\gamma$ is smooth.

**Solution:** We choose $[0,1]$ as the parameter interval and define a path $\gamma$ as follows (see Figure 2.3.2):

$$\gamma(t) = \begin{cases} 4t, & \text{if } 0 \le t \le 1/4; \\ 1 + (4t-1)i, & \text{if } 1/4 \le t \le 1/2; \\ 3 - 4t + i, & \text{if } 1/2 \le t \le 3/4; \\ (4-4t)i, & \text{if } 3/4 \le t \le 1. \end{cases}$$

This is continuous on $[0,1]$ and smooth on each subinterval in the partition $0 < 1/4 < 1/2 < 3/4 < 1$. It traces each side of the square in succession, moving in the counterclockwise direction. On the first interval, $\gamma'$ is the constant 4, on the second it is $4i$, on the third it is $-4$, and on the fourth it is $-4i$.

**Riemann Integral of Complex-Valued Functions.** The integral of a function along a path will be defined in terms of the Riemann integral on an interval. This is the familiar Riemann integral from calculus, except that the functions being integrated will be complex-valued. This difference requires a few comments.

If $f(t) = g(t) + ih(t)$ is a complex-valued function on an interval $[a,b]$, where $g$ and $h$ are real-valued, then we will say that $f$ is *Riemann integrable* on $[a,b]$ if both $g$ and $h$ are Riemann integrable on $[a,b]$ as real-valued functions. We then define the integral of $f$ on $[a,b]$ by

$$\int_a^b f(t)\,dt = \int_a^b g(t)\,dt + i\int_a^b h(t)\,dt. \tag{2.3.2}$$

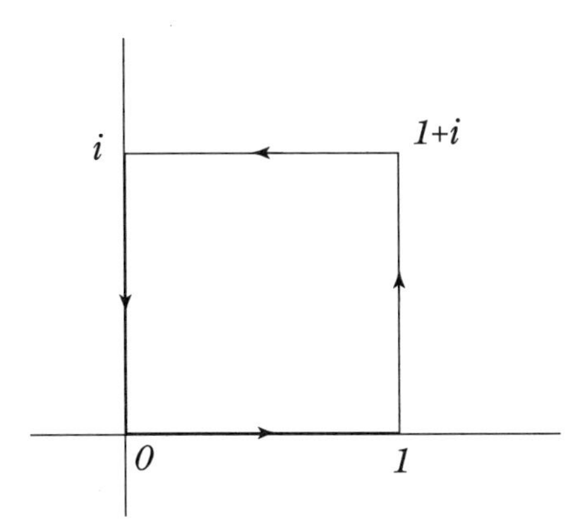


**Figure 2.3.2.** The Path of Example 2.3.4.

This Riemann integral for complex-valued functions has the properties one would expect given knowledge of the Riemann integral for real-valued functions. The next three theorems cover some of these properties.

**Theorem 2.3.5.** *Let $f_1$ and $f_2$ be Riemann integrable functions on $[a,b]$ and $\alpha$ and $\beta$ complex numbers. Then, $\alpha f_1 + \beta f_2$ is integrable on $[a,b]$, and*

$$\int_a^b (\alpha f_1(t) + \beta f_2(t))\, dt = \alpha \int_a^b f_1(t)\, dt + \beta \int_a^b f_2(t)\, dt.$$

**Proof.** That this is true if the constants $\alpha$ and $\beta$ are real follows directly from expressing $f_1$ and $f_2$ in terms of their real and imaginary parts. Thus, to prove the theorem we just need to show that $\int_a^b if(t)\, dt = i\int_a^b f(t)\, dt$ if $f = g + ih$ is an integrable function on $[a,b]$. However,

$$\begin{aligned}\int_a^b i(g(t) + ih(t))\, dt &= \int_a^b (-h(t) + ig(t))\, dt \\ &= -\int_a^b h(t)\, dt + i\int_a^b g(t)\, dt = i\left(\int_a^b (g(t) + ih(t))\, dt\right).\end{aligned}$$

This completes the proof. □

**Theorem 2.3.6.** *If $f$ is a function defined on $[a,b]$ and $c \in (a,b)$, then $f$ is integrable on $[a,b]$ if and only if it is integrable on $[a,c]$ and $[c,b]$. In this case*

$$\int_a^b f(t)\, dt = \int_a^c f(t)\, dt + \int_c^b f(t)\, dt.$$

**Proof.** This follows from the fact that the same things are true of the integrals of the real and imaginary parts $g$ and $h$ of $f$. □

**Theorem 2.3.7.** *If $f$ is an integrable function on $[a,b]$, then*

$$\left|\int_a^b f(t)\, dt\right| \le \int_a^b |f(t)|\, dt.$$

**Proof.** This is proved using a trick. We set $w = \int_a^b f(t)\,dt$. If $w = 0$, there is nothing to prove. If $w \neq 0$, let $u = \overline{w}/|w|$. Then $uw = |w|$ and so

$$\left|\int_a^b f(t)\,dt\right| = u\int_a^b f(t)\,dt = \int_a^b uf(t)\,dt.$$

Since this is a real number, the integral of the imaginary part of $uf$ is zero and we have

$$\left|\int_a^b f(t)\,dt\right| = \int_a^b \operatorname{Re}(uf(t))\,dt \leq \int_a^b |uf(t)|\,dt = \int_a^b |f(t)|\,dt. \qquad \square$$

A complex-valued function which is defined and continuous on an interval $[a, b]$ is clearly Riemann integrable on $[a, b]$, since its real and imaginary parts are continuous, and continuous real-valued functions on closed, bounded intervals are Riemann integrable.

**Integration Along a Path.** If $\gamma$ is a path, then $\gamma'$ exists and is continuous on each interval $[a_{i-1}, a_i]$ in a partition $a = a_0 < a_1 < \cdots < a_n = b$ of the parameter interval $[a, b]$. At the points $a_1, a_2, \cdots, a_{n-1}$ the definition of $\gamma'$ is ambiguous – $\gamma'(a_j)$ has one value from the derivative of $\gamma$ on $[a_{j-1}, a_j]$ and another from the derivative of $\gamma$ on $[a_j, a_{j+1}]$. In order to remove this ambiguity, we choose to define $\gamma'$ so as to be left continuous at these points. That is, at $a_j$, we choose the value for $\gamma'$ that comes from its definition on $[a_{j-1}, a_j]$. Then $\gamma'$ is well defined on $I = [a, b]$.

If $f$ is a complex-valued function defined and continuous on a set $E$ containing $\gamma(I)$, then the function $f(\gamma(t))\gamma'(t)$ is a well-defined function on $I$ which is piecewise continuous in the following sense: It is continuous everywhere on $[a, b]$ except at the partition points $a_1, a_2, \cdots, a_{n-1}$. It is left continuous at these points, and the limit from the right exists and is finite at these points as well. In other words, this function is continuous from the left everywhere on $[a, b]$ and continuous except at finitely many points where it has simple jump discontinuities.

A function of this type is Riemann integrable on $[a, b]$. To see this, first observe that it is Riemann integrable on each subinterval $[a_{j-1}, a_j]$ because, on such an interval, the function agrees with a continuous function except at one point, $a_{j-1}$. A continuous function on a closed interval is Riemann integrable and changing its value at one point does not effect this fact or the value of the integral. Furthermore, by Theorem 2.3.6, if a function is Riemann integrable on two contiguous intervals, then it is integrable on their union. It follows that a function which is integrable on each subinterval in a partition of $[a, b]$ will be integrable on $[a, b]$.

The above discussion settles the question of the Riemann integrability of the integrand in the following definition.

**Definition 2.3.8.** Let $\gamma : [a, b] \to \mathbb{C}$ be a path and let $f$ be a function which is defined and continuous on a set $E$ which contains $\gamma([a, b])$. Then we define the integral of $f$ over $\gamma$ to be

$$\int_\gamma f(z)\,dz = \int_a^b f(\gamma(t))\gamma'(t)\,dt. \tag{2.3.3}$$

One may think of this definition in the following way: the contour integral on the left in (2.3.3) is defined to be the Riemann integral obtained by replacing $z$ by $\gamma(t)$ and $dz$ by $\gamma'(t)dt$ and integrating over the parameter interval for $\gamma$.

In practice, we will calculate contour integrals by breaking the path up into its smooth sections, calculating the integrals over these sections and then adding the results. That this is legitimate follows from the fact that the Riemann integral of a function over the union of two contiguous intervals on the line is the sum of the integrals over the two intervals.

**Examples.**

**Example 2.3.9.** Find $\int_\gamma z\,dz$ if $\gamma$ is the circular path defined in Example 2.3.2.

**Solution:** By Example 2.3.2, we have $\gamma(t) = r\,e^{it}$ for $0 \le t \le 2\pi$ and $\gamma'(t) = ir\,e^{it}$. Thus,

$$\int_\gamma z\,dz = \int_0^{2\pi} r\,e^{it}\,ir\,e^{it}\,dt = ir^2\int e^{2it}\,dt = ir^2\int_0^{2\pi}(\cos 2t + i\sin 2t)\,dt$$
$$= ir^2\int_0^{2\pi}\cos 2t\,dt - r^2\int_0^{2\pi}\sin 2t\,dt = 0.$$

**Example 2.3.10.** Find a path $\gamma$ which traces the straight line from 0 to $i$ followed by the straight line from $i$ to $i+1$. Then calculate $\int_\gamma z^2\,dz$ for this path $\gamma$.

**Solution:** We may choose $\gamma$ to be the path parameterized on $[0,2]$ as follows:

$$\gamma(t) = \begin{cases} it, & \text{if } 0 \le t \le 1; \\ i+t-1, & \text{if } 1 \le t \le 2. \end{cases}$$

We calculate the integrals over each of the two smooth sections of the path. On $[0,1]$ we have $(\gamma(t))^2 = -t^2$ and $\gamma'(t) = i$. Thus, the integral over the first section of the path is

$$\int_0^1 (\gamma(t))^2\gamma'(t)\,dt = \int_0^1 -t^2 i\,dt = -t^3 i/3\Big|_0^1 = -i/3.$$

On $[1,2]$ we have $(\gamma(t))^2 = t^2 - 2t + 2(t-1)i$ and $\gamma'(t) = 1$. Thus, the integral over the second section of the path is

$$\int_1^2 (\gamma(t))^2\gamma'(t)\,dt = \int_0^1 (t^2 - 2t + 2(t-1)i)\,dt = (t^3/3 - t^2 + (t^2 - 2t)i)\Big|_1^2 = -2/3 + i.$$

Thus, $\int_\gamma z^2\,dz = -i/3 - 2/3 + i = -2/3 + 2i/3$.

**Example 2.3.11.** Find a path $\gamma$ which traces once around the triangle with vertices $0, 1, i$ in the counterclockwise direction, starting at 0. For this path $\gamma$, find $\int_\gamma \overline{z}\,dz$.

**Solution:** A path $\gamma$ with the required properties has parameter interval $[0,3]$ and is given by

$$\gamma(t) = \begin{cases} t, & \text{if } 0 \le t \le 1; \\ 2-t+(t-1)i & \text{if } 1 \le t \le 2; \\ (3-t)i & \text{if } 2 \le t \le 3. \end{cases}$$

On the interval $[0,1]$, we have $\overline{\gamma(t)} = t$ and $\gamma'(t) = 1$. Hence,

$$\int_0^1 \overline{\gamma(t)}\gamma'(t)\,dt = \int_0^1 t\,dt = 1/2.$$

On the interval $[1,2]$, we have $\overline{\gamma(t)} = 2 - t - (t-1)i$ and $\gamma'(t) = -1 + i$. Hence,

$$\int_1^2 \overline{\gamma(t)}\gamma'(t)\,dt = \int_1^2 (2t - 3 + i)\,dt = i.$$

On the interval $[2,3]$, we have $\overline{\gamma(t)} = t - 3$ and $\gamma'(t) = -i$. Hence,

$$\int_0^1 \overline{\gamma(t)}\gamma'(t)\,dt = \int_2^3 (3-t)i\,dt = i/2.$$

If we add the contributions of each of these three intervals, the result is

$$\int_\gamma \overline{z}\,dz = 1/2 + i + i/2 = 1/2 + (3/2)i.$$

## Exercise Set 2.3

1. Find $\int_0^\pi e^{it}\,dt$.
2. Find $\int_0^1 \sin(it)\,dt$.
3. Find $\int_0^{2\pi} e^{int}\,e^{imt}\,dt$ for all integers $n$ and $m$.
4. Find a path which traces the straight line joining $2 - i$ to $-1 + 3i$.
5. If $z_0 \in \mathbb{C}$, find a path which traces the circle of radius $r$, centered at $z_0$, (a) once in the counterclockwise direction, (b) once in the clockwise direction, (c) three times in the counterclockwise direction.
6. Prove that if $\gamma(t) = x(t) + iy(t)$ is a curve defined on an interval $I$, with real and imaginary parts $x(t)$ and $y(t)$, and if $c \in I$, then $\gamma'(c)$ exists if and only if $x'(c)$ and $y'(c)$ exist and, in this case, $\gamma'(c) = x'(c) + iy'(c)$.
7. Show that if $f$ is a smooth complex-valued function on an interval $[a,b]$, then $\int_a^b f'(t)\,dt = f(b) - f(a)$.
8. Suppose $\gamma$ is a path with parameter interval $[a,b]$. Use the result of the previous exercise to show that $\int_\gamma 1\,dz = \gamma(b) - \gamma(a)$.
9. Find $\int_\gamma z^2\,dz$ if $\gamma$ traces a straight line from 0 to $w$.
10. Find $\int_\gamma z^{-1}\,dz$ and $\int_\gamma \overline{z}\,dz$ for the circular path $\gamma(t) = 3\,e^{it}$, $0 \le t \le 2\pi$.
11. Find $\int_\gamma \mathrm{Re}(z)\,dz$ if $\gamma$ is the path of Example 2.3.11.
12. With $\gamma$ as in the previous exercise, find $\int_\gamma \mathrm{Im}(z^2)\,dz$.
13. Is it generally true that $\mathrm{Re}(\int_\gamma f(z)\,dz) = \int_\gamma \mathrm{Re}(f(z))\,dz$?

## 2.4. Properties of Contour Integrals

We begin this section with the question of parameter independence. To what extent does the integral of a function along a path depend on how the path is parameterized? The same geometric figure $\gamma(I)$ may be parameterized in many ways. For example, the top third of the unit circle may be parameterized by

$$\begin{aligned} &\gamma_1(t) = -t + i\sqrt{1-t^2}, \quad -\sqrt{3}/2 \le t \le \sqrt{3}/2, \quad \text{or} \\ &\gamma_2(t) = \mathrm{e}^{it} = \cos t + i \sin t, \quad \pi/6 \le t \le 5\pi/6, \end{aligned} \tag{2.4.1}$$

and these are only two of infinitely many possibilities. Does the integral of a function over the upper third of the unit circle depend on which of these parmeterizations is chosen?

**Parameter Changes that Change the Integral.** The following example shows that some changes of parameterization do change the integral.

**Example 2.4.1.** Find $\int_{\gamma_1} 1/z\, dz$ if $\gamma_1(t) = r\,\mathrm{e}^{it}$ on $[0, 2\pi]$ is the circular path of Example 2.3.2. Does the answer change if the circle is traversed in the clockwise direction instead, using the path $\gamma_2(t) = r\,\mathrm{e}^{-it}$ on $[0, 2\pi]$?

**Solution:** From Example 2.3.2 we know that the path $\gamma_1(t) = r\,\mathrm{e}^{it}$ has $\gamma_1'(t) = ir\,\mathrm{e}^{it}$ and so the given integral is

$$\int_{\gamma_1} \frac{dz}{z} = \int_0^{2\pi} \frac{(r\,\mathrm{e}^{it})'}{r\,\mathrm{e}^{it}} dt = \int_0^{2\pi} i\, dt = 2\pi i.$$

On the other hand, the derivative of $\gamma_2 = \mathrm{e}^{-it}$ is $-ir\,\mathrm{e}^{-it}$ and so

$$\int_{\gamma_2} \frac{dz}{z} = \int_0^{2\pi} -i\, dt = -2\pi i.$$

This example shows that the integral along a path depends not only on the geometric figure that is the image $\gamma(I)$ of the path, but also on the direction the path is traversed (at the very least).

Also, traversing a portion of the curve more than once may affect the integral. For example, if we were to go around the circle twice in Example 2.4.1, by choosing $\gamma(t) = \mathrm{e}^{2it}$ on $[0, 2\pi]$, the result would be $4\pi i$ instead of $2\pi i$.

**The Independence of Parameterization Theorem.** There is a degree to which the integral is independent of the parameterization. Certain ways of changing the parameterization do not effect the integral, as the following theorem shows.

**Theorem 2.4.2.** *Let $\gamma_1 : [a, b] \to \mathbb{C}$ be a path and $\alpha : [c, d] \to [a, b]$ a smooth function with $\alpha(c) = a$ and $\alpha(d) = b$. If $\gamma_2$ is the path with parameter interval $[c, d]$ defined by $\gamma_2(t) = \gamma_1(\alpha(t))$, then*

$$\int_{\gamma_2} f(z)dz = \int_{\gamma_1} f(z)dz$$

*for every function $f$ defined and continuous on a set $E$ containing $\gamma_1([a, b]) = \gamma_2([c, d])$.*

**Proof.** We have $\gamma_2(t) = \gamma_1(\alpha(t))$ and, by the chain rule,

$$\gamma_2'(t) = \gamma_1'(\alpha(t))\alpha'(t).$$

Thus,

$$\begin{aligned}\int_{\gamma_2} f(z)\,dz &= \int_c^d f(\gamma_2(t))\gamma_2'(t)\,dt \\ &= \int_c^d f(\gamma_1(\alpha(t)))\gamma_1'(\alpha(t))\alpha'(t)\,dt \\ &= \int_a^b f(\gamma_1(s))\gamma_1'(s)\,ds \\ &= \int_{\gamma_1} f(z)\,dz,\end{aligned}$$

where the third equality follows from the substitution $s = \alpha(t)$. This completes the proof. □

Note that the condition that $\alpha(c) = a$ and $\alpha(d) = b$ is essential in the above theorem. It says that $\alpha$ takes the endpoints of the parameter inverval $[c, d]$ to the endpoints of the parameter interval $[a, b]$ in an order preserving fashion.

**Example 2.4.3.** Are the integrals of a continuous function over the two paths in (2.4.1) necessarily the same?

**Solution:** Yes. If we set $\alpha(t) = -\cos t$, then $\alpha$ is a smooth function mapping the parameter interval $[\pi/6, 5\pi/6]$ to the parameter interval $[-\sqrt{3}/2, \sqrt{3}/2]$ in an order preserving fashion. Furthermore, $\gamma_2 = \gamma_1 \circ \alpha$. Thus, the above theorem insures that the integral of a continuous function over $\gamma_1$ is the same as its integral over $\gamma_2$.

Doesn't Example 2.4.1 contradict Theorem 2.4.2? After all, if $\gamma_1(t) = \mathrm{e}^{it}$ on $[0, 2\pi]$ and $\alpha : [0, 2\pi] \to [0, 2\pi]$ is defined by $\alpha(t) = 2\pi - t$, then $\gamma_2(t) = \gamma_1(\alpha(t)) = \mathrm{e}^{-it}$. By Example 2.4.1 the integrals of $1/z$ over these two curves are different. Doesn't Theorem 2.4.2 say they should be the same? No. The conditions $\alpha(a) = c$ and $\alpha(b) = d$ are not satisfied by this choice of $\alpha$, since $\alpha(0) = 2\pi$ and $\alpha(2\pi) = 0$. In other words, this choice of $\alpha$ reverses the order of the endpoints of the parameter interval rather than preserving that order.

In general, the conditions $\alpha(a) = c$ and $\alpha(b) = d$ guarantee that, overall, $\gamma_2$ traverses the curve in the same direction as $\gamma_1$. If $\alpha'$ were positive on the entire interval, then $\alpha$ would be increasing on this interval and $\gamma_1$ and $\gamma_2$ would be moving in the same direction at each point of the curve. If $\alpha'$ is not positive on all of $[c, d]$, then there may be intervals where one path reverses direction and backtracks, while the other path does not. These things do not affect the integral, because if a curve does backtrack for a time, it has to turn around and recover the same ground in order to catch up to the other curve in the end. This is an intuitive explanation; the actual proof that the integral is unaffected is in the proof of the above theorem.

Theorem 2.4.2 leads to a strategy which, for some paths $\gamma_1$ and $\gamma_2$ with the same image, yields a proof that they determine the same integral: Suppose that the parameter intervals for the two paths can each be partitioned into $n$ subintervals in such a way that for $j = 1, \cdots, n$, $\gamma_1$ on its $j$th subinterval and $\gamma_2$ on its $j$th subinterval are related by a smooth function $\alpha_j$, as in Theorem 2.4.2. If this can be done, then it clearly follows that $\int_{\gamma_1} f(z)\,dz = \int_{\gamma_2} f(z)\,dz$ for any function $f$ which is continuous on a set containing $\gamma_1(I)$. For this reason, Theorem 2.4.2 is sometimes called the *independence of parameterization theorem.*

**Remark 2.4.4.** Since path integrals are essentially independent of the way the path is parameterized, we will often describe a path without specifying a parameterization. Instead, we will just give a description of the geometric object that is traced, the direction, and how many times. For example, we may describe a path as tracing once around the unit circle in the counterclockwise direction, or tracing once around the boundary $\partial\Delta$ of a given triangle $\Delta$ in the counterclockwise direction, or as tracing the straight line path from a complex number $w_1$ to a complex number $w_2$. In the first two cases we may simply write

$$\int_{|z|=1} f(z)\,dz \quad \text{or} \quad \int_{\partial\Delta} f(z)\,dz$$

for the corresponding path integral. In the latter case, we may write

$$\int_{w_1}^{w_2} f(z)\,dz$$

for the path integral along the straight line from $w_1$ to $w_2$.

**Closed Curves.** The curves in Examples 2.3.2 and 2.3.4 both have the property that they begin and end at the same point – that is, they are closed curves. A closed curve $\gamma$ on a parameter interval $[a, b]$ is one that satisfies $\gamma(a) = \gamma(b)$. A closed curve which is a path will be called a *closed path.*

The famous integral theorem of Cauchy states that the integral of an analytic function $f$ around a closed path is 0, provided there is an appropriate relationship between the curve $\gamma$ and the domain $U$ on which $f$ is analytic (roughly speaking, the curve should lie in $U$ but not go around any *holes* in $U$). Since the function $f(z) = z$ is analytic on $\mathbb{C}$ (as is any polynomial in $z$), the next example illustrates this phenomenon.

**Example 2.4.5.** Find $\int_\gamma z\,dz$ if $\gamma$ is the path of Example 2.3.4.

**Solution:** From Example 2.3.4 we know that the path $\gamma(t)$ has values $4t, 1 + (4t-1)i, 3-4t+i, (4-4t)i$ and derivatives $4, 4i, -4$, and $-4i$ on the four subintervals of the partition $0 < 1/4 < 1/2 < 3/4 < 1$. Thus, the integrals over the four smooth pieces of our curve are

$$\int_0^{1/4} 4t \cdot 4\,dt = 8t^2\Big|_0^{1/4} = 1/2,$$

$$\int_{1/4}^{1/2} (1 + (4t-1)i) \cdot 4i\,dt = (4ti - 8t^2 + 4t)\Big|_{1/4}^{1/2} = i - 1/2,$$

$$\int_{1/2}^{3/4} (3 - 4t + i) \cdot (-4)\,dt = (-12t + 8t^2 - 4ti)\Big|_{1/2}^{3/4} = -1/2 - i,$$

$$\int_{3/4}^{1} (4 - 4t)i \cdot (-4i)\,dt = (+16t - 8t^2)\Big|_{3/4}^{1} = 1/2.$$

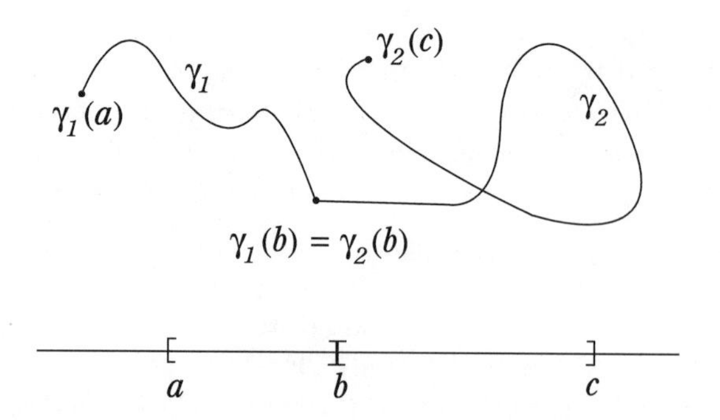


**Figure 2.4.1.** The Join of Two Paths.

Since these add up to 0, we have $\int_\gamma z\,dz = 0$.

The function $1/z$ is also analytic, except at $z = 0$. The circular path of Example 2.4.1 is closed and lies in the domain where $1/z$ is analytic. So why is the integral not 0? Because the path goes around a hole in the domain of $1/z$ – it goes around $\{0\}$.

**Additivity Properties of Contour Integrals.** If $\gamma$ is a path with parameter interval $[a, b]$, then we can use Theorem 2.4.2 to change the parameter interval to any other interval $[c, d]$ with $c < d$, in a way that does not affect the image of $\gamma$ or integrals over $\gamma$. In fact, if we set

$$\alpha(t) = a + \frac{b-a}{d-c}(t-c),$$

then $\alpha$ is smooth, $\alpha([c, d]) = [a, b]$, $\alpha(c) = a$ and $\alpha(d) = b$. Thus, $\gamma_1(t) = \gamma(\alpha(t))$ defines a path $\gamma_1$ with the same image as $\gamma$ and, by Theorem 2.4.2, a path which determines the same integral for continuous functions on its image. Thus, without loss of generality, we may always assume that the parameter interval for a path is any interval we choose.

If $\gamma_1$ and $\gamma_2$ are two paths so that $\gamma_1$ ends where $\gamma_2$ begins, then we can join the two paths to form a single new path $\gamma_1 + \gamma_2$. We do this as follows: If $\gamma_1$ has parameter interval $[a, b]$, we choose a parameter interval of the form $[b, c]$ for $\gamma_2$. The fact that $\gamma_2$ begins where $\gamma_1$ ends means that $\gamma_1(b) = \gamma_2(b)$. We define $\gamma_1 + \gamma_2$ on $[a, c]$ by

$$(\gamma_1 + \gamma_2)(t) = \begin{cases} \gamma_1(t) & \text{if } t \in [a, b], \\ \gamma_2(t) & \text{if } t \in [b, c]. \end{cases} \tag{2.4.2}$$

The path $\gamma_1 + \gamma_2$ is called the *join* of $\gamma_1$ and $\gamma_2$.

In Example 2.4.1, changing the path from one tracing the circle counterclockwise to one tracing the circle clockwise had the effect of changing the sign of the integral. As we shall see, this always happens. If $\gamma : [a, b] :\to \mathbb{C}$ is a path, denote by $-\gamma$ the path defined by

$$-\gamma(t) = \gamma(a + b - t).$$

Then $-\gamma(a) = \gamma(b)$ and $-\gamma(b) = \gamma(a)$. In fact, $-\gamma$ traces the same geometric figure as $\gamma$, but it does so in the opposite direction.

For some closed curves, such as circles, and boundaries of rectangles, triangles, etc., there is clearly a clockwise direction around the curve and a counterclockwise direction. If such a curve is parameterized so that it is traversed in the counterclockwise direction, we will say the resulting closed path has *positive orientation.* If it is traversed in the clockwise direction, we will say it has *negative orientation.* Clearly, if $\gamma$ has positive orientation, then $-\gamma$ has negative orientation.

The common starting and ending point of a closed path can be changed without changing the integral of a function over this path. This is done by representing the closed path as the join of two paths which connect the original starting and ending point to the new one. One then uses part (b) of the next theorem. The details are left to the exercises.

The next theorem states the elementary properties of path integrals having to do with linearity and path additivity. Part (b) follows immediately from the corresponding additivity property of the Riemann integral on the line and we have already used it several times. We leave the proofs of (a) and (c) to the exercises (Exercise 2.4.5).

**Theorem 2.4.6.** *Let $\gamma, \gamma_1, \gamma_2$ be paths with $\gamma_1$ ending where $\gamma_2$ begins, $f$ and $g$ two functions which are continuous on a set $E$ containing the images of these paths, and $a$ and $b$ complex numbers. Then*

(a) $\displaystyle\int_\gamma (af(z) + bg(z))\, dz = a \int_\gamma f(z)\, dz + b \int_\gamma g(z)\, dz;$

(b) $\displaystyle\int_{\gamma_1+\gamma_2} f(z)\, dz = \int_{\gamma_1} f(z)\, dz + \int_{\gamma_2} f(z)\, dz;$

(c) $\displaystyle\int_{-\gamma} f(z)\, dz = -\int_\gamma f(z)\, dz.$

Part (a) of this theorem says that a path integral is a linear function of the integrand, Part (b) says that it is an additive function of the path, while Part (c) shows why the notation $-\gamma$ is appropriate for the curve that is $\gamma$ traversed in the opposite direction.

**Length of a Path.** We define the length $\ell(\gamma)$ of a path $\gamma$ in $\mathbb{C}$ in the same way the length of a curve in $\mathbb{R}^2$ is defined in calculus.

**Definition 2.4.7.** If $\gamma(t) = x(t) + iy(t)$ is a path in $\mathbb{C}$ with parameter interval $[a, b]$, then the length $\ell(\gamma)$ of $\gamma$ is defined to be

$$\ell(\gamma) = \int_a^b |\gamma'(t)|\, dt = \int_a^b \sqrt{x'(t)^2 + y'(t)^2}\, dt.$$

**Example 2.4.8.** Prove that the above definition of length yields the correct length for a path which traces once around a circle of radius $r$.

**Solution:** The path is $\gamma(t) = r\,e^{it}$, with parameter interval $[0, 2\pi]$. The derivative of $\gamma$ is $\gamma'(t) = ir\,e^{it}$ and so $|\gamma'(t)| = r$. Thus, $\ell(\gamma) = \int_0^{2\pi} r\, dt = 2\pi r$.

It will be important in coming sections to be able to obtain good upper bounds on the absolute value of a path integral. The key theorem that produces such upper bounds is the following.

**Theorem 2.4.9.** *Let $\gamma$ be a path in $\mathbb{C}$ and $f$ a function continuous on a set containing $\gamma(I)$. If $|f(z)| \le M$ for all $z \in \gamma(I)$, then*

$$\left| \int_\gamma f(z)\, dz \right| \le M\ell(\gamma).$$

**Proof.** If the parameter interval for $\gamma$ is $[a, b]$, then

$$\begin{aligned}\left| \int_\gamma f(z)\, dz \right| &= \left| \int_a^b f(\gamma(t))\gamma'(t)\, dt \right| \le \int_a^b |f(\gamma(t))\gamma'(t)|\, dt \\ &\le \int_a^b M|\gamma'(t)|\, dt = M \int_a^b |\gamma'(t)|\, dt = M\ell(\gamma).\end{aligned}$$

□

The next example is a typical application of this theorem.

**Example 2.4.10.** Show that if $f$ is a bounded continuous function on $\mathbb{C}$, and $\gamma_R$ is the path $\gamma_R(z) = R\,e^{it}$ for $t \in [0, 2\pi]$, then

$$\lim_{R\to\infty} \int_{\gamma_R} \frac{f(z)}{(z-w)^2}\, dz = 0 \tag{2.4.3}$$

for each $w \in \mathbb{C}$.

**Solution:** The statement that $f$ is bounded means there is an upper bound $M$ for $|f|$. That is, $|f(z)| \le M$ for all $z \in \mathbb{C}$. We also have $|z - w| \ge |z| - |w|$ by the second form of the triangle inequality. If $z \in \gamma(I)$, then $|z| = R$ and so $|z - w| \ge R - |w|$, which implies $|z - w|^{-2} \le (R - |w|)^{-2}$. Thus, for $z \in \gamma(I)$, we have the following bound on the integrand of (2.4.3):

$$\left| \frac{f(z)}{z - w} \right| \le \frac{M}{(R - |w|)^2}.$$

Since $\ell(\gamma_R) = 2\pi R$, Theorem 2.4.9 implies that

$$\left| \int_{\gamma_R} \frac{f(z)}{(z-w)^2}\, dz \right| \le \frac{2\pi MR}{(R - |w|)^2}.$$

The right side of this inequality has limit 0 as $R \to \infty$ and this implies (2.4.3).

## Exercise Set 2.4

1. Compute $\int_\gamma z^2\, dz$ if $\gamma$ is any path which traces once around the circle of radius one in the counterclockwise direction.
2. Compute $\int_\gamma 1/z\, dz$ if $\gamma$ is any path which traces twice around the circle of radius one, centered at 0, in the counterclockwise direction.
3. If $z_0$ and $w_0$ are two points of $\mathbb{C}$, compute $\int_\gamma z\, dz$ if $\gamma$ is any path which traces the straight line from $z_0$ to $w_0$ once.

4. Compute the integral of the previous exercise for any smooth path $\gamma$ which begins at $z_0$ and ends at $w_0$.
5. Prove Parts (a) and (c) of Theorem 2.4.6.
6. Describe a smooth, order preserving function $\alpha$ which takes the parameter interval $[0,1]$ to the parameter interval $[2,5]$.
7. Prove that a parameter change $\gamma \to \gamma \circ \alpha$, like the one in Theorem 2.4.2, does not change the length of a path provided $\alpha$ is an non-decreasing function (has a non-negative derivative).
8. Show that $\left| \int_\gamma \frac{\cos z}{z}\, dz \right| \le 2\pi e$ if $\gamma$ is a path that traces the unit circle once. Hint: Show that $|\cos z| \le e$ if $|z| = 1$.
9. Show that if $\Delta$ is a triangle in the plane of diameter $d$ (length of its longest side), and if $f$ is a continuous function on $\Delta$ with $|f|$ bounded by $M$ on $\Delta$, then

$$\left| \int_{\partial\Delta} f(z)\, dz \right| \le 3Md.$$

10. Prove that $\int_\gamma p(z)\, dz = 0$ if $\gamma(t) = e^{it}$, $0 \le t \le 2\pi$, and $p(z)$ is any polynomial in $z$ (this is a special case of Cauchy's Theorem, but do not assume Cauchy's Theorem in your proof).
11. Let $R(z)$ be the remainder after $n$ terms in the power series for $e^z$. That is,

$$R(z) = e^z - \sum_{k=1}^{n} \frac{z^k}{k!} = \sum_{k=n+1}^{\infty} \frac{z^k}{k!}.$$

Prove that $|R(z)| \le \dfrac{e-1}{(n+1)!}$ if $|z| \le 1$.

12. Prove that $\int_\gamma e^z\, dz = 0$ if $\gamma(t) = e^{it}$, $0 \le t \le 2\pi$ using the previous exercise and Exercise 10.
13. Prove that if $\gamma$ is a closed path with parameter interval $I = [a,b]$ and common starting and ending point $z = \gamma(a) = \gamma(b)$ and $w$ is any other point on $\gamma(I)$, then there is another closed path $\gamma_1$ with $\gamma_1(I) = \gamma(I)$, which determines the same integral, but has $w$ as common starting and ending point. Hint: Use part (b) of Theorem 2.4.6.

## 2.5. Cauchy's Integral Theorem for a Triangle

The core material of any beginning Complex Variables text is the proof of Cauchy's Integral Theorem and the exploration of its consequences. Roughly speaking, Cauchy's Integral Theorem states that the integral of an analytic function around a closed path is zero, provided the path is contained in the open set $U$ on which the function is analytic and does not go around any "holes" in $U$. Part of the problem here is to make sense of the idea of a "hole" in an open set and to decide what it means for a path to go around such a hole.

We take a first step toward the proof of Cauchy's Theorem in this section by proving it in the case where the path is the boundary of a triangle and the function is analytic on an open set containing the triangle.

The proof of this result will make essential use of a couple of properties of compact sets. Thus, we will precede the proof with a discussion of compact sets.

**Compact Sets.** A subset $K$ of $\mathbb{R}^n$ is called *compact* if every open cover of $K$ has a finite subcover. Here, by an open cover of $K$, we mean a collection of open sets whose union contains $K$. A finite subcover is then a finite subcollection of this collection which also has union containing $K$.

Here we will state without proof a number of facts about compact sets. The proofs can be found in any text on Advanced Calculus or undergraduate Real Analysis.

**Theorem 2.5.1** (Heine-Borel Theorem). *A subset of $\mathbb{R}^n$ is compact if and only if it is closed and bounded.*

Another characterization of compact sets is obtained by using complementation to turn the statement about open covers in the definition into a statement about collections of closed sets. The result is the following:

A set $K$ is compact if and only if for every collection of closed subsets of $K$ with empty intersection, there is a finite subcollection with empty intersection. The same thing stated somewhat differently is:

**Theorem 2.5.2.** *A set $K$ is compact if and only if, whenever a collection of closed subsets of $K$ has the property that each finite subcollection has non-empty intersection, then the full collection also has non-empty intersection.*

A consequence of the previous theorem is the following:

**Corollary 2.5.3.** *If*

$$A_1 \supset A_2 \supset \cdots \supset A_n \supset \cdots$$

*is a nested sequence of non-empty compact subsets of $\mathbb{R}^n$, then $\bigcap A_n \neq \emptyset$.*

This is one of the properties of compact sets that we shall need in our proof of Cauchy's Theorem on a triangle. The others are as follows:

**Theorem 2.5.4.** *If $f$ is a continuous function, defined and continuous on a compact subset $K$ of $\mathbb{R}^n$, with values in $\mathbb{R}^m$, then $f(K)$ is also compact.*

A non-empty compact subset of $\mathbb{R}$ contains both a maximal element and a minimal element. If $f$ is a real-valued function, defined and continuous on a compact set $K$, then $f(K)$ is a compact subset of the line and, hence, has maximal and minimal elements. This proves the following corollary of the previous theorem.

**Corollary 2.5.5.** *A continuous real-valued function on a compact set has a maximal value and a minimal value.*

**Antiderivatives.** There is one case in which it is very easy to prove that the integral of a continuous function $f$ around a closed path is 0. This is the case where the function $f$ has an antiderivative.

As with functions of a real variable, an antiderivative for a function $f$, defined on an open set $U$, is a function $g$ such that $g' = f$ on $U$.

**Theorem 2.5.6.** *If $f$ is a continuous function defined on an open set $U$, and if $f$ has an antiderivative $g$ on $U$, then*

$$\int_\gamma f(z)\,dz = g(\gamma(b)) - g(\gamma(a))$$

*if $\gamma$ is any path in $U$ with parameter interval $[a, b]$. If $\gamma$ is a closed path, then this integral is 0.*

**Proof.** Since $g' = f$ on $U$, we have

$$\int_\gamma f(z)\,dz = \int_a^b f(\gamma(t))\gamma'(t)\,dt = \int_a^b g'(\gamma(t))\gamma'(t)\,dt.$$

There is a version of the chain rule which holds for the composition of an analytic function with a path (Exercise 2.5.4). It tells us that

$$(g(\gamma(t)))' = g'(\gamma(t))\gamma'(t).$$

Thus,

$$\int_\gamma f(z)\,dz = \int_a^b (g(\gamma(t)))'\,dt = g(\gamma(b)) - g(\gamma(a)),$$

by the complex version of the Fundamental Theorem of Calculus (Exercise 2.3.7). If the path is closed, then $\gamma(b) = \gamma(a)$ and the integral is 0. □

**Cauchy's Theorem.** The fact that a function $f$ has a complex derivative at $w$ means that, near $w$, it can be approximated by a linear function of $z$ (more precisely, by a polynomial of degree one in $z$). Such a function has a complex antiderivative, as does any polynomial, and so its integral around a closed curve is zero. Thus, a function with a complex derivative at $w$ can be approximated near $w$ by a function whose integral around any closed path is zero. This is the basis for an argument that the integral of a function with a complex derivative at $w$, around a small triangle containing $w$, is much smaller than one would predict (using, for example, the estimate given in Exercise 2.3.9). This is made precise in the next lemma, which is the basis for the proof of Cauchy's Theorem for triangles.

**Lemma 2.5.7.** *Let $f$ be a function which is continuous on a neighborhood of $w \in \mathbb{C}$ and which has a complex derivative at $w$. Then for every $\epsilon > 0$, there is a $\delta > 0$ such that*

$$\left|\int_{\partial\Delta} f(z)\,dz\right| < \epsilon d^2$$

*if $\Delta$ is any triangle containing $w$, of diameter $d \le \delta$.*

**Proof.** Since $f$ is continuous on a neighborhood of $w$, we may choose an $r > 0$ such that $f$ is continuous on the open disc $D_r(w)$.

Since $f$ has a complex derivative at $w$,

$$\lim_{z \to w} \frac{f(z) - f(w)}{z - w} = f'(w)$$

exists. Thus, we may choose a positive $\delta < r$ such that $|z - w| < \delta$ implies that

$$\left| \frac{f(z) - f(w)}{z - w} - f'(w) \right| < \frac{\epsilon}{3}. \tag{2.5.1}$$

If we multiply (2.5.1) by $|z - w|$, the result is

$$|f(z) - f(w) - f'(w)(z - w)| < \frac{\epsilon}{3}|z - w| \quad \text{for all} \quad z \in D_\delta(w). \tag{2.5.2}$$

If $\Delta$ is a triangle of diameter $d \leq \delta$, containing $w$, we set

$$I = \int_{\partial\Delta} f(z)dz.$$

Then

$$I = \int_{\partial\Delta} (f(w) + f'(w)(z - w))dz + \int_{\partial\Delta} (f(z) - f(w) - f'(w)(z - w))dz.$$

However,

$$\int_{\partial\Delta} (f(w) + f'(w)(z - w))dz = 0,$$

because the integrand has a complex antiderivative: $f(w)z + f'(w)(z^2/2 - wz)$ (remember $w$ is a constant – the variable is $z$), and so

$$I = \int_{\partial\Delta} (f(z) - f(w) - f'(w)(z - w))dz. \tag{2.5.3}$$

Then (2.5.2) implies that

$$|I| < d^2\epsilon,$$

since the modulus of the integrand in (2.5.3) is bounded by $\epsilon d/3$ and the length of the path $\partial\Delta$ is no more than $3d$. This completes the proof. □

**Theorem 2.5.8.** *Let $f$ be a function which is analytic in an open set $U$, and suppose that $\Delta$ is a triangle contained in $U$. Let $\partial\Delta$ denote the boundary of $\Delta$, considered as a closed path with positive orientation. Then*

$$\int_{\partial\Delta} f(z)\,dz = 0.$$

**Proof.** We set

$$I = \int_{\partial\Delta} f(z)\,dz.$$

Our objective is to prove that $I = 0$. We will do this by showing that $|I| < \epsilon$ for every positive number $\epsilon$. Thus, let $\epsilon$ be an arbitrary positive number.

We subdivide the triangle $\Delta$ into four smaller triangles by joining the midpoints of the sides of $\Delta$. The resulting four triangles are all similar to $\Delta$ with sides exactly half as long as the corresponding sides of $\Delta$ (see Figure 2.5.1).

We now apply Parts (b) and (c) of Theorem 2.4.6. The sum of the integrals around each of these smaller triangles is a sum of integrals along their edges, with each of the edges interior to the original triangle occurring twice – once going one

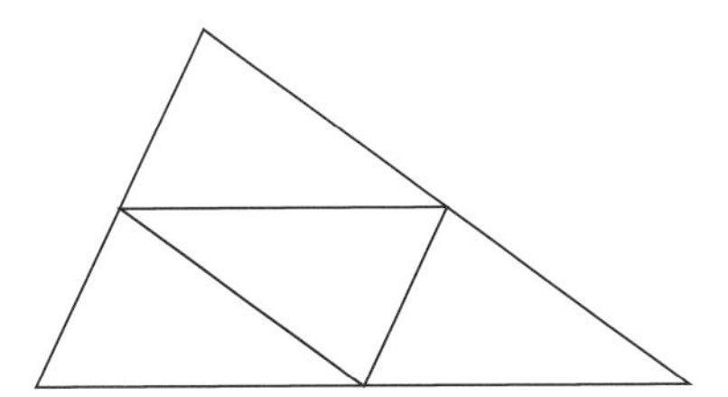

**Figure 2.5.1.** Subdividing the Triangle $\Delta$ in Theorem 2.5.8.

direction and once going the opposite direction. Thus, the contributions of these interior edges cancel, leaving only the contributions from the edges which lie along the boundary of the original triangle. It follows that the sum of the integrals of $f$ around the boundaries of these four smaller triangles is $I$, and so one of these integrals must have modulus at least $|I|/4$. Let $\Delta_1$ denote the corresponding triangle. In other words, $\Delta_1$ is chosen from the four subtriangles so that

$$|I_1| \geq |I|/4 \quad \text{where} \quad I_1 = \int_{\partial\Delta_1} f(z)\, dz.$$

Note also that, if $h$ is the diameter of $\Delta$ (which is the length of its longest side), then $\Delta_1$ has diameter $h_1 = h/2$.

We now repeat the above construction with $\Delta$ replaced by $\Delta_1$. That is, we subdivide $\Delta_1$ into four similar triangles and choose one of them, call it $\Delta_2$, with the property that

$$|I_2| \geq |I_1|/4 \geq |I|/4^2 \quad \text{where} \quad I_2 = \int_{\partial\Delta_2} f(z)\, dz,$$

and with diameter $h_2 = h_1/2 = h/2^2$.

Proceeding by induction, we may choose for each $n$ a triangle $\Delta_n$, of diameter $h_n$, so that

$$\Delta_n \subset \Delta_{n-1},$$

$$|I_n| \geq |I|/4^n \quad \text{where} \quad I_n = \int_{\partial\Delta_n} f(z) dz \tag{2.5.4}$$

and

$$h_n = h/2^n. \tag{2.5.5}$$

The collection $\{\Delta_n\}$ is a nested sequence of closed bounded non-empty sets in the plane (see Figure 2.5.2) and, hence, by Corollary 2.5.3, there is a point $w$ in the intersection $\bigcap_n \Delta_n$.

We now apply the previous lemma with $\epsilon/h^2$ replacing the $\epsilon$ of the lemma. We conclude there is a $\delta > 0$ such that the integral of $f$ around any triangle containing $w$, of diameter $d \leq \delta$, is less than $d^2\epsilon/h^2$.

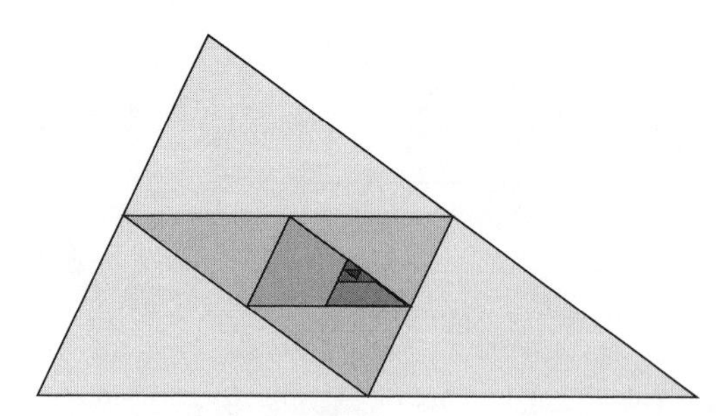

**Figure 2.5.2.** The Nested Sequence of Triangles in Theorem 2.5.8.

We may then choose $n$ large enough that $h_n = h/2^n < \delta$. Since $h_n$ is the diameter of $\Delta_n$ and $w \in \Delta_n$, we have

$$|I_n| < \frac{h_n^2}{h^2}\epsilon.$$

Putting this together with (2.5.5), we conclude

$$|I_n| < \frac{\epsilon}{4^n}.$$

Combined with (2.5.4), this yields

$$|I| \le 4^n |I_n| < \epsilon.$$

Since $\epsilon$ was an arbitrary positive number, we conclude that $I = 0$. This completes the proof. □

We will need a slightly stronger version of this theorem in which we allow the possibility that there is one point in $\Delta$ where $f$ may not have a complex derivative, but where $f$ is continuous.

**Theorem 2.5.9.** *Let $f, U$, and $\Delta$ be as in the previous theorem except that we assume that $f$ is continuous on $U$ and analytic on $U \setminus \{c\}$ for some exceptional point $c \in \Delta$. Then we still have*

$$\int_{\partial\Delta} f(z)dz = 0.$$

**Proof.** If $c$ is a vertex of $\Delta$, then, given $\epsilon > 0$, we may subdivide $\Delta$ into smaller triangles in such a way that the one containing $c$ has circumference less than $\epsilon/M$, where $M$ is the maximum value of $|f|$ on $\Delta$ (Figure 2.5.3(a)). The integral of $f$ over the boundary of this triangle will then be less than or equal to $\epsilon$. The integrals of $f$ over the boundaries of other triangles in the subdivision are all 0 by the previous theorem (none of them contains $c$). As before, the integral of $f$ over $\partial\Delta$ is the sum of the integrals over the boudaries of the triangles in the subdivision and, hence, has modulus less than or equal to $\epsilon$. Since $\epsilon$ was arbitrary, we conclude that the integral of $f$ over $\partial\Delta$ is zero.

If $c$ is in $\Delta$ but is not a vertex, then the triangle can be subdivided into triangles which do contain $c$ as a vertex (Figure 2.5.3(b)). The integral around the boundary

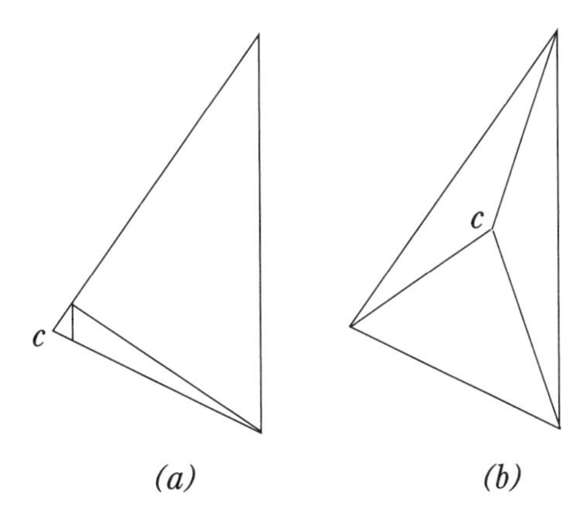


**Figure 2.5.3.** Dealing with an Exceptional Point $c$.

of each of these is zero, so the integral around $\partial\Delta$ is also 0. This completes the proof. □

## Exercise Set 2.5

1. Prove the Bolzano-Weierstrass Theorem: If $K$ is a compact subset of $\mathbb{R}^n$, then every sequence in $K$ has a subsequence which converges to an element of $K$.
2. Use Corollary 2.5.5 to show that if $K$ is a compact subset of $\mathbb{C}$ and $f$ is a continuous complex-valued function on $K$, then the modulus $|f(z)|$ of $f$ takes on a maximal value at some point of $K$.
3. Show that if $K$ is a compact subset of $\mathbb{C}$, then there is a point $z_0 \in K$ of minimum modulus – that is, a point $z_0 \in K$ such that
$$|z_0| \leq |z| \quad \text{for all} \quad z \in K.$$
4. Prove that if $g$ is analytic on an open subset $U$ of $\mathbb{C}$ and $\gamma : [a,b] \to U$ is a path in $U$, then
$$(g(\gamma(t)))' = g'(\gamma(t))\gamma'(t)$$
for $t \in [a,b]$. Hint: The proof is very similar to the proof of Theorem 2.2.7.
5. Calculate $\int_\gamma z^n\,dz$ if $n$ is a non-negative integer and $\gamma$ is a path in the plane joining the point $z_0$ to the point $w_0$. Hint: Use Theorem 2.5.6.
6. Show that $\int_\gamma p(z)\,dz = 0$ if $\gamma$ is any closed path in the plane and $p$ is any polynomial.
7. Calculate $\int_\gamma 1/z\,dz$ if $\gamma$ is any path in $\mathbb{C}$ joining $-i$ to $i$ which does not cross the half-line $(-\infty, 0]$ on the real axis. Hint: Use the result of Example 2.2.11 and Theorem 2.5.6.

8. Using the same hint as in the previous exercise, show that
$$\int_\gamma \frac{1}{z}\,dz = 0$$
if $\gamma$ is any closed path contained in the complement of the set of non-positive real numbers. Compare this with Example 2.4.1.

9. If $\sqrt{z}$ is defined by $\sqrt{z} = \mathrm{e}^{(\log z)/2}$ for the branch of the log function defined by the condition $-\pi/2 \le \arg(z) \le 3\pi/2$, find an antiderivative for $\sqrt{z}$ and then find $\int_\gamma \sqrt{z}\,dz$, where $\gamma$ is any path from $-1$ to $1$ which lies in the upper half-plane.

10. Prove that if $f$ is analytic in an open set containing a rectangle $R$, then the path integral of $f$ around the boundary of this rectangle is 0.

11. Let $\gamma$ be the path which traces the straight line from $1$ to $1+i$, then the straight line from $1+i$ to $i$ and then the straight line from $i$ to $0$. Calculate $\int_\gamma z^n\,dz$.

12. Let $\Delta$ be the triangle with vertices $1-i, i,$ and $-1-i$ and $S$ be the square with vertices $1-i, 1+i, -1+i,$ and $-1-i$. If $f$ is any function which is analytic on $\mathbb{C}\setminus\{0\}$, prove that
$$\int_{\partial\Delta} f(z)\,dz = \int_{\partial S} f(z)\,dz,$$
where $\partial\Delta$ and $\partial S$ are traversed in the counterclockwise direction.

13. For any pair of points $a, b$ in $\mathbb{C}$, denote the integral of a function $f$ along the straight line segment joining $a$ to $b$ by $\int_a^b f(z)\,dz$, as in Remark 2.4.4. Suppose $f$ is analytic in an open set containing the triangle with vertices $a, b, c$. Show that
$$\int_a^c f(z)\,dz - \int_a^b f(z)\,dz = \int_b^c f(z)\,dz.$$

14. Show that Theorem 2.5.9 can be strengthened to conclude that the integral of $f$ around any triangle in $U$ is 0 if $f$ is continuous on $U$ and analytic on $U\setminus I$, where $I$ is an interval contained in $U$. Hint: First consider the case where one side of the triangle lies along the interval $I$.

15. If $f$ is analytic on an open set $U$, then the integral of $f$ around the boundary of any triangle in $U$ is 0 (Theorem 2.5.8), as is its integral around the boundary of any rectangle in $U$ (Exercise 2.5.10). What other geometric figures have this property? What is the most general theorem along these lines you can think of?

## 2.6. Cauchy's Theorem for a Convex Set

A convex set $C$ in $\mathbb{C}$ is a set with the property that if $a$ and $b$ are points in $C$, then the line segment joining $a$ and $b$ is also contained in $C$.

**Existence of Antiderivatives.** The strategy for proving Cauchy's Theorem for convex sets is to prove that every analytic function on a convex set has an antiderivative and then apply Theorem 2.5.6. The first step in this program is accomplished with the following theorem.

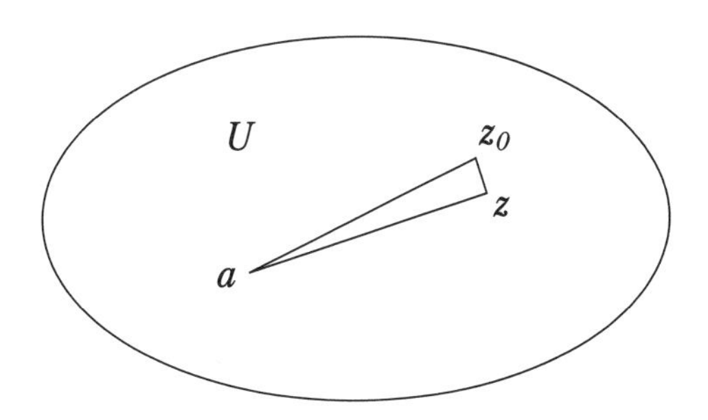


**Figure 2.6.1.** The Triangle $\Delta$ of Theorem 2.6.1.

**Theorem 2.6.1.** *Let $U$ be a convex open set and suppose $f$ is a function which is continuous on $U$ and has the property that its integral around the boundary of any triangle in $U$ is zero. If $a \in U$ is fixed and $F(z)$ is defined for all $z \in U$ by*

$$F(z) = \int_a^z f(w)\, dw,$$

*then $F'(z) = f(z)$ for all $z \in U$.*

**Proof.** Let $[a, z]$ denote the line segment joining $a$ to $z$, considered as a path from $a$ to $z$. For $z \in U$, this line segment lies entirely in $U$ and so we may define a function $F(z)$ by

$$F(z) = \int_a^z f(w)\, dw,$$

where by this we mean the path integral of $f$ along the path $[a, z]$, as in Remark 2.4.4. We will show that $F'(z) = f(z)$. To do this, we let $z$ and $z_0$ be points of $U$ and consider the triangle $\Delta$ with vertices $a, z, z_0$ (see Figure 2.6.1). The fact that $U$ is convex implies that $\Delta \subset U$.

Let $\partial\Delta$ denote the boundary of $\Delta$, considered as a contour which goes from $a$ to $z$ to $z_0$ and then to $a$ again. Since, by hypothesis, the integral of $f$ around the boundary of any triangle in $U$ is 0, we have

$$\begin{aligned}
0 &= \int_{\partial\Delta} f(w)\, dw \\
&= \int_a^z f(w)\, dw + \int_z^{z_0} f(w)\, dw + \int_{z_0}^a f(w)\, dw \\
&= F(z) - F(z_0) - \int_{z_0}^z f(w)\, dw.
\end{aligned}$$

We conclude that

$$F(z) - F(z_0) = \int_{z_0}^z f(w)\, dw.$$

If we add and subtract the number $f(z_0)$ in the integrand of this integral, we obtain

$$\begin{aligned} F(z) - F(z_0) &= \int_{z_0}^{z} f(z_0)\,dw + \int_{z_0}^{z} (f(w) - f(z_0))\,dw \\ &= f(z_0)(z - z_0) + \int_{z_0}^{z} (f(w) - f(z_0))\,dw. \end{aligned}$$

If we divide by $z - z_0$ and subtract the first term on the right from both sides, we get

$$\frac{F(z) - F(z_0)}{z - z_0} - f(z_0) = \frac{1}{z - z_0} \int_{z_0}^{z} (f(w) - f(z_0))\,dw. \tag{2.6.1}$$

Thus, to finish the proof that $F'(z_0) = f(z_0)$, we just need to show that the expression on the right in (2.6.1) has limit 0 as $z \to z_0$.

Given $\epsilon > 0$, we may choose a $\delta > 0$ such that $|f(w) - f(z_0)| < \epsilon$ when $|w - z_0| < \delta$. This follows from the fact that $f$ is continuous on $U$. If $|z - z_0| < \delta$, then $|w - z_0| < \delta$ for every $w$ on the line segment $[z_0, z]$ and so $|f(w) - f(z_0)| < \epsilon$ for every $w$ on this line segment. Then

$$\left| \int_{z_0}^{z} (f(w) - f(z_0))\,dw \right| < \epsilon |z - z_0|$$

and so

$$\left| \frac{1}{z - z_0} \int_{z_0}^{z} (f(w) - f(z_0))\,dw \right| < \epsilon$$

whenever $|z - z_0| < \delta$. This shows that

$$\lim_{z \to z_0} \frac{1}{z - z_0} \int_{z_0}^{z} (f(w) - f(z_0))\,dw = 0,$$

which completes the proof. □

**Cauchy's Integral Theorem.** We now have all the tools in place to prove Cauchy's Integral Theorem for convex sets. This is not the most general form of the theorem – that will come later – but it is sufficiently general to allow us to derive a wealth of surprising consequences.

**Theorem 2.6.2.** *Let $U$ be a convex open set and suppose $f$ is a function which is analytic on $U$, except possibly at one point, where it is at least continuous. Then*

$$\int_{\gamma} f(z)\,dz = 0$$

*for every closed path in $U$.*

**Proof.** By the previous theorem, and Theorem 2.5.8, $f$ has a complex antiderivative $F$ in $U$. Then Theorem 2.5.6 implies that the integral of $f$ around any closed path is zero. □

One of the obvious applications of Cauchy's Theorem is in proving independence of path results for path integrals.

**Corollary 2.6.3.** *If $U$ is a convex set and $f$ is analytic on $U$, and $a, b \in U$, then $\int_{\gamma} f(z)\,dz$ is the same for all paths $\gamma$ in $U$ which begin at $a$ and end at $b$.*

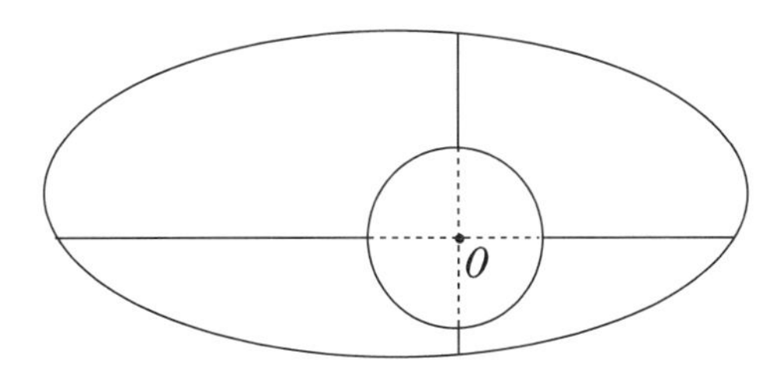


**Figure 2.6.2.** The Picture for Example 2.6.4.

The proof is left as an exercise. Another kind of path independence is illustrated by the following example.

**Example 2.6.4.** Without doing any calculating, prove that the integral of $1/z$ around any positively oriented ellipse with 0 inside is $2\pi i$.

**Solution:** From Example 2.4.1 we know that the integral of $1/z$ around a positively oriented circle centered at 0 is $2\pi i$. Choose such a circle, with radius small enough that the circle lies inside the ellipse. Then join the circle to the ellipse with four line segments, two of which lie along the $x$-axis, and two of which lie along the $y$-axis. This creates four closed paths, each of which consists of a path along a piece on the circle followed by a line segment, followed by a piece of the ellipse followed by a line segment leading back to the original point. Each of these closed paths is contained in a convex open set on which $1/z$ is analytic and so the integral of $1/z$ around each of them is 0. However, the sum of these integrals is also the difference between the integral of $1/z$ around the circle and its integral around the ellipse, because the contributions of the line segments cancel. Thus the integral of $1/z$ around the circle is the same as that around the ellipse and we conclude that the latter is $2\pi i$.

**Index of a Path around a Point.** Having proved Cauchy's Integral Theorem for a convex set, we can now prove a companion result – Cauchy's Integral Formula on a convex set. There are several versions of this result. The one we will present here allows the integral to take place over a very general path, but requires that we know how many times the path goes around a given point. The tool that measures this is described in the following definition.

**Definition 2.6.5.** Let $\gamma : I \to \mathbb{C}$ be any closed path in $\mathbb{C}$ and let $z$ be a point of $\mathbb{C}$ which does not lie on $\gamma(I)$. We set

$$\text{Ind}_\gamma(z) = \frac{1}{2\pi i} \int_\gamma \frac{dw}{w - z}.$$

This is called the *index* of $z$ with respect to $\gamma$. It is also sometimes called the *winding number* of $\gamma$ around $z$.

**Theorem 2.6.6.** *If $\gamma$ is a closed path in $\mathbb{C}$ with parameter interval $I = [a, b]$, then $\text{Ind}_\gamma(z)$ is an integer-valued function of $z$ defined on the complement of $\gamma(I)$.*

**Proof.** Let $z_0$ a point in the complement of $\gamma(I)$. Since $\gamma$ is a closed path, we have $\gamma(a) = \gamma(b)$. We define a complex-valued function $\lambda(t)$ on the interval $a \leq t \leq b$ by

$$\lambda(t) = \int_a^t \frac{\gamma'(s)}{\gamma(s) - z_0}\, ds.$$

Then $\lambda(a) = 0$ and $\lambda(b) = 2\pi i \operatorname{Ind}_\gamma(z)$. If we can show that $\mathrm{e}^{\lambda(b)} = 1$, then the proof will be complete, since the only numbers $w$ with $\mathrm{e}^w = 1$ are the numbers $2\pi i\, n$ where $n$ is an integer.

By the Fundamental Theorem of Calculus,

$$\lambda'(t) = \frac{\gamma'(t)}{\gamma(t) - z_0},$$

while the derivative of $\mathrm{e}^{\lambda(t)}$ is

$$(\mathrm{e}^{\lambda(t)})' = \mathrm{e}^{\lambda(t)}\, \lambda'(t) = \mathrm{e}^{\lambda(t)}\, \frac{\gamma'(t)}{\gamma(t) - z_0}.$$

It follows that

$$\left(\frac{\mathrm{e}^{\lambda(t)}}{\gamma(t) - z_0}\right)' = \frac{1}{(\gamma(t) - z_0)^2}\left(\mathrm{e}^{\lambda(t)}\, \frac{\gamma'(t)}{\gamma(t) - z_0}(\gamma(t) - z_0) - \mathrm{e}^{\lambda(t)}\, \gamma'(t)\right) = 0.$$

Hence, $\mathrm{e}^{\lambda(t)} / (\gamma(t) - z_0)$ is a constant. We conclude that

$$\frac{\mathrm{e}^{\lambda(t)}}{\gamma(t) - z_0} = \frac{\mathrm{e}^{\lambda(a)}}{\gamma(a) - z_0} = \frac{1}{\gamma(a) - z_0}$$

for every $t \in [a, b]$. If we set $t = b$, this gives us

$$\mathrm{e}^{\lambda(b)} = \frac{\gamma(b) - z_0}{\gamma(a) - z_0}.$$

Since $\gamma(a) = \gamma(b)$, it follows that $\mathrm{e}^{\lambda(b)} = 1$. This completes the proof. □

**Cauchy's Integral Formula for Convex Sets.**

**Theorem 2.6.7.** *Let $U$ be a convex open set, $f$ a function which is analytic on $U$ and $\gamma : I \to U$ a closed path in $U$. Then*

$$\operatorname{Ind}_\gamma(z) f(z) = \frac{1}{2\pi i} \int_\gamma \frac{f(w)}{w - z} dw,$$

*for every point $z \in U$ which does not lie on $\gamma(I)$.*

**Proof.** Consider the function $g(z, w)$ defined for $z, w \in U$ by

$$g(z, w) = \begin{cases} \dfrac{f(w) - f(z)}{w - z} & \text{if } w \neq z, \\ f'(z), & \text{otherwise.} \end{cases}$$

For each fixed $z \in U$, this function is analytic in $w$ everywhere on $U$ except possibly at $w = z$, but it is at least continuous at $w = z$. Since Theorem 2.6.2 holds even if the function is not analytic at some point but is continuous there, it follows that

$$0 = \int_\gamma g(z,w)dw = \int_\gamma \frac{f(w)}{w-z}dw - \int_\gamma \frac{f(z)}{w-z}dw$$
$$= \int_\gamma \frac{f(w)}{w-z}dw - 2\pi i \,\mathrm{Ind}_\gamma(z)f(z),$$

as long as $z$ is not on the contour $\gamma$ (note that this is required in order to write the integral in the first line above as the difference of two integrals, since otherwise these two integrals might not exist individually). We conclude that

$$\mathrm{Ind}_\gamma(z)f(z) = \frac{1}{2\pi i}\int_\gamma \frac{f(w)}{w-z}dw,$$

as required. This completes the proof. □

This is a striking result, for it says that the values of an analytic function at points "inside" a closed path are determined by its values at points on the path. Here, a point is considered inside the path if the path has non-zero index at the point.

**Corollary 2.6.8.** *If $U$ is a convex open set, $z \in U$ and $\gamma$ is a closed path in $U$ with $\mathrm{Ind}_\gamma(z) = 1$, then*

$$f(z) = \frac{1}{2\pi i}\int_\gamma \frac{f(w)}{w-z}dw$$

*for every function $f$ analytic on $U$.*

Intuitively, the meaning of the hypothesis $\mathrm{Ind}_\gamma(z) = 1$ in the above corollary is that the closed path $\gamma$ goes around $z$ once and does so in the positive direction.

Cauchy's Integral Theorem and Cauchy's Integral Formula have a wealth of applications. We will begin exploring these in the next chapter.

However, in order for Cauchy's Integral Theorem, in the above form, to be usable, we need to be able to easily compute the index of a curve around a given point. The last section of this chapter is devoted to developing the essential properties of the index function which make this possible.

## Exercise Set 2.6

1. Prove that a function which has complex derivative identically 0 on a convex open set $U$ is constant on $U$.
2. Calculate $\int_\gamma (z^2-4)^{-1}\,dz$ if $\gamma$ is the unit circle traversed once in the positive direction.
3. Calculate $\int_\gamma (1-e^z)^{-1}\,dz$ if $\gamma$ is the circle $\gamma(t) = 2i + e^{it}$.
4. Calculate $\int_\gamma 1/z\,dz$ if $\gamma$ is any circle which does not pass through 0. Note that the answer depends on $\gamma$.
5. Find $\int_\gamma 1/z^2\,dz$ if $\gamma$ is any closed path in $\mathbb{C} \setminus \{0\}$.

6. Show that the principal branch of the log function can be described by the formula $\log(z) = \int_1^z 1/w\, dw$ for $z \notin (-\infty, 0]$.
7. Prove Corollary 2.6.3.
8. Without doing any calculating, show that the integral of $1/z$ around the boundary of the triangle with vertices $i, 1-i, -1-i$ is $2\pi i$.
9. Let $f$ be a function which is analytic on $\mathbb{C}\backslash\{z_0\}$. Show that the contour integral of $f$ around a circle of radius $r > 0$, centered at $z_0$, is independent of $r$.
10. Calculate $\mathrm{Ind}_\gamma(z_0)$ if $\gamma(t) = z_0 + e^{int}, t \in [0, 2\pi]$ and $n$ is any integer.
11. Calculate $\mathrm{Ind}_\gamma(1+i)$ if $\gamma$ is the path which traces the line from 0 to 2, then proceeds counterclockwis around the circle $|z| = 2$ from 2 to $2i$ and then traces the line from $2i$ to 0. What is the answer if this path is traversed in the opposite direction?
12. Use Cauchy's Integral Formula to calculate $\displaystyle\int_{|z|=1} \frac{e^z}{z}\, dz.$
13. Use Cauchy's Formula to show that
$$\int_{|z-1|=1} \frac{1}{z^2-1}\, dz = \pi i, \quad \int_{|z+1|=1} \frac{1}{z^2-1}\, dz = -\pi i.$$
14. Show that
$$\int_{|z|=3} \frac{1}{z^2-1}\, dz = 0.$$
Hint: Use the result of the preceding exercise.
15. Use Cauchy's Integral Formula to prove that if $f$ is a function which is analytic in an open set containing the closed unit disc $\overline{D}_1(0)$, and if $T = \{z : |z| = 1\}$ is the unit circle, then $|f(0)| \le M$, where $M$ is the maximum value of $|f|$ on $T$.
16. Show that if $\gamma$ is a path from $z_1$ to $z_2$ which does not pass through the point $z_0$, then
$$\int_\gamma \frac{1}{w-z_0}\, dw = \log\left(\frac{z_2-z_0}{z_1-z_0}\right),$$
for some branch of the log function. Note that, in the case where $z_1 = z_2$, this is just Theorem 2.6.6.

## 2.7. Properties of the Index Function

If $\gamma$ is a closed path, then removing $\gamma(I)$ from the plane results in a set which is divided into a number of connected pieces. These are open sets called the *connected components* of the complement of $\gamma(I)$. We will prove that $\mathrm{Ind}_\gamma(z)$ is constant on each of these components. Thus, to calculate $\mathrm{Ind}_\gamma(z)$ on a given component, one only needs to calculate it at one point of the component.

Before proving this, we need to have a firm idea of what a connected component is. This leads to a discussion of connected sets.

### Connected Sets.

**Definition 2.7.1.** A set $E \subset \mathbb{C}$ is *separated* if there exists a pair $A, B$ of open subsets of $\mathbb{C}$ such that $E \subset A \cup B$, $A \cap E \neq \emptyset$, $B \cap E \neq \emptyset$ and $A \cap B = \emptyset$. The pair $A, B$ is then said to *separate* $E$. If $E$ is not separated, then it is said to be *connected.*

The union of a family of connected sets with a point in common is also connected (Exercise 2.7.1). It follows that, if $z \in E$, then the union of all connected subsets of $E$ containing $z$ is itself connected. This implies that each point of $E$ is contained in a maximal connected subset of $E$. A maximal connected subset of $E$ is called a *connected component* of $E$ or simply a *component* of $E$. Two components of $E$ are either disjoint or identical, since, otherwise, their union would be a connected set larger than one of them. Thus, the components of $E$ form a pairwise disjoint family of subsets of $E$ whose union is $E$.

In this section we are primarily concerned with open sets and their components. If $E$ is open, then the sets $A \cap E$ and $B \cap E$ of Definition 2.7.1 are open subsets of $E$. It follows that an open set $E$ is separated if and only if it is the union of two disjoint non-empty open subsets of itself. It is connected if this is not the case. The next theorem states the essential facts regarding connected open sets that we will need in this section.

An open set $U$ is said to be *path connected* if every two points in $U$ can be connected by a path which lies entirely in $U$.

**Theorem 2.7.2.** *Let $U$ be an open subset of $\mathbb{C}$. Then*

(a) *each component of $U$ is also open;*

(b) *$U$ is connected if and only if it is path connected.*

**Proof.** We prove (b). The proof of (a) is left as an exercise.

Suppose $U$ is connected. Given $z \in U$, let $V_z$ be the set of points of $U$ that are connected to $z$ by a path in $U$, and let $w$ be some other point of $U$. There is an open disc $D$, centered at $w$, which is contained in $U$. Since any two points in $D$ are connected by a line segment, either all points of $D$ are in $V_z$ or all points of $D$ are in $U \setminus V_z$. Hence, $V_z$ and $U \setminus V_z$ are open subsets of $U$ whose union is $U$. Since $U$ is connected, one of them must be empty. Since it contains $z$, $V_z$ is not empty, and so $U \setminus V_z$ must be empty. This means $V_z = U$ and every point of $U$ is connected to $z$ by a path in $U$. Hence, $U$ is path connected.

Conversely, suppose $U$ is path connected. If $U = A \cup B$, where $A$ and $B$ are disjoint non-empty open sets, then the function $f$ which is 1 on $A$ and 0 on $B$ is a continuous function on $U$, since the inverse image of any open subset of $\mathbb{R}$ under $f$ is $A$, $B$, $U$, or $\emptyset$, and these are all open. Now since $U$ is path connected, there is a path $\gamma$ connecting a point of $A$ to a point of $B$. Then $f \circ \gamma$ is a continuous function on an interval $I$ which takes on the values 1 and 0 and only these values. This is impossible, by the Intermediate Value Theorem. The resulting contradiction shows that there is no pair $A$, $B$ as above, and so $U$ is connected. □

By the above theorem, if $K$ is a compact subset of $\mathbb{C}$ (such as the image $\gamma(I)$ of a closed path $\gamma$), then $\mathbb{C} \setminus K$ is the union of its connected components, each of which is an open, path connected set.

**Theorem 2.7.3.** *If $K$ is a compact subset of $\mathbb{C}$, then $\mathbb{C} \setminus K$ has exactly one unbounded component.*

**Proof.** If $K$ is a compact set, then $K$ is closed and bounded. Since it is bounded, it is contained in some closed disc $\overline{D}_r(0)$. Then its complement $\mathbb{C} \setminus K$ is open and contains the complement of $\overline{D}_r(0)$. Since the complement of $\overline{D}_r(0)$ is connected, it is contained in one of the components of $\mathbb{C} \setminus K$. This means all the other components of $\mathbb{C} \setminus K$ are contained in $\overline{D}_r(0)$ and, hence, are bounded. □

**Example 2.7.4.** What are the components of $\mathbb{C} \setminus T$, where $T$ is the unit circle?

**Solution:** Clearly the sets $\{z \in \mathbb{C} : |z| < 1\}$ and $\{z \in \mathbb{C} : |z| > 1\}$ are path connected and, hence, connected. These two connected open sets have $\mathbb{C} \setminus T$ as union and so they must be the components of $\mathbb{C} \setminus T$. One of them is bounded and the other is unbounded.

**Index is Constant on Components.** We can now establish the result alluded to at the beginning of this section.

**Theorem 2.7.5.** *If $\gamma : I \to \mathbb{C}$ is a closed path, then $\operatorname{Ind}_\gamma(z)$ is constant on each component of $\mathbb{C} \setminus \gamma(I)$, and is zero in the unbounded component of $\mathbb{C} \setminus \gamma(I)$.*

**Proof.** The set $\gamma(I)$ is the image of a compact set under a continuous function and so it is compact, hence, closed. Its complement $\mathbb{C} \setminus \gamma(I)$ is, therefore, open. Thus, if $z_0 \in \mathbb{C} \setminus \gamma(I)$, there is an open disc, centered at $z_0$, and contained in $\mathbb{C} \setminus \gamma(I)$. Let $R$ be the radius of one such disc. We will show that, on some smaller disc, centered at $z_0$, $\operatorname{Ind}_\gamma(z)$ is constant.

Suppose $r$ is a positive number less than $R$ and $z \in D_r(z_0)$. Then

$$\begin{aligned} \operatorname{Ind}_\gamma(z) - \operatorname{Ind}_\gamma(z_0) &= \frac{1}{2\pi i}\int_\gamma \frac{dw}{w-z} - \frac{1}{2\pi i}\int_\gamma \frac{dw}{w-z_0} \\ &= \frac{1}{2\pi i}\int_\gamma \frac{z-z_0}{(w-z)(w-z_0)}\,dw. \end{aligned} \tag{2.7.1}$$

Furthermore, every point $w$ of $\gamma(I)$ is at least a distance $R$ away from $z_0$ and a distance $R-r$ away from $z$. That is,

$$|w - z_0| \geq R \quad \text{and} \quad |w - z| \geq R - r.$$

Since $|z - z_0| < r$, this implies that the integrand of the last integral in (2.7.1) is less than or equal to $r/R(R-r)$ and, hence, that

$$|\operatorname{Ind}_\gamma(z) - \operatorname{Ind}_\gamma(z_0)| \leq \frac{r\ell(\gamma)}{2\pi R(R-r)}.$$

We can make the right side of this inequality as small as we want by choosing $r$ sufficiently small. In particular, we can make it less than 1. However, $\operatorname{Ind}_\gamma(z)$ and $\operatorname{Ind}_\gamma(z_0)$ are both integers. If they differ by less than 1, then they are the same. Thus, if $r$ is chosen small enough, $\operatorname{Ind}_\gamma(z)$ is constant on $D_r(z_0)$.

Let $A$ be a component of $\mathbb{C} \setminus \gamma(I)$, and for each integer $n$ let $V_n$ be the set of points of $A$ on which $\text{Ind}_\gamma(z) = n$. The above argument shows that each $V_n$ is an open subset of $A$. So is the union of all the sets $V_m$ for which $m \neq n$. These two sets separate $A$ unless one of them is empty. Since $A$ is connected, one of them must be empty. This means that if $V_n$ is not empty, then it is all of $A$. Thus, only one of the sets $V_n$ can be non-empty and this means that $\text{Ind}_\gamma(z)$ is constant on $A$.

It remains to show that $\text{Ind}_\gamma(z) = 0$ on the unbounded component of $\mathbb{C} \setminus \gamma(I)$. Let $D$ be an open disc containing $\gamma(I)$ and let $z_0$ be a point outside this disc. Then $z_0$ is in the unbounded component of $\mathbb{C} \setminus \gamma(I)$. Furthermore,

$$\text{Ind}_\gamma(z_0) = \frac{1}{2\pi i} \int_\gamma \frac{dw}{w - z_0} = 0$$

by Cauchy's Integral Theorem, since $D$ is a convex set containing $\gamma(I)$ and $\dfrac{1}{w - z_0}$ is analytic on $D$. Since $\text{Ind}_\gamma(z)$ is constant on each component of $\mathbb{C} \setminus \gamma(I)$, it must be identically 0 on the unbounded component. □

**Example 2.7.6.** Calculate $\text{Ind}_\gamma(z)$ for the path $\gamma$ which traces $n$ times around a circle of radius $r$ centered at $z_0$, where, if $n$ is positive, this means $\gamma$ traces the circle in the counterclockwise direction, and, if $n$ is negative, it means $\gamma$ traces the circle in the clockwise direction.

**Solution:** A parameterization for such a $\gamma$ is $\gamma(t) = z_0 + r\,e^{int}$ with parameter interval $I = [0, 2\pi]$. Here $\gamma'(t) = irn\,e^{int}$ and so

$$\text{Ind}_\gamma(z_0) = \frac{1}{2\pi i} \int_0^{2\pi} \frac{\gamma'(t)dt}{\gamma(t) - z_0} = \frac{1}{2\pi i} \int_0^{2\pi} in\,dt = n.$$

We use the previous theorem to find the index at other points $z$. Since the interior of the circle traced by $\gamma$ is a component of $\mathbb{C} \setminus \gamma(I)$, $\text{Ind}_\gamma(z)$ must be constant on it. Therefore, it has the value $n$ at every $z$ with $|z - z_0| < r$. The other component of $\mathbb{C} \setminus \gamma(I)$ is the unbounded component $\{z : |z - z_0| > r\}$. On it, $\text{Ind}_\gamma(z) = 0$.

Thus, in this example, $\text{Ind}_\gamma(z)$ is an integer $n$ which is the number of times the path goes around $z$ if the path has positive orientation (goes in the counterclockwise direction), and is the negative of this number if the path has negative orientation. This includes the case where the path does not go around $z$ at all, because $z$ lies outside the circle. Then $n = 0$.

**Crossing a Path.** For most paths, the index function can easily be computed using the principle that if a path is crossed from right to left at a "simple" point of the path, then the index increases by 1. We will make this statement precise below and outline its proof. Most of the details are left to the exercises.

**Definition 2.7.7.** Let $\gamma$ be a path with parameter interval $I = [a, b]$ and let $D$ be an open disc in the plane. We will say that $\gamma$ *simply splits* $D$ if

(a) $J = \gamma^{-1}(D)$ is a non-empty open subinterval of $I$, or (in the case where the path is closed and $\gamma(a) = \gamma(b) \in D$) a union of two half-open subintervals $[a, c)$ and $(d, b]$; and

(b) $D \setminus \gamma(J)$ has exactly two components.

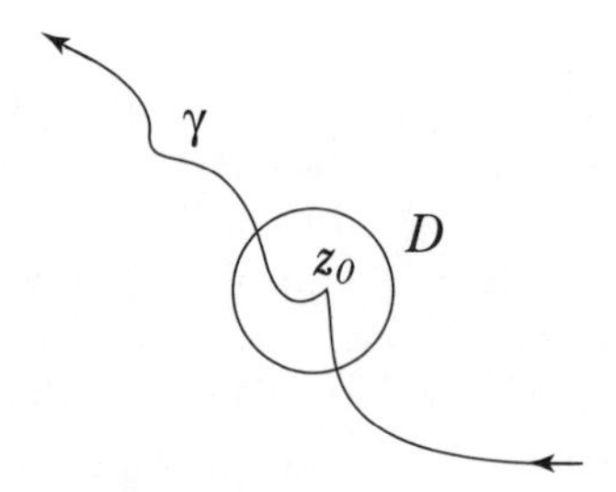


**Figure 2.7.1.** A Disc Simply Split by a Path.

Roughly speaking, a disc $D$ is simply split by a path if the path passes just once through $D$ and cuts it into exactly two connected components.

In this situation, there is a way to make sense of which of the two components is to the left of the path and which is to the right (Exercise 2.7.3).

**Theorem 2.7.8.** *Let $\gamma$ be a closed path which simply splits a disc $D$. Then*

$$\mathrm{Ind}_\gamma(z) = 1 + \mathrm{Ind}_\gamma(w)$$

*if $z$ is in the left and $w$ in the right component of $D \setminus \gamma(J)$.*

*Thus, if $z_0$ is a point of $\gamma$ which is the center of a disc which is simply split by $\gamma$, then $\mathrm{Ind}_\gamma(z)$ increases by 1 as $z$ crosses $\gamma$ from right to left at $z_0$.*

The proof is left to the exercises.

The next example illustrates how to use this to compute $\mathrm{Ind}_\gamma(z)$ in specific situations.

**Example 2.7.9.** Suppose $\gamma$ is a path which traces the circle $|z| = 2$ once in the counterclockwise direction beginning at 2 and then traces the circle $|z-1| = 1$ once in the counterclockwise direction. Find $\mathrm{Ind}_\gamma(z)$ for each $z$ that is not in $\gamma(I)$.

**Solution:** The components of $\mathbb{C} \setminus \{\gamma(I)\}$ are

$$\begin{aligned} A &= \{z \in \mathbb{C} : |z| > 2\}; \\ B &= \{z \in \mathbb{C} : |z| < 2,\ |z-1| > 1\}; \\ C &= \{z \in \mathbb{C} : |z-1| < 1\}. \end{aligned}$$

Since $A$ is the unbounded component, $\mathrm{Ind}_\gamma(z) = 0$ if $z \in A$. Crossing $\gamma$ from right to left at $2i$ takes us from points of $A$ to points of $B$. Hence, by the preceding theorem, $\mathrm{Ind}_\gamma(z) = 1$ for $z \in B$. Crossing $\gamma$ from right to left at $1+i$ takes us from points of $B$ to points of $C$ and so, again by the preceding theorem, $\mathrm{Ind}_\gamma(z) = 2$ if $z \in C$.

If $\gamma$ is a path with parameter interval $I = [a, b]$, and $t_0 \in (a, b)$, then the derivatives of $\gamma$ as a function on $[a, t_0]$ and $\gamma$ as a function on $[t_0, b]$ both exist at $t_0$ (see Definition 2.3.1 and the discussion preceding it). They are equal if $t_0$ is a smooth point of the path. If $t_0$ is not a smooth point, then $\gamma$ has two derivatives at $t_0$ – a left derivative $D_\ell\gamma$ and a right derivative $D_r\gamma$. At the endpoint $a$ of $I$, $\gamma$ has only a right derivative, while at the endpoint $b$ it has only a left derivative.

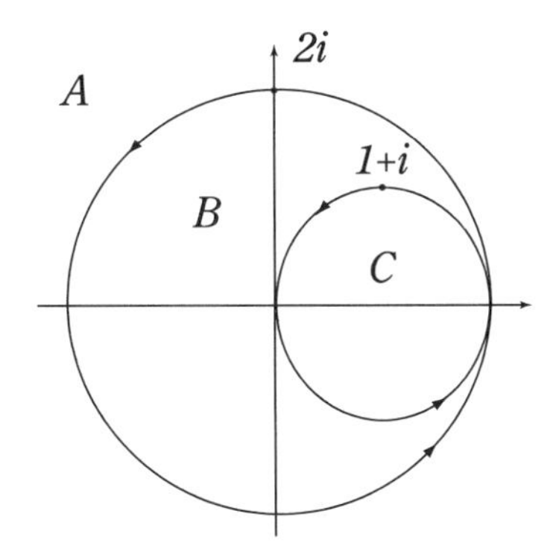


**Figure 2.7.2.** The Path for Example 2.7.9.

**Definition 2.7.10.** If $\gamma$ is a path with parameter interval $I = [a, b]$, a point $z_0$ on $\gamma(I)$ is said to be a *simple point* if $z_0 = \gamma(t_0)$ for exactly one parameter value $t_0 \in (a, b)$ or for the two values $a$ and $b$ (in this case, $\gamma$ is closed), and if the left and right derivatives of $\gamma$ at $t_0$ (at $a$ and $b$ if $z_0 = \gamma(z) = \gamma(b)$) are both non-zero.

In other words, a point is a simple point of a path if the path passes through the point just once and it both approaches the point from a definite direction with a positive speed and leaves the point in a definite direction with a positive speed. This leads to a very useful criterion for a point on a path to be the center of a disc that is simply split by the path.

**Theorem 2.7.11.** *If $z_0$ is a simple point of the path $\gamma$, then there is an open disc, centered at $z_0$, which is simply split by $\gamma$.*

Exercises 2.7.11 through 2.7.14 are devoted to proving this theorem in the case where the point $z_0$ is actually a smooth simple point of $\gamma$. In Exercise 2.7.15, the reader is asked to modify this argument so as to prove the theorem in general.

Note that in attempting to prove the above theorem, we may assume that $z_0 = \gamma(t_0)$ for some interior point $t_0$ of the parameter interval (otherwise we can just reparameterize to make this the case). Also, we may assume that $z_0 = 0$ since, otherwise, we can just translate the curve to make this the case. Both of these assumptions are made in Exercises 2.7.11 to 2.7.15.

Theorems 2.7.11 and 2.7.8 imply that if a closed path $\gamma$ is crossed from right to left at a simple point, then $\text{Ind}_\gamma$ increases by 1. For most of the paths $\gamma$ that we shall encounter, all but finitely many points of $\gamma$ are simple points. Exceptions to this rule are paths which retrace parts of themselves, cross themselves infinitely often, or come to a dead stop over some segment of the parameter interval.

## Exercise Set 2.7

1. Prove that the union of a family of connected sets with a point in common is also a connected set.

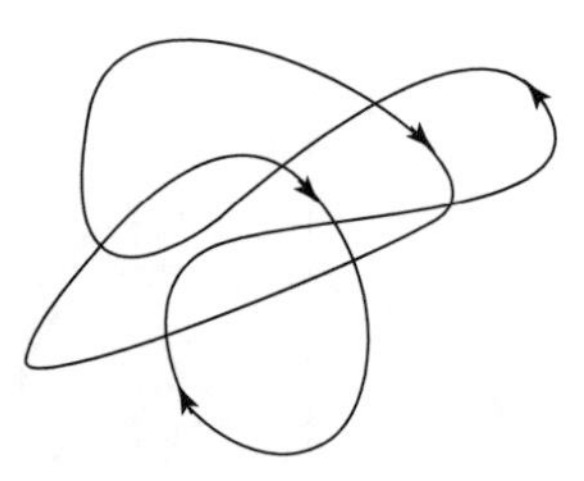

**Figure 2.7.3.** The Path for Exercise 2.7.9.

2. Prove Part (a) of Theorem 2.7.2. That is, prove that each component of an open set is open.
3. Let $D$ be an open disc and $\gamma : I \to \mathbb{C}$ be a closed path which simply splits $D$. Argue that, if we think of the positive direction along the curve through $D$ as being "up", then it makes sense to think of one of the components into which $\gamma$ splits $D$ as the "left" one and the other as the "right" one. Describe how to tell which is which.
4. Suppose a closed path $\gamma$ simply splits a disc $D$ as in Figure 2.7.1. Define two new paths $\gamma_1$ and $\gamma_2$ as follows: The curve $\gamma_1$ agrees with $\gamma$ until $\gamma$ enters $D$. It then departs from $\gamma$ and instead traces the boundary of $D$ in the counterclockwise direction until it rejoins $\gamma$. It agrees with $\gamma$ from that point on. The path $\gamma_2$ does the same thing except it traces the boundary of $D$ in the clockwise direction. For which points $z$ inside $D$ does $\operatorname{Ind}_\gamma(z) = \operatorname{Ind}_{\gamma_1}(z)$? For which points $z$ inside $D$ does $\operatorname{Ind}_\gamma(z) = \operatorname{Ind}_{\gamma_2}(z)$? What is $\operatorname{Ind}_{\gamma_1}(z) - \operatorname{Ind}_{\gamma_2}(z)$ if $z$ is any point inside $D$? Hint: Use Cauchy's Integral Theorem and Example 2.7.6.
5. Use the results of the previous exercise to prove Theorem 2.7.8.
6. Prove that if $\gamma$ is a closed path whose complement has just two components and if $\gamma$ has at least one simple point, then $\operatorname{Ind}_\gamma(z) = \pm 1$ on the bounded component.
7. In Example 2.7.9 how would the answers differ if the inner circle is traced in the clockwise direction rather than the counterclockwise direction?
8. If a path $\gamma$ traces a figure eight once, what are the possibilities for $\operatorname{Ind}_\gamma(z)$ in the two bounded components of the complement of the figure eight?
9. Determine the value of $\operatorname{Ind}_\gamma(z)$ in each of the components of $\mathbb{C} \setminus \gamma(I)$ if $\gamma$ is the curve of Figure 2.7.3.
10. Suppose $\gamma_1$ and $\gamma_2$ are closed paths and $z$ is not on either path. Show that $\operatorname{Ind}_{\gamma_1+\gamma_2}(z) = \operatorname{Ind}_{\gamma_1}(z) + \operatorname{Ind}_{\gamma_2}(z)$ and $\operatorname{Ind}_{-\gamma_1} = -\operatorname{Ind}_{\gamma_1}(z)$. Hint: Use Theorem 2.4.6.
11. Let $\gamma(t) = x(t) + iy(t)$ be a path with parameter interval $I = [a, b]$ and $t_0$ a point in $(a, b)$ at which the path is smooth and simple. For further simplicity, assume $\gamma(t_0) = 0$ (we can always achieve this by translating the path). Prove that, even though $\gamma''(t_0)$ may not exist, the second derivative of the function
$$h(t) = |\gamma(t)|^2 = x^2(t) + y^2(t)$$
does exist at $t_0$ and equals $2|\gamma'(t_0)|^2$.

12. Let $\gamma$ and $t_0$ be as in the previous exercise. Prove that there is an interval $(c, d) \subset (a, b)$, containing $t_0$, such that $|\gamma(t)|$ is strictly decreasing on $[c, t_0]$ and strictly increasing on $[t_0, d]$. Show that this implies that, if $\delta = \min\{|\gamma(c)|, |\gamma(d)|\}$, then $\gamma(t)$ crosses each circle of radius less than $\delta$, centered at 0, exactly once for $t \in (c, t_0)$ and exactly once for $t \in (t_0, d)$.
13. Let $\gamma$, $t_0$ and $(c, d)$ be as in the previous exercise. Prove that there is an open disc $D$, centered at 0, such that $\gamma^{-1}(D)$ is an open subinterval $J$ of $(c, d)$. Hint: Begin by showing you can choose $D$ small enough that it contains no points $\gamma(t)$ for $t \notin (c, d)$; then use the result of the previous exercise.
14. With $\gamma$, $t_0$, $(c, d)$, and $D$ as in the previous exercise, prove that $D$ is simply split by $\gamma$. This proves Theorem 2.7.11 in the case of a smooth simple point.
15. How would the argument outlined in the previous four exercises need to be modified to prove Theorem 2.7.11 for a point which is a simple point of $\gamma$, but not a smooth point? Note that, in this case, $\gamma(t)$ will have two derivatives at $t_0$ – one from the left and one from the right.

Chapter 3

# Power Series Expansions

In this section we present several striking applications of Cauchy's theorems. The first of these is the existence of local power series expansions for analytic functions. This leads to a number of other results, including the Fundamental Theorem of Algebra and detailed information about the zeroes and singularities of analytic functions.

Before we show that analytic functions have power series expansions we need to develop a deeper understanding of convergence issues for power series. The first section of the chapter is devoted to this task.

## 3.1. Uniform Convergence

We would like to be able to integrate and differentiate power series term by term. This is shown to be legitimate in the case of real power series in the typical advanced calculus or foundations of analysis course. The key to doing this is to show that power series *converge uniformly* on certain sets. We will give a brief development of these ideas in the context of complex-valued functions of a complex variable.

**Definition 3.1.1.** Let $\{f_n\}$ be a sequence of functions defined on a set $S \subset \mathbb{C}$. Then

(a) the sequence $\{f_n\}$ converges *pointwise* to the function $f$ on $S$ if, for each $z \in S$, the sequence of numbers $\{f_n(z)\}$ converges to the number $f(z)$ – that is, for each $z \in S$ and each $\epsilon > 0$, there is an $N$ such that

$$|f_n(z) - f(z)| < \epsilon \quad \text{for all} \quad n \geq N;$$

(b) the sequence $\{f_n\}$ converges *uniformly* on $S$ if for each $\epsilon > 0$ there exists an $N$ such that

$$|f_n(z) - f(z)| < \epsilon \quad \text{for all} \quad n \geq N \quad \text{and all} \quad z \in S.$$

There is a subtle but crucial difference between statements (a) and (b) in the above definition: In (b), given $\epsilon$, there must be an $N$ that works for all $z \in S$. In (a), for each $z$ there must be an $N$, but $N$ depends on $z$, in general, and there may not be an $N$ that works simultaneously for all $z \in S$.

The importance of uniform convergence stems primarily from two facts that are proved below and are extensively used thereafter: (1) The limit of a uniformly convergent sequence of continuous functions is also continuous; and (2) the integral along a path of the limit of a uniformly convergent sequence of continuous functions is the limit of their integrals.

**Theorem 3.1.2.** *If $E$ is a subset of $\mathbb{C}$ and $\{f_n\}$ is a sequence of continuous functions on $E$ which converges uniformly on $E$ to a function $f$, then $f$ is also continuous on $E$.*

**Proof.** Let $z_0$ be a point of $E$. Given $\epsilon > 0$, we choose $N$ such that

$$n \geq N \quad \text{implies} \quad |f(z) - f_n(z)| < \frac{\epsilon}{3} \quad \text{for all} \quad z \in E.$$

We can do this because $\{f_n\}$ converges uniformly to $f$ on $E$.

We next choose a $\delta > 0$ such that

$$|f_N(z) - f_N(z_0)| < \frac{\epsilon}{3} \quad \text{whenever} \quad z \in E \quad \text{and} \quad |z - z_0| < \delta.$$

We can do this because $f_N$ is continuous on $E$.

Then $|z - z_0| < \delta$ and $z \in E$ imply

$$\begin{aligned} |f(z) - f(z_0)| &\leq |f(z) - f_N(z)| + |f_N(z) - f_N(z_0)| + |f_N(z_0) - f(z_0)| \\ &< \frac{\epsilon}{3} + \frac{\epsilon}{3} + \frac{\epsilon}{3} = \epsilon. \end{aligned}$$

We conclude that $\lim_{z \to z_0} f(z) = f(z_0)$, and so $f$ is continuous at $z_0$. Since $z_0$ was a general point of $E$, $f$ is continuous on $E$. □

**Theorem 3.1.3.** *If $\gamma : I \to \mathbb{C}$ is a path and $\{f_n\}$ is a sequence of continuous functions on $\gamma(I)$ which converges uniformly on $\gamma(I)$ to $f$, then*

$$\lim_{n \to \infty} \int_\gamma f_n(z)\, dz = \int_\gamma f(z)\, dz. \tag{3.1.1}$$

**Proof.** Given $\epsilon > 0$, we choose $N$ such that

$$|f(z) - f_n(z)| < \epsilon / \ell(\gamma) \quad \text{for all} \quad n \geq N, \ z \in \gamma(I).$$

Then Theorem 2.4.9 implies that, for $n \geq N$,

$$\left| \int_\gamma f(z)\, dz - \int_\gamma f_n(z)\, dz \right| = \left| \int_\gamma (f(z) - f_n(z))\, dz \right| \leq \frac{\epsilon}{\ell(\gamma)} \ell(\gamma) = \epsilon$$

and this proves (3.1.1). □

**Example 3.1.4.** If $f_n(z) = |z|^n$, then prove that the sequence $\{f_n\}$ converges pointwise on $\overline{D}_1(0)$ but not uniformly. Show that it does converge uniformly on any disc $\overline{D}_r(0)$ with $r < 1$.

**Solution:** If $|z| < 1$, then $|z|^n \to 0$. If $|z| = 1$, then $|z|^n \to 1$. Thus, the sequence converges pointwise and the limit function $f(z)$ is 0 if $|z| < 1$ and 1 if $|z| = 1$. The convergence is not uniform because the limit function is not continuous.

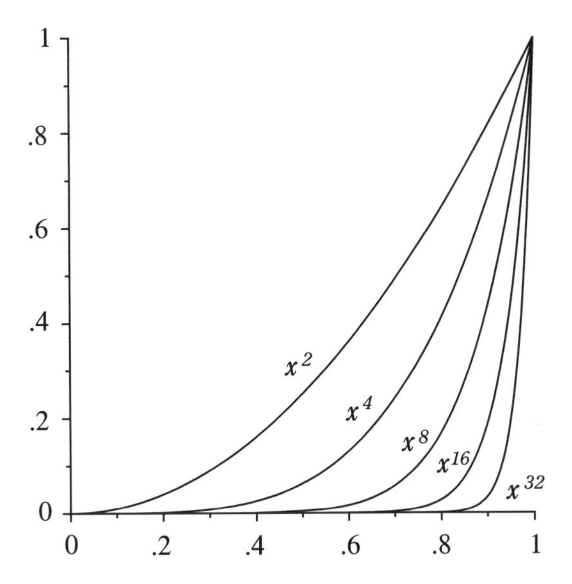


**Figure 3.1.1.** The Sequence $\{x^n\}$ does not Converge Uniformly on $[0, 1]$.

The fact that the convergence is not uniform can also be seen directly: No matter how large $n$ is chosen, we can always find a $z$ with $|z| < 1$ such that $|f_n(z) - f(z)| = |z|^n \geq 1/2$. In fact, $(1/2)^{1/n}$ is such a $z$. Thus, the condition for uniform convergence fails to hold for $\epsilon = 1/2$.

On the other hand, if $z \in \overline{D}_r(0)$ with $r < 1$, then $|z| \leq r$ and $|z|^n \leq r^n$. Given $\epsilon > 0$, if $N$ is chosen larger than $\log \epsilon / \log r$, then $n \geq N$ implies

$$|f_n(z) - f(z)| = |z|^n \leq r^n < \epsilon.$$

Since $N$ was chosen independent of $z$, the convergence is uniform on $\overline{D}_r(0)$.

The real analogue of the above example is the sequence $\{x^n\}$ on $[0, 1]$, which is illustrated in Figure 3.1.1.

**Uniform Convergence of Series.** We say that an infinite series $\sum_{k=0}^{\infty} f_k(z)$ of functions, defined on a set $E$, converges uniformly on $E$ if the sequence of partial sums $\{s_n\}$ converges uniformly on $E$, where we recall that

$$s_n(z) = \sum_{k=0}^{n} f_k(z).$$

There is a very useful criterion which insures uniform convergence of such a sequence. This is the *Weierstrass M-test*:

**Theorem 3.1.5** (Weierstrass $M$-Test). *Let*

$$\sum_{k=0}^{\infty} f_k(z) \tag{3.1.2}$$

*be an infinite series of functions defined on a set $E$. If there is a convergent series of non-negative numbers*

$$\sum_{k=0}^{\infty} M_k \tag{3.1.3}$$

*such that $|f_k(z)| \le M_k$ for all $k$ and all $z \in E$, then* (3.1.2) *converges uniformly on $E$.*

**Proof.** The comparison test, comparing (3.1.2) to (3.1.3), shows that, for each $z \in E$, the series (3.1.2) converges. Let $s(z)$ be the number it converges to, and let $s_n(z)$ denote the $n$th partial sum of (3.1.2). Then

$$|s(z) - s_n(z)| \le \sum_{k=n+1}^{\infty} |f_k(z)| \le \sum_{k=n+1}^{\infty} M_k. \tag{3.1.4}$$

Since the series (3.1.3) converges, given $\epsilon > 0$, we can choose $N$ such that the right side of (3.1.4) is less than $\epsilon$ for all $n \ge N$. Then (3.1.4) implies that $|s(z)-s_n(z)| < \epsilon$ for all $n \ge N$ and all $z \in E$. Since $N$ was chosen independently of $z$, this shows that the convergence is uniform. □

**Example 3.1.6.** Show that the series $\sum_{k=1}^{\infty} z^k/k^2$ converges uniformly on the closed unit disc $\overline{D}_1(0)$.

**Solution:** We have $|z^k/k^2| \le 1/k^2$ if $|z| \le 1$. Furthermore, $\sum_{k=1}^{\infty} 1/k^2$ converges, because it is a $p$-series with $p = 2$. Hence, by the Weierstrass $M$-test, the series $\sum_{k=1}^{\infty} z^k/k^2$ converges uniformly on $\overline{D}_1(0)$.

**Uniform Convergence of Power Series.** In Section 1.2 we stated without proof that a complex power series converges on a certain open disc and diverges at all points in the complement of the corresponding closed disc. The radius of this disc is called the *radius of convergence* of the power series. We are now prepared to prove this result and give a formula for the radius of convergence.

The theorem that does this uses the notion of lim sup of a sequence. This is defined as follows.

If $S$ is a non-empty set of real numbers, then an *upper bound* for $S$ is a number $M$ such that $s \le M$ for every $s \in S$. If there is an upper bound for $S$, then $S$ is said to be *bounded above.* The completeness axiom for the real number system states that each non-empty set of real numbers $S$ that is bounded above, has a least upper bound $L$. This means $L$ has two properties: (1) it is greater than or equal to each element of $S$, and (2) it is less than every other number with this property. We will denote the least upper bound of a non-empty set $S$ which is bounded above by $\sup(S)$. If the set is non-empty, but not bounded above, we set $\sup(S) = \infty$. The notion of inf or *greatest lower bound* is defined analogously, but the inequalities are all reversed.

**Definition 3.1.7.** If $\{a_k\}$ is a sequence of real numbers, then $\limsup\{a_k\}$ is the limit of the non-increasing sequence $\{u_n\}$ defined by

$$u_n = \sup\{a_k : k \ge n\}.$$

The sequence $\{u_n\}$ of this definition is non-increasing because as $n$ increases, the set of numbers $\{a_k : k \geq n\}$ gets smaller. The sequence $\{u_n\}$ may not, however, be bounded below and so $\limsup\{a_k\}$ may be $-\infty$. Also, the $u_n$ could all be $+\infty$, in which case $\limsup\{a_k\} = +\infty$.

Of course, there is an analogous notion, lim inf, which uses inf instead of sup in the above definition.

The lim sup and lim inf of a sequence $\{a_k\}$ always exist (but may be infinite), even though $\lim a_k$ itself may not exist. In fact the sequence has a limit (which may be infinite) if and only if $\limsup a_k = \liminf a_k$. In this case $\lim a_k$ is this common value.

**Theorem 3.1.8.** *Given a power series $\sum_{k=0}^\infty c_k(z-z_0)^k$, let*

$$R = \left(\limsup |c_k|^{1/k}\right)^{-1}.$$

*Then the series converges absolutely if $z \in D_R(z_0)$ and diverges if $z \notin \overline{D}_R(z_0)$. Furthermore, it converges uniformly on each closed disc $\overline{D}_r(z_0)$ with $r < R$. Thus, $R$ is the radius of convergence of the given power series.*

**Proof.** Since we can always make a change of variables which replaces $z - z_0$ by $z$, we may as well assume that $z_0 = 0$.

By definition, $\limsup |c_k|^{1/k}$ is the limit of the non-increasing sequence $\{u_n\}$ where

$$u_n = \sup\{|c_k|^{1/k} : k \geq n\}.$$

If $r < R$, we choose a number $t$ with $r < t < R$. Then $t^{-1} > R^{-1} = \lim u_n$. This implies that for large enough $n$ the numbers $u_n$ are less than $t^{-1}$. If $n$ is one such integer, then, for all $k \geq n$,

$$|c_k|^{1/k} < t^{-1} \quad \text{and, hence,} \quad |c_k| < t^{-k}.$$

Now if $|z| \leq r$, then this implies that

$$|c_k z^k| < \left(\frac{r}{t}\right)^k \quad \text{for all} \quad k \geq n. \tag{3.1.5}$$

Since $|r/t| < 1$, the geometric series $\sum_{k=n}^\infty (r/t)^k$ converges. Then the Weierstrass $M$-test implies that the series $\sum_{k=n}^\infty c_k z^k$ converges uniformly in the disc $\overline{D}_r(0)$. The same is true of the original series $\sum_{k=0}^\infty c_k z^k$, since the convergence or uniform convergence of a series is unaffected by the first $n$ terms if $n$ is fixed.

Since the series converges on each closed disk $\overline{D}_r(0)$, with radius less than $R$, it converges at each point $z$ in the open disc $D_R(0)$.

It remains to prove that the series diverges at each $z$ with $|z| > R$. Given such a $z$, we have

$$|z|^{-1} < \lim u_n.$$

This implies that, for each $n$, there is a $k > n$ with

$$|z|^{-1} < |c_k|^{1/k}, \quad \text{so that} \quad |c_k z^k| > 1.$$

But this means that there is a subsequence of the sequence of terms $\{c_k z^k\}$ consisting of numbers with modulus greater than 1. Since the sequence of terms does not converge to 0, the series diverges by the term test (Exercise 1.2.9). □

The above theorem has the following corollary, the proof of which is left as an exercise.

**Corollary 3.1.9.** *If $f$ has a power series expansion about $z_0$ which converges on the disc $D_R(z_0)$, then $f$ is continuous on this disc.*

**Example 3.1.10.** Prove that if $\sum_{k=0}^{\infty} c_k z^k$ is a power series with radius of convergence $R$, then the power series $\sum_{k=1}^{\infty} k c_k z^{k-1}$ also has radius of convergence $R$.

**Solution:** If we multiply the second series by $z$, the set on which the series converges does not change, and so its radius of convergence does not change. The resulting series is $\sum_{k=0}^{\infty} k c_k z^k$. Let $R_1$ be its radius of convergence. By the previous theorem,

$$R = \left(\limsup |c_k|^{1/k}\right)^{-1},$$

and

$$R_1 = \left(\limsup |kc_k|^{1/k}\right)^{-1} = \left(\limsup k^{1/k} |c_k|^{1/k}\right)^{-1}.$$

The sequence $k^{1/k}$ has limit 1 (Exercise 3.1.11), and so the factor $k^{1/k}$ does not effect the lim sup (Exercise 3.1.12). Hence, $R_1 = R$.

Theorem 3.1.8 and Theorem 3.1.3 combine to prove that it is legitimate to integrate a power series term by term.

**Theorem 3.1.11.** *Let $f(z) = \sum_{k=0}^{\infty} c_k (z - z_0)^k$, where the radius of convergence of this power series is $R$. Then*

$$\int_{z_0}^{z} f(w)\, dw = \sum_{k=0}^{\infty} \frac{c_k}{k+1} (z - z_0)^{k+1} \tag{3.1.6}$$

*for all $z \in D_R(z_0)$.*

**Proof.** If we set $s_n(z) = \sum_{k=0}^{n} c_k (z - z_0)^k$ for each positive integer $n$, then

$$\int_{z_0}^{z} s_n(w)\, dw = \sum_{k=0}^{n} \frac{c_k}{k+1} (z - z_0)^{k+1}$$

because the integral is linear and we know how to integrate $(z - z_0)^k$. To finish the proof of (3.1.6), we just need to take the limit of both sides and use Theorem 3.1.3 to bring the limit inside the integral on the left. Of course, we need to know that the convergence of $\{s_n\}$ to $f$ is uniform on $[z_0, z]$. This, however, follows from Theorem 3.1.8, since the interval $[z_0, z]$ is inside the closed disc $\overline{D}_r(z_0)$, where $r = |z| < R$. □

## Exercise Set 3.1

1. Show that the sequence $\{1/(nz)\}$ converges uniformly to 0 on every set of the form $\{z : |z| \geq r\}$ for fixed $r > 0$, but it does not converge uniformly on $\{z : z \neq 0\}$.

2. Show that the sequence $\{\sin(x/n)\}$ converges uniformly to 0 on any interval of the form $[0, k]$, but it does not converge uniformly on $[0, \infty)$.
3. Show that the sequence $\{\arctan(nx)\}$ converges pointwise but not uniformly on $\mathbb{R}$.
4. Use the Weierstrass $M$-test (but not Theorem 3.1.8) to show that the series $\sum_{k=1}^{\infty} \frac{z^k}{k!}$ converges uniformly on $D_R(0)$ for each $R > 0$.
5. Use the Weierstrass $M$-test to show that the series $\sum_{k=1}^{\infty} \frac{k+z}{k^3+1}$ converges uniformly on $\overline{D}_1(0)$.
6. Show that for each $r > 0$ the series
$$\sum_{k=0}^{\infty} \frac{1}{k^2 - z}$$
converges uniformly on the set
$$E_r = \{z : |z| \leq r, \; z \neq k^2 \text{ for } k = 0, 1, 2, \cdots\}.$$
7. Prove that the series $\sum_{k=1}^{\infty} k^{-z}$ converges uniformly on each set of the form $\{z \in \mathbb{C} : \mathrm{Re}(z) > s\}$, with $s > 1$. The function to which it converges is called the Riemann Zeta Function.
8. If the series of the previous exercise is differentiated term by term, does the resulting series still converge uniformly on $\{z \in \mathbb{C} : \mathrm{Re}(z) > s\}$ if $s > 1$?
9. For each $n$ find $\sup\{1 + (-1)^k + 1/k : k \geq n\}$.
10. Find the radius of convergence of the power series $\sum_{k=0}^{\infty} (2 + (-1)^k)^k z^k$.
11. Prove that $\lim k^{1/k} = 1$.
12. Prove that if $\{a_k\}$ and $\{b_k\}$ are two sequences of non-negative numbers with $\lim a_k = a$ and $\limsup b_k = b$, then $\limsup a_k b_k = ab$.
13. Prove Corollary 3.1.9.
14. Can a power series of the form $\sum_{k=0}^{\infty} c_k (z-1)^k$ converge at $z = 3$ and diverge at $z = 0$? Why?
15. Using the power series expansion $\frac{1}{1+w} = \sum_{k=0}^{\infty} (-1)^k w^k$, find a power series expansion for $\int_0^z \frac{1}{1+w}\, dw$ about 0. What is the radius of convergence of this power series? What function does it converge to?
16. If a function $E(z)$ is defined on $\mathbb{C}$ by
$$E(z) = \int_0^z e^{-w^2}\, dw,$$
find a power series expansion for $E(z)$ about 0. Where does this power series converge?

## 3.2. Power Series Expansions

A function of a real variable can be differentiable, even infinitely differentiable on an interval and still not have a convergent power series expansion in that interval. For functions of a complex variable, the situation is quite different. We will prove that every analytic function has convergent power series expansions about each point of its domain. In fact, we will prove that a function $f$ is analytic if and only if it has such expansions. First, we show that a function which has a power series expansion on a disc is analytic on that disc. This involves showing that we can differentiate power series term by term.

**Differentiating Power Series.** The next theorem and its corollaries concern a function $f$ defined by a convergent power series

$$f(z) = \sum_{n=0}^{\infty} c_n (z - z_0)^n \tag{3.2.1}$$

with radius of convergence $R$.

**Theorem 3.2.1.** *If $f$ is defined as above, then $f$ is analytic on $D_R(z_0)$ and $f'$ has a convergent power series expansion*

$$f'(z) = \sum_{n=1}^{\infty} n c_n (z - z_0)^{n-1},$$

*which converges to $f'$ on $D_R(z_0)$.*

**Proof.** If $f$ has a power series expansion (3.2.1) with radius of convergence $R$, let $g$ be the function which we hope turns out to be the derivative of $f$ – that is, we set

$$g(z) = \sum_{n=1}^{\infty} n c_n (z - z_0)^{n-1}.$$

This series has the same radius of convergence as the series for $f$ (see Example 3.1.10) and so it converges on $D_R(z_0)$ and converges uniformly on any smaller closed disc. By Corollary 3.1.9, $g$ is continous on $D_R(z_0)$, and by Theorem 3.1.11,

$$\int_{z_0}^{z} g(w)\, dw = \sum_{n=1}^{\infty} c_n z^n = f(z) - f(z_0)$$

on $D_R(z_0)$. Theorem 2.6.1 tells us this function is an antiderivative for $g(z)$. Since $f(z_0)$ is a constant, this means $f' = g$ on $D_R(z_0)$, as required. □

**Example 3.2.2.** Find a power series expansion for the principal branch of the log function about the point $z_0 = 1$.

**Solution:** We know that the derivative of $\log(z)$ is $1/z$ (Exercise 2.2.14). We also know that, since

$$\frac{1}{z} = \frac{1}{1 - (1 - z)},$$

the power series expansion of $1/z$ about $z_0 = 1$ is

$$\frac{1}{z} = \sum_{n=0}^{\infty}(1-z)^n = \sum_{n=0}^{\infty}(-1)^n(z-1)^n. \tag{3.2.2}$$

Since we can differentiate term by term, a series which has this as its derivative is the series

$$\sum_{n=0}^{\infty}(-1)^n\frac{(z-1)^{n+1}}{n+1} = -\sum_{n=1}^{\infty}(-1)^{n-1}\frac{(z-1)^n}{n}. \tag{3.2.3}$$

The series (3.2.2) and (3.2.3) both have radius of convergence 1. If $f(z)$ denotes the sum of series (3.2.3), then $f'(z) = 1/z = \log'(z)$. It follows that $f$ and log differ by a constant. Since they both have the value 0 at $z = 1$, they are the same. Therefore, (3.2.3) is the power series expansion of $\log(z)$ about 1.

Using Theorem 3.2.1, the following can be proved in the same way as its real variable counterpart. The details are left to the exercises.

**Corollary 3.2.3.** *If $f$ has a power series expansion* (3.2.1) *about $z_0$, with radius of convergence $R$, then it has derivatives of all orders on $D_R(z_0)$. Its $k$th derivative is*

$$f^{(k)}(z) = \sum_{n=k}^{\infty}\frac{n!\,c_n}{(n-k)!}(z-z_0)^{n-k}.$$

*In particular, its $k$th derivative at $z_0$ is given by*

$$f^{(k)}(z_0) = k!\,c_k.$$

This immediately implies:

**Corollary 3.2.4.** *If $f$ has a power series expansion* (3.2.1) *about $z_0$, with positive radius of convergence, then it has only one such expansion. In fact, the coefficients $\{c_n\}$ for such an expansion are uniquely determined by the equations*

$$c_n = \frac{f^{(n)}(z_0)}{n!}.$$

**Power Series Expansions of Analytic Functions.** We are now ready to present the most important application of Cauchy's theorems – the proof of the existence of local power series expansions of analytic functions.

**Theorem 3.2.5.** *Let $f$ be analytic in an open set $U$ and suppose $D_r(z_0)$, $r > 0$, is an open disc contained in $U$. Then there is a power series expansion for $f$*

$$f(z) = \sum_{n=0}^{\infty} c_n(z-z_0)^n,$$

*which converges to $f(z)$ on $D_r(z_0)$. Furthermore, the coefficients of this power series are the numbers*

$$c_n = \frac{1}{2\pi i}\int_{|w-z_0|=s}\frac{f(w)}{(w-z_0)^{n+1}}\,dw, \tag{3.2.4}$$

*where $s$ is any number with $0 < s < r$.*

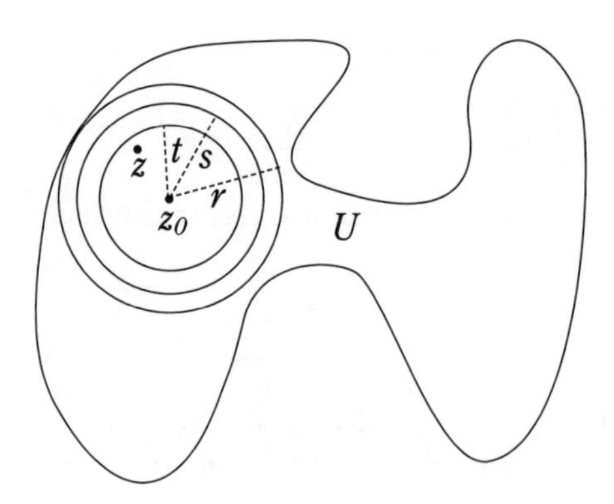


**Figure 3.2.1.** Setup for the Proof of the Existence of Power Series Expansions.

**Proof.** If $0 < t < s < r$, $|w - z_0| = s$, and $|z - z_0| \le t$, then

$$\left|\frac{z - z_0}{w - z_0}\right| \le \frac{t}{s} < 1,$$

and so we have

$$\frac{w - z_0}{w - z} = \left(1 - \frac{z - z_0}{w - z_0}\right)^{-1} = \sum_{n=0}^{\infty} \left(\frac{z - z_0}{w - z_0}\right)^n, \tag{3.2.5}$$

where the geometric series on the right is dominated by the constant geometric series $\sum_{n=0}^{\infty}(t/s)^n$. By Theorem 3.1.5, this implies that (3.2.5) converges uniformly as a function of $z \in D_t(z_0)$ and also as a function of $w \in \partial D_s(z_0)$.

If we multiply (3.2.5) by $f(w)/(w - z_0)$ and integrate around the circle of radius $s$, then, since the series converges uniformly in $w$, we may integrate term by term. Using Cauchy's Integral Formula, this yields

$$\begin{aligned} f(z) &= \frac{1}{2\pi i} \int_{|z_0 - w| = s} \frac{f(w)}{w - z}\, dw \\ &= \frac{1}{2\pi i} \sum_{n=0}^{\infty} \left(\int_{|w - z_0| = s} \frac{f(w)}{(w - z_0)^{n+1}}\, dw\right) (z - z_0)^n. \end{aligned}$$

This gives us a power series expansion of $f$ on $D_s(z_0)$ with coefficients given by (3.2.4).

Now given any $z \in D_r(z_0)$, we may choose $s$ such that $|z - z_0| < s < r$. With this choice of $s$, the above series is defined and converges at $z$. However, it follows from Cauchy's Integral Theorem that the integrals defining the coefficients of this series do not depend on the choice of $s$ (see Exercise 2.6.9). Thus, we have a power series expansion for $f$ which converges on $D_r(z_0)$, with coefficients given by (3.2.4). This completes the proof. □

**Corollary 3.2.6.** *If $f$ is analytic on an open set $U$, then $f$ has derivatives of all orders on $U$ and they are all analytic.*

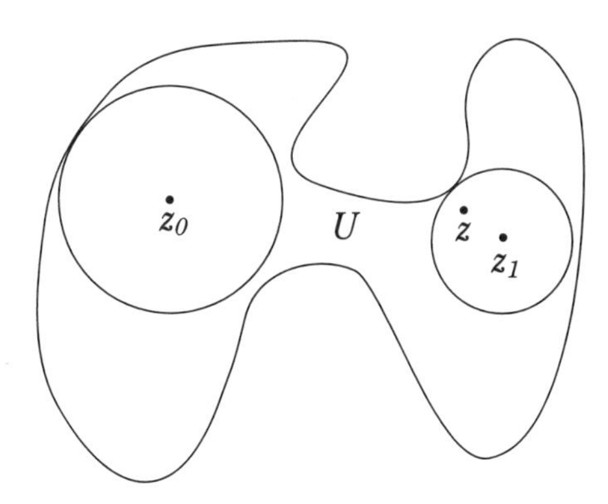


**Figure 3.2.2.** Shift to an Expansion About a Different Point of $U$.

**Proof.** Given any point $z_0 \in U$, there is a disc $D_r(z_0)$ centered at $z_0$ which is contained in $U$. On this disc, $f$ has a convergent power series expansion. By Corollary 3.2.3 $f$ has derivatives of all orders on this disc. These derivatives are analytic because each of them has a complex derivative. Therefore, $f$ has analytic derivatives of all orders on all of $U$. □

It is important to emphasize that the power series expansion of an analytic function $f$ about a point $z_0$ converges in the largest open disc, centered at $z_0$, that is contained in the domain $U$ of $f$. In general, it will not converge at other points of $U$. To obtain a power series expansion of $f$ that converges in a neighborhood of a point $z$ outside this disc, we have to shift to a power series expansion about a different point of $U$ – a point $z_1$ with the property that the largest disc centered at $z_1$ and contained in $U$ contains the point $z$ (see Figure 3.2.2). For example, the power series expansion of $\log z$ about $z = 1$ given in Example 3.2.2 has radius of convergence 1, and so it does not converge at $z = i$. There is a power series expansion of $\log z$ about $z = i$ (Exercise 3.2.7), but it is a different power series than the one about $z = 1$.

**Example 3.2.7.** Show that if $f$ is a function which is analytic in an open set $U$, $z_0 \in U$, and $f(z_0) = 0$, then $g(z) = f(z)/(z - z_0)$ can be given a value at $z_0$ which makes it analytic on $U$.

**Solution:** By the above theorem $f$ has a power series expansion about $z_0$ which converges in the largest open disc $D$ which is centered at $z_0$ and contained in $U$. Since $f(z_0) = 0$, the constant term of this series is 0. Hence, this expansion has the form

$$f(z) = c_1(z - z_0) + c_2(z - z_0)^2 + \cdots + c_n(z - z_0)^n + \cdots .$$

If we give the function $g(z)$ the value $c_1$ at $z = z_0$, then it agrees on $D$ with the sum of the power series

$$c_1 + c_2(z - z_0) + c_3(z - z_0)^2 + \cdots + c_n(z - z_0)^{n-1} + \cdots .$$

Since it has a power series expansion on $D$, $g$ is analytic on $D$. Also, $g$ is analytic in $U \setminus \{z_0\}$ since $f$ is analytic, and $z - z_0$ is analytic and non-vanishing on this set.

A function, defined on the union of two open sets and analytic on each of them, is clearly also analytic on their union; and so $g$ is analytic on $U$.

**Cauchy's Estimates.** We now know that a function which is analytic in an open disc $D_r(z_0)$ has a power series expansion about $z_0$ which converges in that disc. We also have integral formulas (3.2.4) for the coefficients of this power series. However, we also have the formulas

$$c_n = \frac{f^{(n)}(z_0)}{n!} \tag{3.2.6}$$

for these coefficients from Corollary 3.2.4. Combining these yields integral formulas for the derivatives of an analytic function.

**Theorem 3.2.8.** *Let $f$ be analytic in an open set containing the closed disc $\overline{D}_R(z_0)$. Then,*

$$f^{(n)}(z_0) = \frac{n!}{2\pi i} \int_{|w-z_0|=R} \frac{f(w)}{(w-z_0)^{n+1}}\, dw \tag{3.2.7}$$

*for $n = 0, 1, 2, \cdots$.*

**Proof.** We just need to observe that if $\overline{D}_R(z_0) \subset U$, with $U$ open, then there is an open disc $D_r(z_0)$ with $\overline{D}_R(z_0) \subset D_r(z_0) \subset U$ (Exercise 3.2.4). We can then apply Theorem 3.2.5 with $s = R$ to obtain a power series expansion of $f$ with coefficients given by (3.2.4). Then Corollary 3.2.3 relates these coefficients to the derivatives of $f$ at $z_0$. □

This leads to a very powerful tool. By estimating the size of the integrands in this formula, we can get estimates on the size of the derivatives of $f$. These estimates are called *Cauchy's estimates.*

**Theorem 3.2.9** (Cauchy's Estimates). *If $f$ is analytic on an open set containing the closed disc $\overline{D}_R(z_0)$, and if $|f(z)| \le M$ on the boundary of this disc, then*

$$|f^{(n)}(z_0)| \le \frac{n!M}{R^n} \tag{3.2.8}$$

*for $n = 0, 1, 2, \cdots$.*

**Proof.** We use the previous theorem. Since $|w - z_0| = R$ and $|f(w)| \le M$ for $w$ on the path $|w - z_0| = R$, the integrand of (3.2.7) is bounded by $M/R^{n+1}$. The length of the path is the circumference of a circle of radius $R$ and so it is $2\pi R$. Thus, Theorem 2.4.9 implies that

$$|f^{(n)}(z_0)| \le \frac{n!}{2\pi} \frac{M}{R^{n+1}} 2\pi R = \frac{n!M}{R^n}$$

for each non-negative integer $n$. □

This theorem will provide the crucial step in the proof of Liouville's Theorem in the next section.

**Example 3.2.10.** Find upper bounds on the derivatives at 0 of a function $f$ which is analytic on the unit disc $D_1(0)$ and has modulus bounded by one on this disc. Also find bounds on the moduli of coefficients in the power series expansion of this function about 0.

**Solution:** We apply Cauchy's estimates. If $r < 1$, then $f$ is analytic in the open set $D_1(0)$, which contains $\overline{D}_r(0)$. Since $|f(z)| \leq 1$ on $D_1(0)$, we may choose $M = 1$ and $R = r$ in Cauchy's estimates. We conclude

$$|f^{(n)}(0)| \leq \frac{n!}{r^n}.$$

However, since $r$ was any positive number less than 1, we may pass to the limit as $r \to 1$ and conclude that

$$|f^{(n)}(0)| \leq n!.$$

By (3.2.6), the corresponding estimate on the power series coefficients is

$$|c_n| \leq 1.$$

**Morera's Theorem.** This is a very handy tool for showing that a function is analytic.

**Theorem 3.2.11** (Morera's Theorem). *Let $f$ be a continuous function defined on an open set $U$. If the integral of $f$ is 0 around the boundary of every triangle that is contained in $U$, then $f$ is analytic in $U$.*

**Proof.** Theorem 2.6.1 says that a function $f$ that is continuous on a convex open set $U$, and has the property that its integral around any triangle in $U$ is 0, has a complex antiderivative $g$ in $U$. However, the fact that $g' = f$ on $U$ means, in particular, that $g$ is analytic on $U$. But then $f$ is analytic on $U$ by Corollary 3.2.6.

The hypothesis that $U$ is convex in the above argument is not necessary, since every open set is a union of convex open sets (open discs, in fact). Thus, we can apply the argument of the previous paragraph to each open disc contained in $U$ and conclude that $f$ is analytic on each of them. In particular, it has a derivative at each point of $U$ and, hence, is analytic on $U$. □

An example of how Morera's Theorem is used is provided by the following theorem.

**Theorem 3.2.12.** *Let $\{f_n\}$ be a sequence of analytic functions on an open set $U$ and suppose this sequence converges uniformly to $f$ on each compact subset of $U$. Then $f$ is analytic on $U$.*

**Proof.** Since $f_n \to f$ uniformly on each compact subset of $U$, $f$ is continuous on $U$. The convergence is uniform, in particular, on $\partial\Delta$ for every triangle $\Delta$ contained in $U$. Given such a triangle $\Delta$, Theorem 3.1.3 implies that

$$\int_{\partial\Delta} f(z)\,dz = \lim_{n\to\infty} \int_{\partial\Delta} f_n(z)\,dz.$$

Since each $f_n$ is analytic, Theorem 2.5.8 implies

$$\int_{\partial\Delta} f_n(z)\,dz = 0$$

for each $n$. We conclude that

$$\int_{\partial\Delta} f(z)\, dz = 0,$$

for every triangle $\Delta \subset U$. By Morera's Theorem, $f$ is analytic on $U$. □

## Exercise Set 3.2

1. Use the power series expansion for $\dfrac{1}{1-z}$ about 0 to find the power series expansion of $\dfrac{1}{(1-z)^2}$ about 0.
2. Find a power series expansion of $\sqrt{1+z}$ about 0, where the square root function is defined in terms of the principal branch of the log function. What is the radius of convergence of this series?
3. Prove Corollary 3.2.3.
4. Prove that if $\overline{D}_R(z_0)$ is a closed disc contained in an open set $U$, then there is an open disc $D_r(z_0)$ such that $\overline{D}_R(z_0) \subset D_r(z_0) \subset U$.
5. Show that if $f$ is analytic on an open set $U$, then, as a function on $\mathbb{R}^2$, it is $\mathcal{C}^\infty$ – that is, its partial derivatives of all orders exist and are continuous.
6. The function $\dfrac{1}{\cos z}$ has a power series expansion about $z_0 = 0$. Without finding the series, show that its radius of convergence is $\pi/2$.
7. Find the power series expansion of the principal branch of the log function about the point $z = i$. There are several ways to do this, one of which is really easy (see Example 3.2.2).
8. Use power series methods to show that the function which is $\dfrac{\sin z}{z}$ when $z \neq 0$ and 1 when $z = 0$ is analytic on the whole complex plane.
9. If $f$ is analytic and not identically 0 on a disc $D_r(z_0)$, show that there is a non-negative integer $k$ and a function $g$, which is also analytic in $D_r(z_0)$, such that $f(z) = (z - z_0)^k g(z)$ and $g(z_0) \neq 0$.
10. Prove that if $f$ is analytic on the disc $D_R(z_0)$ and $|f(z)| \leq M$ on $D_R(z_0)$, then $|f'(z_0)| \leq M/R$.
11. Suppose $p(z) = a_3z^3 + a_2z^2 + a_1z + a_0$ is a polynomial of degree 3. If $|p(z)| \leq 1$ on the unit circle $\{z : |z| = 1\}$, then show that $|a_3| \leq 1$.
12. Suppose $f$ is analytic in an open set $U$. Also, suppose $z \in U$ and the distance from $z$ to the complement of $U$ is $d$. If $|f(w)| \leq M$ for all $w \in U$, find estimates, similar to Cauchy's estimates, on the size of $|f^{(n)}(z)|$ in terms of $M$ and $d$.
13. Suppose $f$ is analytic on a disc $D_r(z_0)$ and unbounded (there is no $M$ such that $|f(z)| \leq M$ on $D_r(z_0)$). Then prove that the radius of convergence of the power series expansion of $f$ about $z_0$ is $r$.

14. Use Morera's Theorem to show that if $f$ is continuous on an open set $U$ and analytic on $U \setminus E$, where $E$ is either a point or a line segment, then $f$ is actually analytic on all of $U$.
15. Use Cauchy's estimates to prove that if $\{f_n\}$ is a sequence of analytic functions on an open set $U$, converging uniformly to $f$ on each compact subset of $U$, then $\{f_n^{(k)}\}$ converges uniformly to $f^{(k)}$ on each compact subset of $U$.
16. Use Morera's Theorem to prove that if $U$ is an open subset of $\mathbb{C}$, $I = [a, b]$ is an interval on the real line, and $g(z, t)$ is a continuous function on $U \times I$ which is analytic in $z$ for each $t \in I$, then the function

$$f(z) = \int_a^b g(z, t)\, dt$$

is analytic in $U$.

## 3.3. Liouville's Theorem

Liouville's Theorem is simple to state, very easy to prove (given what we know at this point), and extremely powerful. It concerns entire functions, where an *entire function* is a function which is analytic in the entire complex plane. It also concerns bounded functions, where a function is bounded on a set $E$ if there is a positive constant $M$ such that $|f(z)| \le M$ for every $z \in E$. If $f$ is bounded on its domain, we simply say it is *bounded*. Thus, a bounded entire function is a function which is analytic and bounded on $\mathbb{C}$.

**Theorem 3.3.1** (Liouville's Theorem). *The only bounded entire functions are the constant functions.*

**Proof.** The reader who did Exercise 3.2.10 of the preceding section has nearly completed the proof of Liouville's Theorem. The exercise states a simple consequence of the Cauchy estimates: If a function $f$ is analytic on a disc of radius $R$, centered at $z_0$, and if $|f(z)| \le M$ for all $z$ in this disc, then

$$|f'(z_0)| \le \frac{M}{R}. \tag{3.3.1}$$

If $f$ is bounded and entire, then $f(z)$ is analytic on the entire plane and $|f(z)|$ is bounded by some number $M$ on the entire plane. This implies that (3.3.1) holds for all positive numbers $R$ and all $z_0 \in \mathbb{C}$. If we take the limit as $R \to \infty$ in (3.3.1), we conclude that $|f'(z_0)| = 0$ for all $z_0 \in \mathbb{C}$. In other words, the derivative of $f$ is identically zero. This implies $f$ is a constant (see Exercise 2.6.1). □

One has to see the consequences of this theorem to appreciate its power. In the remainder of this section and the exercises we will introduce a number of these. Others will occur later.

**The Fundamental Theorem of Algebra.** At the very beginning of the text, we promised that we would prove that every non-constant polynomial with complex coefficients has a complex root. This is the *Fundamental Theorem of Algebra.* Before proving this theorem, it will be convenient to introduce limits at infinity.

**Definition 3.3.2.** If $f$ is a function defined on an unbounded set $E$ (so $E$ contains points of arbitrarily large modulus) and $L \in \mathbb{C}$, then we say

$$\lim_{z \to \infty} f(z) = L$$

if, for every $\epsilon > 0$, there is an $R > 0$ such that $|f(z) - L| < \epsilon$ whenever $|z| > R$ and $z \in E$.

This concept satisfies the same basic rules as other kinds of limits: limit of the sum is sum of the limits, limit of the product is product of the limits, limit of the quotient is quotient of the limits if the denominator does not have limit zero, etc. These facts, as well as the fact that

$$\lim_{z \to \infty} \frac{1}{z} = 0$$

(Exercise 3.3.1), are used in the proof of the Fundamental Theorem of Algebra. Before proving that theorem, we prove the following simple result.

**Theorem 3.3.3.** *If a function $f$ is defined and continuous on the entire plane and if $\lim_{z \to \infty} f(z)$ exists, then $f$ is bounded on $\mathbb{C}$.*

**Proof.** If $\lim_{z \to \infty} f(z) = L$, there exists an $R > 0$ such that $|f(z) - L| < 1$ whenever $|z| > R$. By the triangle inequality, this implies

$$|f(z)| < |L| + 1 \quad \text{if} \quad |z| > R.$$

Since $f$ is continuous on $\mathbb{C}$ and $\overline{D}_R(0)$ is closed and bounded, hence compact, $f$ is bounded on $\overline{D}_R(0)$. Since $f$ is bounded on $\overline{D}_R(0)$ and on its exterior, it is bounded on all of $\mathbb{C}$. □

**Theorem 3.3.4** (Fundamental Theorem of Algebra). *Every non-constant complex polynomial has a complex root.*

**Proof.** Let $p(z) = a_n z^n + a_{n-1} z^{n-1} + \cdots + a_1 z + a_0$ be a non-constant complex polynomial of degree $n$. Then $n \geq 1$ and $a_n \neq 0$. We will show the assumption that $p$ has no root leads to a contradiction.

If $p$ has no root, then $p(z) \neq 0$ for every $z \in \mathbb{C}$. This implies that $1/p$ is an entire function. We will show that it is also bounded.

If we define a function $h(z)$ by

$$h(z) = \frac{z^n}{p(z)} = \frac{1}{a_n + a_{n-1} z^{-1} + \cdots + a_1 z^{n-1} + a_0 z^{-n}},$$

then

$$\frac{1}{p(z)} = \frac{h(z)}{z^n} \tag{3.3.2}$$

for $z \neq 0$. Furthermore

$$\lim_{z \to \infty} h(z) = 1/a_n$$

and so

$$\lim_{z\to\infty} \frac{1}{p(z)} = \lim_{z\to\infty} \frac{h(z)}{z^n} = 0.$$

Since the limit of $1/p$ exists at infinity, the previous theorem implies that $1/p$ is bounded on all of $\mathbb{C}$. So Liouville's Theorem implies that it is a constant. In fact, the constant must be zero, since $1/p(z)$ has limit 0 at $\infty$. This is clearly a contradiction, since $1/p(z)$ cannot take on the value 0 on $\mathbb{C}$. We conclude that $p(z)$ must have a root. □

The Fundamental Theorem of Algebra has a number of important consequences. We will discuss only a few of them.

**Factoring Polynomials.** A polynomial is said to factor completely if it can be written as a product

$$p(z) = b(z - z_1)(z - z_2) \cdots (z - z_n)$$

of linear factors. Here, $n$ is the degree of $p$ and the numbers $z_1, z_2, \cdots, z_n$ are the roots of $p$. The roots need not all be distict. If a root occurs $k$ times in this factorization, it is said to be a root of *mutiplicity* $k$. Necessarily, the number of roots is the degree of the polynomial if each root is counted as many times as its multiplicity.

**Corollary 3.3.5.** *Each complex polynomial factors completely.*

**Proof.** The proof is by induction on the degree $n$ of the polynomial. Obviously a polynomial of degree 0 or 1 factors completely. If every polynomial of degee $n$ factors completely and $p$ is a polynomial of degree $n+1$, then we use the previous theorem to assert that $p$ has a root – call it $z_{n+1}$. It follows that $p$ factors as

$$P(z) = (z - z_{n+1})q(z),$$

where $q$ is a polynomial of degree $n$. By assumption, $q$ factors as

$$q(z) = b(z - z_1)(z - z_2) \cdots (z - z_n).$$

Then $p$ factors as

$$p(z) = b(z - z_1)(z - z_2) \cdots (z - z_n)(z - z_{n+1}).$$

This completes the induction step and finishes the proof of the corollary. □

**Eigenvalues of Matrices.** An eigenvalue for an $n \times n$ matrix $A$ is a complex number $\lambda$ such that the matrix $\lambda I - A$ is singular (has no inverse). Here, $I$ is the $n \times n$ identity matrix.

**Corollary 3.3.6.** *An $n \times n$ matrix $A$ has at least one complex eigenvalue.*

**Proof.** The determinant $\det(\lambda I - A)$ of $\lambda I - A$ is a polynomial in $\lambda$ of degree $n$. This is the *characteristic polynomial* of $A$. By Kramer's Rule, $\lambda I - A$ is singular for a given $\lambda$ if and only if $\det(\lambda I - A) = 0$ – that is, if and only if $\lambda$ is a root of the charcteristic polynomial. By the Fundamental Theorem of Algebra, the characteristic polynomial has a root. Thus, $A$ has an eigenvalue. □

The above result is the essential ingredient in the proof that every complex matrix can be put in upper triangular form (Exercise 3.3.14).

**Example 3.3.7.** Find the eigenvalues of the matrix $\begin{pmatrix} 1 & 2 \\ -1 & 3 \end{pmatrix}$.

**Solution:** The characteristic polynomial of this matrix is

$$\det \begin{pmatrix} \lambda - 1 & -2 \\ 1 & \lambda - 3 \end{pmatrix} = \lambda^2 - 4\lambda + 5.$$

By the Quadratic Formula, the roots of this polynomial are $2 \pm 2i$. Thus, the eigenvalues of the above matrix are $2 + 2i$ and $2 - 2i$.

**Differential Equations.** The Fundamental Theorem of Algebra also has applications to Differential Equations. A homogeneous constant-coefficient differential equation is an equation of the form

$$a_n y^{(n)} + a_{n-1} y^{(n-1)} + \cdots + a_1 y' + a_0 y = 0. \tag{3.3.3}$$

**Example 3.3.8.** Prove that each homogeneous constant-coefficient differential equation has a solution of the form $y = \mathrm{e}^{\lambda x}$, where $\lambda$ is a complex number. Which numbers $\lambda$ yield solutions?

**Solution:** We simply plug $y = \mathrm{e}^{\lambda x}$ into the equation (3.3.3) and see if there are values of $\lambda$ that yield solutions. Since $(\mathrm{e}^{\lambda x})' = \lambda\, \mathrm{e}^{\lambda x}$, the result is

$$(a_n \lambda^n + a_{n-1} \lambda^{n-1} + \cdots + a_1 \lambda + a_0)\, \mathrm{e}^{\lambda x} = 0.$$

The polynomial in parentheses is called the *auxiliary polynomial* for (3.3.3). Clearly $\mathrm{e}^{\lambda x}$ is a solution of (3.3.3) if and only if $\lambda$ is a root of this polynomial. Since every non-constant polynomial has a root, equation (3.3.3) has a solution of the form $y = \mathrm{e}^{\lambda x}$.

If the auxiliary polynomial has distinct roots $\lambda_1, \cdots, \lambda_n$, then the general solution to equation (3.3.3) is a linear combination of the solutions $\mathrm{e}^{\lambda_j x}$.

If the coefficients of (3.3.3) are real numbers and we are looking only for real solutions to this differential equation, then we exploit the fact that the non-real roots of a polynomial with real coefficients occur in conjugate pairs (Exercise 3.3.10). If $\lambda = \alpha + i\beta$ and $\overline{\lambda} = \alpha - i\beta$ are roots of the auxilliary equation for (3.3.3), then

$$\mathrm{e}^{\alpha x} \cos \beta x = \frac{\mathrm{e}^{\lambda x} + \mathrm{e}^{\overline{\lambda} x}}{2} \quad \text{and}$$

$$\mathrm{e}^{\alpha x} \sin \beta x = \frac{\mathrm{e}^{\lambda x} - \mathrm{e}^{\overline{\lambda} x}}{2i}$$

both give real solutions to (3.3.3), as does any linear combination

$$C\, \mathrm{e}^{\alpha x} \cos \beta x + D\, \mathrm{e}^{\alpha x} \sin \beta x = \mathrm{e}^{\alpha x}(C \cos \beta x + D \sin \beta x),$$

where $C$ and $D$ are arbitrary real constants.

**Example 3.3.9.** Find all solutions to the differential equation $y'' - 2y' + 5y = 0$. Then find all real solutions.

**Solution:** The auxiliary equation is $\lambda^2 - 2\lambda + 5 = 0$, which has solutions $\lambda = 1 \pm 2i$. Thus, the general solution to the differential equation is

$$y = A\,\mathrm{e}^{(1+2i)x} + B\,\mathrm{e}^{(1-2i)x},$$

where $A$ and $B$ are complex constants. The general real solution is

$$y = \mathrm{e}^{x}(C\cos 2x + D\sin 2x),$$

where $C$ and $D$ are real constants.

**Characterization of Polynomials.** In the proof of Liouville's Theorem, we only use Cauchy's estimate on the first derivative of an analytic function on a disc. Here we present a generalization of Liouville's Theorem that uses Cauchy's estimates on higher derivatives.

**Theorem 3.3.10.** *An entire function $f$ is a polynomial of degree at most $n$ if and only if there are positive constants $A$ and $B$ such that*

$$|f(z)| \le A + B|z|^n \tag{3.3.4}$$

*for all $z \in \mathbb{C}$.*

**Proof.** We will prove that every polynomial satisfies an inequality of the form (3.3.4). We leave the converse as an exercise in the application of Cauchy's estimates.

Suppose $p(z) = a_n z^n + a_{n-1}z^{n-1} + \cdots + a_1 z + a_0$ is a polynomial of degree at most $n$. Then

$$\lim_{z\to\infty} \frac{p(z)}{z^n} = a_n.$$

If we apply the definition of limit, using $\epsilon = 1$, we conclude that there is an $R > 0$ such that $|z| > R$ implies

$$\left|\frac{p(z)}{z^n} - a_n\right| < 1.$$

It follows from the triangle inequality that

$$\frac{|p(z)|}{|z|^n} < |a_n| + 1 \quad \text{if} \quad |z| > R.$$

If we set $B = |a_n| + 1$, then

$$|p(z)| < B|z|^n \quad \text{if} \quad |z| > R.$$

This gives a bound on $|p|$ on the complement of the closed disc $\overline{D}_R(0)$. A closed disc of finite radius is a compact set and so $p$ is bounded on $\overline{D}_R(0)$ – say, $|p(z)| \le A$ on this disc. If we combine this with our bound on the complement of $\overline{D}_R(0)$, we conclude that

$$|p(z)| \le A + B|z|^n$$

on all of $\mathbb{C}$, as required. □

## Exercise Set 3.3

1. Prove that $\lim_{z\to\infty} \frac{1}{z} = 0$.
2. We say $\lim_{z\to\infty} f(z) = \infty$ if for each $K > 0$, there is an $M > 0$ such that $|f(z)| > K$ whenever $|z| > M$. Prove that $\lim_{z\to\infty} f(z) = \infty$ if and only if $\lim_{z\to\infty} 1/f(z) = 0$.
3. Show that if $f$ is an entire function and $\lim_{z\to\infty} f(z) = \infty$, then $f$ must have a zero somewhere in $\mathbb{C}$. Hint: See the previous exercise.
4. Prove that if $f$ is an entire function which satisfies $|f(z)| \geq 1$ on the entire plane, then $f$ is constant.
5. Prove that if an entire function has real part which is bounded above, then the function is constant.
6. Prove that if an entire function $f$ is not constant, then its range $f(\mathbb{C})$ is dense in $\mathbb{C}$ – meaning that the closure of $f(\mathbb{C})$ is $\mathbb{C}$. Hint: If $f(\mathbb{C})$ is not dense, then there is a point $z_0$ and a disc $D_r(z_0)$, centered at $z_0$, such that $f(\mathbb{C}) \cap D_r(z_0) = \emptyset$.
7. Show that if $f$ is an entire function and $|f(z)| \leq K|z|$ for all $z \in \mathbb{C}$, where $K$ is a positive constant, then $f(z) = Cz$ for some constant $C$.
8. Show that if $f$ is an entire function and $|f(z)| \leq K|\,e^z\,|$ for all $z \in \mathbb{C}$, where $K$ is a positive constant, then $f(z) = C\,e^z$ for some constant $C$.
9. Finish the proof of Theorem 3.3.10 by using Cauchy's estimates to prove that the only entire functions $f$ that satisfy an inequality of the form $|f(z)| \leq A + B|z|^n$ for all $z \in \mathbb{C}$ are polynomials of degree at most $n$.
10. Show that if $p$ is a polynomial with real coefficients, then the non-real roots of $p$ occur in conjugate pairs. That is, show that if $w$ is a root, then so is $\overline{w}$.
11. Find the roots of the polynomial $p(z) = z^2 - 2z + 2$.
12. Find the eigenvalues of the matrix $\begin{pmatrix} 1 & 1 \\ -1 & 1 \end{pmatrix}$.
13. Find all solutions of the differential equation $y'' - 4y' + 13y = 0$. Then find all real solutions.
14. Use Corollary 3.3.6 to prove that if $A$ is an $n \times n$ matrix, then $A$ is conjugate to an upper triangular matrix.
15. Use the result of Exercise 10 to prove that every polynomial with real coefficients factors as a product of polynomials of degree at most 2, also with real coefficients.
16. Are there any non-constant entire functions $f$ that satisfy an inequality of the form

$$|f(z)| \leq A + B\log|z| \quad \text{for all } z \text{ with } |z| \geq 1,$$

where $A$ and $B$ are positive constants?

## 3.4. Zeroes and Singularities

The existence of power series expansions leads to a great deal of information about the local structure of analytic functions. For example, Exercise 3.2.9 makes an assertion that is important enough to be expanded and restated as a theorem.

**Theorem 3.4.1.** *If $f$ is a function which is analytic in a open set $U$, then for each $z_0 \in U$, exactly one of the following statements is true:*

(1) *there is an open disc, centered at $z_0$, on which $f$ is identically 0;*

(2) *there is a non-negative integer $k$, an open disc $D_r(z_0)$, and a function $g$, analytic on $U$, such that*

$$f(z) = (z - z_0)^k g(z) \quad \textit{for all} \quad z \in D_r(z_0) \tag{3.4.1}$$

*and $g(z)$ is non-zero at each point of $D_r(z_0)$.*

**Proof.** The function $f$ has a power series expansion

$$f(z) = \sum_{n=0}^{\infty} c_n (z - z_0)^n \tag{3.4.2}$$

which converges in a disc $D_R(z_0)$ of positive radius $R$. Some of the coefficients $c_n$ may be zero. If they are all zero, then $f$ is identically zero on $D_R(z_0)$ and so (1) holds in this case.

If the coefficients $c_n$ are not all zero, there is a smallest $n$ for which $c_n \neq 0$ – call it $k$. Then (3.4.2) becomes

$$f(z) = \sum_{n=k}^{\infty} c_n (z - z_0)^n = (z - z_0)^k \sum_{n=0}^{\infty} c_{n+k} (z - z_0)^n, \tag{3.4.3}$$

where $c_k \neq 0$. We may then define $g$ by

$$g(z) = \begin{cases} \displaystyle\sum_{n=0}^{\infty} c_{n+k} (z - z_0)^n, & \text{if } z \in D_R(z_0); \\ \dfrac{f(z)}{(z - z_0)^k}, & \text{if } z \in U \setminus \{z_0\}. \end{cases}$$

This is a consistent definition, since on $D_R(z_0) \setminus \{z_0\}$ the functions $f(z)/(z - z_0)^k$ and $\sum_{n=0}^{\infty} c_{n+k}(z - z_0)^n$ are equal.

Since $g$ is continuous and $g(z_0) = c_k \neq 0$, there is a positive $r < R$ such that $|g(z) - g(z_0)| < |c_k|$ whenever $|z - z_0| < r$. This implies $g(z) \neq 0$ if $|z - z_0| < r$. In other words, $g(z) \neq 0$ for every $z \in D_r(z_0)$. In this case, (2) holds. □

**Zeroes of Analytic Functions.** By a *zero* of a function, we mean a point in its domain at which it vanishes (has value 0). The previous theorem leads to an important and somewhat surprising result about the zeroes of an analytic function $f$ on a connected open set: Unless $f$ is identically zero, each zero of $f$ has no neighboring zeroes. This is made precise in Part (b) of the following theorem.

**Theorem 3.4.2.** *Let $f$ be a function which is analytic on a connected open set $U$ and is not identically zero. Then*

(a) *for each* $z_0 \in U$*, there is a non-negative integer* $k$*, an* $r > 0$*, and a function* $g$*, analytic in* $U$*, such that*

$$f(z) = (z - z_0)^k g(z) \quad \textit{for all} \quad z \in U,$$

*and* $g(z)$ *has no zeroes on* $D_r(z_0)$*;*

(b) *each* $z_0 \in U$ *is the center of an open disc* $D_r(z_0) \subset U$ *in which there are no zeroes of* $f$ *except possibly* $z_0$ *itself;*

(c) *the set of zeroes of* $f$ *is at most countable.*

**Proof.** Theorem 3.4.1 shows that, at each point $z_0$ of $U$, there are two possibilities: (1) there is a disc centered at $z_0$ in which $f$ is identically 0, and (2) there is a disc centered at $z_0$ in which $f$ has no zeroes except possibly at $z_0$ itself. Let $V_j$, $j = 1, 2$ be the set of points $z_0$ for which the $j$th possibility is the one which occurs. Obviously, $V_1$ and $V_2$ are both open sets, $V_1 \cap V_2 = \emptyset$, and $V_1 \cup V_2 = U$. Thus, either $V_1$ or $V_2$ is empty, since, otherwise, they would separate the connected set $U$. If $V_2 = \emptyset$, then $f$ is identically 0 on $U$. Since $f$ is not identically 0, we conclude that $V_1 = \emptyset$. Since the second of the possibilities in Theorem 3.4.1 is the only one that occurs in this situation, we conclude that (a) and (b) above both hold at every $z_0 \in U$.

We can modify the disc $D_r(z_0)$ in statement (a) of the theorem so that it has a center (which may no longer be $z_0$) with rational coordinates and a rational radius and still has the property that it contains $z_0$ and has no zeroes other than possibly $z_0$. We simply choose a point $z_0'$ with rational coordinates, and a positive rational number $\rho$ such that $|z_0 - z_0'| < \rho < r/2$. Then $z_0 \in D_\rho(z_0') \subset D_r(z_0)$ and so $D_\rho(z_0')$ contains $z_0$ but no other zeroes of $f$. Since we may do this for each zero of $f$ in $U$ and since there are only countably many discs with rational centers and rational radii, we conclude that $f$ can have only countably many zeroes in $U$. □

Recall that an *isolated point* of a set $E$ is a point which is contained in an open disc which contains no other points of $E$. Thus, if $Z(f)$ denotes the set of zeroes of an analytic function $f$ on its domain $U$, then Part (b) of the above theorem implies that each point of $Z(f)$ is an isolated point of $Z(f)$. Actually it says something much stronger. It says that every point $z_0$ of $U$ (whether in $Z(f)$ or not) is the center of a disc containing no points of $Z(f)$ other than $z_0$. There is a term to describe subsets with this property:

**Definition 3.4.3.** Let $U$ be an open subset of $\mathbb{C}$ and $E$ a subset of $U$. We say that $E$ is a *discrete subset* of $U$ if every point $z_0$ of $U$ has a neighborhood which contains no points of $E$ other than possibly $z_0$ itself.

Thus, Part (b) of the above theorem says that the zero set $Z(f)$ of a non-constant analytic function on a connected open set $U$ is a discrete subset of $U$.

It turns out that a subset of an open set $U$ is discrete if and only if no sequence of distinct points of $E$ converges to a point of $U$ (see Exercise 3.4.1). Of course, there may be sequences in $E$ which converge to points not in $U$.

Theorem 3.4.2 has the following easy but important consequence. We leave the proof as an exercise (Exercise 3.4.2).

**Theorem 3.4.4** (Identity Theorem). *Suppose $f$ and $g$ are two analytic functions with domain a connected open set $U$. If $f(w) = g(w)$ at each point $w$ of a non-discrete subset $E$ of $U$, then $f(z) = g(z)$ at every point of $U$.*

If the $f$ of Theorem 3.4.2 actually has a zero at $z_0$, then the integer $k$ is positive. We call it the *order* of the zero of $f$ at $z_0$.

**Example 3.4.5.** What is the order of the zero of the function $f(z) = \cos z - 1$ at 0? What is the function $g$ of Part (a) of Theorem 3.4.2 in this case if $z_0 = 0$?

**Solution:** If we subtract 1 from the power series expansion of $\cos z$ about 0, we obtain

$$\begin{aligned} \cos z - 1 &= -\frac{z^2}{2!} + \frac{z^4}{4!} + \cdots (-1)^n \frac{z^{2n}}{2n!} + \cdots \\ &= z^2 \left( -\frac{1}{2!} + \frac{z^2}{4!} + \cdots (-1)^n \frac{z^{2n-2}}{2n!} + \cdots \right). \end{aligned}$$

We conclude that the order of the zero of $\cos z - 1$ at 0 is 2 and the function $g$ of Part (a) of Theorem 3.4.2 is the function given by the power series in parentheses above. This power series has infinite radius of convergence by the ratio test.

**Theorem 3.4.6.** *If an analytic function $g$ is not zero at a point $z_0$ in its domain, then in some neighborhood $V$ of $z_0$ there is an analytic function $h$ such that $g(z) = e^{h(z)}$ in $V$.*

**Proof.** To find such an $h$, we simply choose a branch of the log function that does not have $g(z_0)$ on its cut line. Then the set on which this branch of the log function is analytic is an open set $W$ which contains $g(z_0)$. If we set $V = g^{-1}(W)$ and define $h$ on $V$ by

$$h(z) = \log(g(z)),$$

then $V$ is a neighborhood of $z_0$ and $g(z) = e^{h(z)}$ on $V$. □

If we combine this result with Theorem 3.4.2, we have:

**Theorem 3.4.7.** *Let $f$ be an analytic function on an open set $U$ and let $z_0$ be a point of $U$ such that $f$ is not identically zero in a neighborhood of $z_0$. Then there exist a non-negative integer $k$, a neighborhood $V \subset U$ of $z_0$, and an analytic function $h$ on $V$, such that*

$$f(z) = (z - z_0)^k \, e^{h(z)}$$

*for all $z \in V$.*

**Isolated Singularities.** If $U$ is an open set, $z_0 \in U$, and $f$ is a function which is analytic on $U \setminus \{z_0\}$, then $f$ is said to have an *isolated singularity* at $z_0$. If $f$ can be given a value at $z_0$ which makes it analytic everywhere on $U$, then the singularity is said to be *removable*.

**Theorem 3.4.8.** *If $f$ has an isolated singularity at $z_0$ and is bounded in some deleted neighborhood of $z_0$, then $z_0$ is a removable singularity of $f$.*

**Proof.** Suppose $f$ is analytic and bounded on $U \setminus \{z_0\}$. If we define a function $g$ by $g(z) = (z - z_0)^2 f(z)$ for $z \neq z_0$ and $g(z_0) = 0$, then $g$ is differentiable at $z_0$. In fact,

$$g'(z_0) = \lim_{z \to z_0} \frac{g(z) - g(z_0)}{z - z_0} = \lim_{z \to z_0} (z - z_0) f(z) = 0.$$

That this limit is 0 is proved as follows. Let $M$ be a bound on $|f(z)|$ on $U \setminus \{z_0\}$. Then $|(z - z_0)f(z)| \leq M|z - z_0|$ on $U$. Since $\lim_{z \to z_0} M|z - z_0| = 0$, it follows that $\lim_{z \to z_0}(z - z_0)f(z) = 0$ as well.

The function $g$ is differentiable at every point of $U \setminus \{z_0\}$ and, by the above, is also differentiable at $z_0$. Thus, it is analytic on all of $U$. Since $g(z_0) = g'(z_0) = 0$, the first two terms of its power series expansion about $z_0$ are 0. It follows that we may factor $(z - z_0)^2$ out of every term of its power series expansion and write $g(z) = (z - z_0)^2 h(z)$ for an analytic function $h$ defined by a power series in a disc centered at $z_0$. Clearly $h$ and $f$ are the same in this disc, except at $z_0$, where $f$ is not defined. Thus, setting $f(z_0) = h(z_0)$ serves to define $f$ at $z_0$ in such a way that it becomes analytic on all of $U$. □

**Example 3.4.9.** Show that $f(z) = \dfrac{\cos z - 1}{z^2}$ has a removable singularity at 0.

**Solution:** This follows immediately from the factorization of $\cos z - 1$ obtained in Example 3.4.5. If $f$ is given the value $-1/2$ at $z = 0$, it becomes analytic on the entire plane.

There are two types of isolated singularities that are not removable. A function $f$ defined on $U \setminus \{z_0\}$ of the form

$$f(z) = \frac{g(z)}{(z - z_0)^k},$$

where $g$ is analytic on $U$, $g(z_0) \neq 0$, and $k$ is a positive integer, is said to have a *pole of order k*. If the order of the pole is 1, then it is called a *simple pole*. An isolated singularity which is not a pole and is not a removable singularity is called an *essential singularity*.

**Example 3.4.10.** If $U$ is an open set and $z_0 \in U$, then analyze the singularities of a function of the form $f/g$, where $f$ and $g$ are analytic on $U$.

**Solution:** The singularities of $f/g$ in $U$ are all isolated because the zeroes of $g$ are isolated. If we factor $f$ and $g$ as in Theorem 3.4.2, then

$$f(z) = (z - z_0)^j p(z) \quad \text{and} \quad g(z) = (z - z_0)^k h(z),$$

where $j$ and $k$ are the orders of the zeroes of $f$ and $g$ at $z_0$ and $p$ and $h$ are analytic in $U$ and non-vanishing in some neighborhood of $z_0$. Then $f(z)/g(z) = (z - z_0)^{j-k} p(z)/h(z)$ with $p(z)/h(z)$ analytic and non-vanishing in a neighborhood of $z_0$. The point $z_0$ is a removable singularity for $f/g$ if $j \geq k$ and, otherwise, is a pole of order $k - j$.

**Example 3.4.11.** Analyze the singularities of the function $f(z) = \dfrac{1}{1 - e^z}$.

**Solution:** The denominator of this fraction has zeroes at the points $\{2\pi k i\}$ for $k$ an integer. Each of these is a zero of order 1 because the derivative of $1 - e^z$

is $-e^z$ and this is non-zero for every $z$ and, in particular, is non-zero at the points $\{2\pi ki\}$. It follows from the preceding example that each of these points is a simple pole for $f$.

Essential singularities are quite wild. In fact, the Big Picard Theorem states that a function with an essential singularity at $z_0$ takes on every complex number but one as a value in every open disc centered at $z_0$. We will prove the Big Picard Theorem and its little brother – the Little Picard Theorem – in Chapter 7. The following is a very much weaker statement than the Big Picard Theorem, but it is still enough to show that an analytic function behaves very wildly near an essential singularity.

**Theorem 3.4.12.** *If $f$ is analytic in $U\setminus\{z_0\}$ and has an essential singularity at $z_0$, then for every open disc $D$, centered at $z_0$ and contained in $U$, the set $f(D\setminus\{z_0\})$ has closure equal to the entire complex plane.*

**Proof.** Suppose there is a disc $D$, centered at $z_0$ and contained in $U$ and there is some complex number $w$ which is not in the closure of $f(D\setminus\{0\})$. Then there is an $r>0$ such that $f(D\setminus\{z_0\})\cap D_r(w)=\emptyset$. This means $|f(z)-w|\geq r$ for every $z\in D\setminus\{z_0\}$. Then

$$\left|\frac{1}{f(z)-w}\right|\leq\frac{1}{r}\quad\text{for all}\quad z\in D\setminus\{z_0\}.$$

Thus, $1/(f(z)-w)$ is analytic and bounded on $D\setminus\{z_0\}$ and, hence, has a removable singularity at $z_0$. In other words, there is an analytic function $h$ on $D$ such that

$$\frac{1}{f(z)-w}=h(z)$$

on $D\setminus\{z_0\}$. If $h$ has no zero at $z_0$, set $k=0$; otherwise, let $k$ be the order of the zero of $h$ at 0. Then

$$h(z)=(z-z_0)^k g(z)$$

for some analytic function $g$ on $D$ which does not have a zero at $z_0$. Solving for $f$, we find

$$f(z)=w+\frac{1}{h(z)}=\frac{w(z-z_0)^k+1/g(z)}{(z-z_0)^k}$$

in some disc containing $z_0$ where $g$ is non-vanishing. Since the numerator is analytic in this disc, $f$ has a pole of order $k$ at $z_0$ (or a removable singularity if $k=0$). Since neither of these things is true, we conclude that there is no $w$ which fails to be in the closure of $f(D)$. This completes the proof. □

The following theorem gives an easy way to identify whether a given isolated singularity is removable, a pole, or essential.

**Theorem 3.4.13.** *Let $f$ be an analytic function with an isolated singularity at $z_0$. Then*

(a) *$f$ has a removable singularity at $z_0$ if and only if $\lim_{z\to z_0} f(z)$ exists and is finite;*
(b) *$f$ has a pole at $z_0$ if and only if $\lim_{z\to z_0} f(z)=\infty$;*
(c) *$f$ has an essential singularity at $z_0$ if and only if $\lim_{z\to z_0} f(z)$ does not exist, even as an infinite limit.*

**Proof.** The function $f$ has a removable singularity at $z_0$ if and only if it can be given a value $w$ at $z_0$ which makes it analytic in a neighborhood of $z_0$. In this case, $\lim_{z\to z_0} f(z) = w$, since $f$ with its new value at $z_0$ must be continuous at $z_0$.

On the other hand, if $f$ has a finite limit as $z \to z_0$, then $f$ is bounded in a neighborhood of $z_0$ and, by Theorem 3.4.8, the singularity at $z_0$ is removable. This proves (a).

If $f$ has a pole at $z_0$, then $f(z) = \dfrac{g(z)}{(x - z_0)^k}$ for some $k > 0$ and some $g$ analytic in a neighborhood of $z_0$ with $g(z_0) \neq 0$. Then $\lim_{z\to z_0} g(z) = g(z_0) \neq 0$ and $\lim_{z\to z_0}(z - z_0)^k = 0$. It follows that $\lim_{z\to z_0} f(z) = \infty$.

On the other hand, if $\lim_{z\to z_0} f(z) = \infty$ the singularity is not removable since $f$ does not have a finite limit as $z \to z_0$. It cannot be essential either, since the previous theorem says that, if it were essential, $f$ would take on values arbitrarily close to any given complex number in every open disc centered at $z_0$. Thus, $f$ could not have limit $\infty$ at $z_0$ if the singularity were essential. This proves (b).

If $\lim_{z\to z_0} f(z)$ does not exist, even as a finite limit, then by (a) and (b) $f$ does not have a removable singularity or a pole. Thus, it has an essential singularity. Conversely, if $f$ has an essential singularity at $z_0$, then it does not have a removable singularity or a pole. Then, also by (a) and (b), $f$ does not have a finite limit at $z_0$ nor does it have limit $\infty$. This proves (c). □

**Example 3.4.14.** Analyze the singularity of the function $f(z) = \mathrm{e}^{1/z}$ at $z = 0$.

**Solution:** This is an essential singularity. The function $f(z) = \mathrm{e}^{1/z}$ takes on the value 1 at all points of the form $z = (2\pi n)^{-1}$ and the value $\mathrm{e}^n$ at all points of the form $z = n^{-1}$. Thus, $f(z)$ approaches 1 as $z$ approaches 0 along one sequence of points and it approaches $\infty$ as $z$ approaches 0 along another sequence of points. Thus, $f$ certainly does not have a limit, finite or infinite, as $z \to 0$.

**Meromorphic Functions.** For many reasons, it is important to study functions which are analytic on an open set $U$ except on a subset of $U$ consisting of points where poles occur. Necessarily, such a subset is a discrete subset of $U$.

**Definition 3.4.15.** Let $U$ be an open set and $E$ a discrete subset of $U$. If $f$ is a function which is analytic on $U \setminus E$ and has a removable singularity or a pole at each point of $E$, then $f$ is called a meromorphic function on $U$.

One reason the set of meromorphic functions on $U$ is interesting is that it is a field, if $U$ is a connected open set. Obviously, we can add and multiply meromorphic functions and the results are still meromorphic. Just as obviously, the appropriate commutative, associative and distributive laws hold and there are zero and identity elements and additive inverses. All of these things are also true of the class of analytic functions on $U$. However, in the class of meromorphic functions we also have the last field axiom satisfied: every non-zero element has a multiplicative inverse.

**Theorem 3.4.16.** *If $U$ is a connected open set and $f$ is a meromorphic function on $U$ which is not identically zero, then $1/f$ is also meromorphic.*

**Proof.** By Theorem 3.4.2 the set $Z(f)$ of zeroes of $f$ is a discrete subset of $U$. By definition, the set $P(f)$ of poles of $f$ is a discrete subset of $U$. It follows that the set $E = Z(f) \cup P(f)$ is also a discrete subset of $U$. The function $1/f$ is analytic on $U \setminus E$ and so the only thing we need to establish is that at points of $E$ it has only removable singularities or poles.

Let $z_0$ be a point of $E$. Since, $E$ is discrete, there is a disc $D$, centered at $z_0$ in which $z_0$ is the only point of $E$. If $z_0 \in E$, then $f$ is analytic everwhere in $D$ except at $z_0$ where is has either a zero or a pole.

If $f$ has a zero of order $k$ at $z_0$, then it factors as $(z - z_0)^k g(z)$ where $g$ is analytic and non-vanishing in $D$. Obviously then $1/f(z) = (z - z_0)^{-k}/g(z)$ has a pole of order $k$ at $z_0$.

If $f$ has a pole of order $k$ at $z_0$, then $f$ factors as $f(z) = (z - z_0)^{-k}h(z)$ where $h$ is analytic and non-vanishing in $D$. Then $1/f(z) = (z - z_0)^k/h(z)$ has a zero of order $k$ at $z_0$ and, hence, a removable singularity at $z_0$. □

## Exercise Set 3.4

1. Prove that a set $E$ is a discrete subset of an open set $U$ if and only if no sequence of distinct points of $E$ converges to a point in $U$.
2. Prove Theorem 3.4.4.
3. Is there a function, analytic on $\mathbb{C}$, which is 0 on the set of points $\{1/n\}$ for $n$ a positive integer and not identically 0? Justify your answer. What if the function is only required to be analytic on $\mathbb{C} \setminus \{0\}$?
4. If $f(z) = \sin z - z$, find the order of the zero of $f$ at 0. Then give the factorization of $f$ at 0, as in Theorem 3.4.1.
5. If $f(z) = \cos z - 1 + z^2/2$, find the order of the zero and the factorization at 0 as in the previous exercise.
6. Give an example to show that Theorem 3.4.2 no longer holds if we drop the hypothesis that $U$ is connected.
7. Prove that if $f$ is an analytic function on a connected open set $U$ and if $K$ is a compact subset of $U$, then there can be at most finitely many zeroes of $f$ in $K$.
8. Show that if $f$ is an analytic function with a zero of order $k$ at $z_0$, then there is a neighborhood $V$ of $z_0$ and an analytic function $g$ on $V$ such that $f = g^k$ on $V$ and $g'(z_0) \neq 0$.
9. Suppose $f$ and $g$ are analytic functions on an open set $U$, $z_0 \in U$, and $f(z_0) = g(z_0) = 0$. Show that
$$\lim_{z\to z_0} \frac{f(z)}{g(z)} = \lim_{z\to z_0} \frac{f'(z)}{g'(z)}$$
and that this limit exists if $\infty$ is allowed as a possible value.
10. Suppose $U$ is a connected open set and $z_0 \in U$. Prove that if $f$ is a non-constant analytic function on $U \setminus \{z_0\}$ and $f$ takes on a certain value $c$ at least once in every deleted neighborhood of $z_0$, then $f$ has an essential singularity at $z_0$.

11. Prove that if $f$ is a function which is analytic in the exterior of the closed disc $\overline{D}_r(0)$ and if $\lim_{z\to\infty} f(z) = 0$, then $f$ has a power series expansion of the form

$$f(z) = \sum_{n=1}^{\infty} a_n z^{-n},$$

which converges on the set $\{z \in \mathbb{C} : |z| > r\}$. Hint: Consider the function $g(z) = f(1/z)$.

In the next five problems, analyze each singularity $z_0$ of $f$. Is it removable, a pole, or essential? If it is a pole, what is its order? If it is removable, what value should you give the function at $z_0$ to make it analytic?

12. $f(z) = \dfrac{1}{z - z^3}$.
13. $f(z) = \sin 1/z$.
14. $f(z) = \dfrac{e^z - 1 - z}{z^2}$.
15. $f(z) = \dfrac{1}{e^z - 1} - \dfrac{1}{z}$.
16. $f(z) = \dfrac{\log z}{(1-z)^2}$, where log is the principal branch of the log function.

## 3.5. The Maximum Modulus Principle

This is another important application of Cauchy's theorems. Before stating it, we state and prove a technical lemma that will be used in its proof. The lemma states that if the average value of a continuous function on an interval is as large, in modulus, as all values of the function on the interval, then the function must be constant.

**Lemma 3.5.1.** *Let $f$ be a continuous complex-valued function defined on an interval $I = [a, b]$ on the real line. If $|f(t)| \le M$ on $I$, where*

$$\left| \frac{1}{b-a} \int_a^b f(t)\, dt \right| = M, \tag{3.5.1}$$

*then $f$ is a constant of modulus $M$ on $I$.*

**Proof.** If we choose a complex number $u$ of modulus 1 such that

$$u \int_a^b f(t)\, dt = \left| \int_a^b f(t)\, dt \right|,$$

then (3.5.1) may be written as

$$\int_a^b (M - uf(t))\, dt = 0. \tag{3.5.2}$$

Let $uf = g + ih$, where $g$ and $h$ are real-valued continuous functions on $I$. Note that $|f(t)| \le M$ implies that $g(t) \le M$ and, hence, that $M - g(t) \ge 0$. Equation (3.5.2) implies

$$\int_a^b (M - g(t))\, dt = 0.$$

However, $\int_a^x (M - g(t))\, dt$ is a differentiable function of $x$ with non-negative derivative $M - g(x)$ on $I$ and, hence, is a non-decreasing function. Since it is 0 at $a$ and $b$, it must be identically zero. This implies that its derivative $M - g(x)$ is identically 0.

We now have that the real part of $uf = g + ih$ is the constant $M$. However, we also have that

$$M^2 \ge |f(t)|^2 = |uf(t)|^2 = g^2(t) + h^2(t) = M^2 + h^2(t),$$

and this implies that $h(t) \equiv 0$ on $I$. Thus, $f$ is the constant $u^{-1}M$, which has modulus $M$. □

**The Maximum Modulus Theorem.**

**Theorem 3.5.2.** *If $f$ is analytic on a connected open set $U$ and $|f|$ has a local maximum at $z_0 \in U$, then $f$ is constant on $U$.*

**Proof.** If $f$ has a local maximum at $z_0 \in U$, then we may choose an $r > 0$ such that $\overline{D}_r(z_0) \subset U$ and such that $|f(z_0)|$ is a maximum for $|f(z)|$ on $\overline{D}_r(z_0)$. Then Cauchy's Integral Theorem tells us that

$$\begin{aligned} f(z_0) &= \frac{1}{2\pi i} \int_{|z - z_0| = r} \frac{f(z)}{z - z_0}\, dz \\ &= \frac{1}{2\pi} \int_0^{2\pi} f(z_0 + r\, e^{it})\, dt \end{aligned} \tag{3.5.3}$$

Since $|f(z_0 + r\, e^{it})| \le |f(z_0)|$ for each $t$, by our choice of $r$, we have the hypotheses of the previous lemma satisfied with $M = |f(z_0)|$, $[a, b] = [0, 2\pi]$, and $f(z_0 + r\, e^{it})$ playing the role of the function $f$ of the lemma. Based on the lemma, we conclude that $f$ is a constant $c$ on the circle $z_0 + r\, e^{it}$. Since this circle is a non-discrete subset of $U$, it follows from the Identity Theorem (Theorem 3.4.4) that $f$ is the constant $c$ on all of $U$. □

Recall that a subset of the plane is compact if it is both closed and bounded. Suppose $U$ is an open set which is bounded. Then its closure $\overline{U}$ is both closed and bounded and, hence, is compact. Suppose $U$ is connected. If $f$ is a continuous complex-valued function on $\overline{U}$, then the continuous real valued function $|f(z)|$ has a maximum value on $\overline{U}$. If $f$ is also analytic on $U$ and non-constant, then the previous theorem implies that this maximum cannot occur at a point of $U$. This means it must occur only on the boundary $\partial U$. This proves the following corollary of Theorem 3.5.2.

**Corollary 3.5.3.** *Suppose $U$ is a connected, bounded, open subset of $\mathbb{C}$. If $f$ is a function which is continuous on $\overline{U}$, analytic on $U$, and non-constant, then the maximum value of $|f(z)|$ on $\overline{U}$ is attained on $\partial U$ and nowhere else.*

The typical example of a set $U$ of the type described in the above corollary is an open disc $\overline{D}_r(z_0)$ of finite radius.

**Example 3.5.4.** Find where the function $f(z) = z^2 - z$ attains its maximum modulus on the closed unit disc.

**Solution:** By Corollary 3.5.3, the maximum occurs only on the unit circle $\{z = e^{it} : t \in [0, 2\pi]\}$. Thus, we need to find where the maximum modulus of the function $|f(e^{it})|$ occurs for $t \in [0, 2\pi]$. This is equivalent to finding where the square of the function has a maximum. Thus, we wish to maximize the function

$$h(t) = |\,e^{2it} - e^{it}\,|^2 = |\,e^{it} - 1|^2 = 2 - 2\cos t.$$

Clearly, the maximum occurs at $t = \pi$ and only there. Thus, the maximum of $|z^2 - z|$ on the closed unit disc is 2 and it occurs only at $z = -1$.

**Schwarz's Lemma.** Schwarz's Lemma is a nice application of the Maximum Modulus Theorem. It will be quite useful later in the theory of conformal mappings.

**Lemma 3.5.5** (Schwarz's Lemma). *Let $f$ be analytic on $D_1(0)$, with $f(0) = 0$ and $|f(z)| \le 1$ for every $z \in D_1(0)$. Then,*

$$|f(z)| \le |z| \quad \text{for all} \quad z \in D_1(0) \tag{3.5.4}$$

*and $|f'(0)| \le 1$. If $|f'(0)| = 1$, then there is a constant $c$ of modulus one such that $f(z) = cz$.*

**Proof.** Since $f(0) = 0$, Theorem 3.4.2 implies that $f(z) = zg(z)$, where $g$ is also analytic in $D_1(0)$. Since $|f(z)| \le 1$ on $D_1(0)$, it follows that

$$|g(z)| \le \frac{1}{r} \quad \text{on the circle} \quad |z| = r$$

for each $r < 1$. The Maximum Modulus Theorem implies that this inequality also holds inside the disc of radius $r$. Since this is true of each $r < 1$, we conclude that $|g(z)| \le 1$ on all of $D_1(0)$ and this implies (3.5.4).

The inequality $|f'(0)| \le 1$ follows from (3.5.4), since

$$f'(0) = \lim_{z \to 0} \frac{f(z)}{z} = \lim_{z \to 0} g(z) = g(0).$$

If $|f'(0)| = 1$, then $|g(0)| = 1$ and this is the maximum value of $g$ on $D_1(0)$. The Maximum Modulus Theorem says this cannot happen unless $g$ is a constant $c$. Then $f(z) = cz$. □

We now give a simple application of Schwarz's Lemma, which is a precursor of its later use in conformal mapping theory. Let $U$ and $V$ be open sets in $\mathbb{C}$. A *bi-analytic map* from $U$ to $V$ is an analytic function $f : U \to V$ with an analytic inverse $f^{-1} : V \to U$. That is, $f^{-1} \circ f(z) = z$ for every $z \in U$ and $f \circ f^{-1}(w) = w$ for every $w \in V$.

**Theorem 3.5.6.** *The only bi-analytic maps from the unit disc to itself that take 0 to 0 are of the form $f(z) = cz$ for a constant $c$ of modulus 1. That is, the only bi-analytic maps of the unit disc onto itself which fix 0 are the rotations.*

**Proof.** If $f : D_1(0) \to D_1(0)$ is bi-analytic with inverse $f^{-1}$, then both $f$ and $f^{-1}$ satisfy the hypotheses of Schwarz's Lemma. Thus,

$$|f'(0)| \leq 1 \quad \text{and} \quad |(f^{-1})'(0)| \leq 1. \tag{3.5.5}$$

However, it follows from the chain rule, applied to the composition $f^{-1} \circ f(z) = z$, that

$$(f^{-1})'(0) = \frac{1}{f'(0)}.$$

Combining this with (3.5.5), we conclude that $|f'(0)| = 1$. By Schwarz'e Lemma, $f(z) = cz$ for some constant $c$ of modulus 1. □

**Harmonic Functions.** Theorems about analytic functions, like Cauchy's Formula and the Maximum Modulus Theorem, imply things about harmonic functions. This is due to the fact that, locally at least, each harmonic function is the real part of an analytic function:

**Theorem 3.5.7.** *Let $u$ be a function which is of class $\mathcal{C}^2$ and harmonic on a convex open set $U$. Then $u$ has a harmonic conjugate $v$ on $U$. That is, there is a harmonic function $v$ on $U$ such that the function $f = u + iv$ is analytic on $U$.*

**Proof.** Consider the function $g = u_x - iu_y$. It is $\mathcal{C}^1$, therefore differentiable, and satisfies the Cauchy-Riemann equations, since

$$u_{xx} = -u_{yy} \quad \text{and} \quad u_{xy} = u_{yx}.$$

Thus, $g$ is analytic in $U$. Since $U$ is convex, Theorem 2.5.6 implies that $g$ has an antiderivative $h$ in $U$. This means $h$ is analytic in $U$ and $h' = g$. If $h = w + iv$, with $w$ and $v$ real, then

$$u_x - iu_y = g = h' = w_x + iv_x = w_x - iw_y.$$

On equating real and imaginary parts in this equation, we conclude that

$$u_x = w_x,$$
$$u_y = w_y.$$

It follows that $w = u + c$ for some real constant $c$. Thus, $f = h - c = u + iv$ is an analytic function in $U$ with real part $u$. □

This theorem leads to two important results about harmonic functions: a maximum principle for harmonic functions and an integral formula.

**Theorem 3.5.8.** *If $u$ is a harmonic function on a connected open set $U$ and $u$ has a local maximum at some point $z_0 \in U$, then $u$ is constant on $U$.*

**Proof.** Let $V$ be a convex neighborhood of $z_0$ with $V \subset U$. The previous theorem implies that $u$ has a harmonic conjugate $v$ on $V$. Then, $f = u + iv$ is analytic on $V$, as is the function

$$g(z) = e^{f(z)}.$$

Since $|g(z)| = e^{u(z)}$, if $u$ has a local maximum at $z_0$, then so does $|g(z)|$. By the Maximum Modulus Theorem, this implies that $g$ is constant on $V$. But if $g$ is constant on the connected open set $V$, then $f$ is also constant on $V$ (Exercise 3.5.10). Hence $u$ is constant on $V$.

The completion of the proof involves showing that if a harmonic function on a connected open set $U$ is constant on a non-empty open subset of $U$, then it is constant on all of $U$. We leave this as an exercise (Exercise 3.5.12). □

The next theorem shows that a harmonic function $u$ has the mean value property: the value of $u$ at a point is equal to its mean value over any circle centered at the point.

**Theorem 3.5.9.** *If $u$ is harmonic on an open set $U$ and $\overline{D}_r(z_0) \subset U$, then*

$$u(z_0) = \frac{1}{2\pi} \int_0^{2\pi} u(z_0 + r\,e^{it})\, dt. \tag{3.5.6}$$

**Proof.** Choose $R > r$ such that $D_R(z_0) \subset U$. Then Theorem 3.5.7 implies that $u$ is the real part of an analytic function $f$ on $D_R(z_0)$. The Cauchy Integral Formula applied to $f$ and the path $\gamma(t) = z_0 + r\,e^{it}$ yields

$$f(z) = \frac{1}{2\pi} \int_0^{2\pi} f(z_0 + r\,e^{it})\, dt.$$

Equation (3.5.6) follows from this by equating real parts. □

**Example 3.5.10.** Find a harmonic congugate for the harmonic function $x^2 - y^2$.

**Solution:** We recognize $x^2 - y^2$ as the real part of the analytic function $z^2 = (x + iy)^2$. The imaginary part of this function is $2xy$ and so $2xy$ is a harmonic conjugate of $x^2 - y^2$.

**Example 3.5.11.** Prove that

$$u(x, y) = \frac{x}{x^2 + y^2}$$

is harmonic on $\mathbb{C} \setminus \{0\}$ and find a harmonic conjugate for it.

**Solution:** We simply observe that $u(x, y)$ is the real part of

$$\frac{x - iy}{x^2 + y^2} = \frac{1}{z}.$$

Therefore, $u$ is harmonic and has

$$\frac{-y}{x^2 + y^2}$$

as a harmonic conjugate.

## Exercise Set 3.5

1. Find where the function $z^2 - 1$ attains its maximum modulus on the closed unit disc.
2. Find where the function $|\,e^z\,|$ attains its maximum value on the closed unit disc.
3. Find where the function $(z - 1)^2$ attains its maximum modulus on the triangle with vertices at $0, 1 + i, 1 - i$.

4. Show that if $f$ is a non-constant analytic function on a connected open set $U$ and if $f$ has no zeroes on $U$, then there are no points of $U$ where $|f(z)|$ has a local minimum.
5. Show that if $f$ is a non-constant, continuous function on $\overline{D}_1(0)$, which is analytic on $D_1(0)$ and $|f(z)| = 1$ for all $z$ on the unit circle, then $f$ has a zero somewhere in $D_1(0)$.
6. Prove that if $f$ is a non-constant entire function, and $c > 0$ is a constant such that the closed set $K = \{z \in \mathbb{C} : |f(z)| \leq c\}$ is bounded, then the open set $U = \{z \in \mathbb{C} : |f(z)| < c\}$ contains at least one zero of $f$.
7. Supppose $f$ is an analytic function on the unit disc $D_1(0)$. Prove that if $|f(z)| \leq 1$ on $D_1(0)$ and $f$ has a zero of order 2 at 0, then $|f(z)| \leq |z|^2$ for all $z \in D_1(0)$.
8. If $f$ is a harmonic function $u$ on a connected open set $U$, prove that any two harmonic conjugates for $u$ must differ by a constant.
9. Prove that if $U$ is a connected open set with compact closure $\overline{U}$, and $u$ is a continuous function on $\overline{U}$ which is harmonic on $U$, then $u$ attains its maximum and minimum values on $\partial U$. Also, prove that if $u$ attains either its maximum or its minimum at a point of $U$, then it is constant.
10. Prove that if $f$ is a continuous function on a connected open set $V$ and if $e^{f(z)}$ is constant on $V$, then $f$ is also constant on $V$.
11. Show that $f(z) = \dfrac{2z-1}{z-2}$ is a bi-analytic map from the unit disc onto itself which takes 0 to 1/2. Hint: Show that $|f(z)| = 1$ on the unit circle and conclude from this that $f$ maps the unit disc into itself. Then show it has an inverse function which also maps the unit disc into itself by directly solving the equation $w = f(z)$ for $z$ as a function of $w$.
12. Prove that if a harmonic function on a connected open set $U$ is constant on a non-empty open subset of $U$, then it is constant on all of $U$.
13. Find a harmonic conjugate for $u(x, y) = e^x \cos y$ on $\mathbb{C}$.
14. Find a harmonic conjugate for $u(x, y) = 1/2 \log(x^2 + y^2)$ on the complement of the non-positive real axis in $\mathbb{C}$.
15. Give an example of an open subset $U$ of $\mathbb{C}$ and a harmonic function on $U$ which has no harmonic conjugate on $U$.

Chapter 4

# The General Cauchy Theorems

In this chapter we extend Cauchy's theorems in two ways: (1) we remove the condition that the open set $U$ be convex; and (2) we replace the path $\gamma$ by a more general type of object called a *cycle*.

In making these improvements, we finally come to grips with an issue which we have avoided so far. This issue concerns the *topology* of the plane. Specifically, there is a phenomenon that occurs in the plane which does not occur in the line. On the line, every connected open set is an open interval, which certainly has no gaps or holes. However, in the plane, there are connected open sets which do have holes. An example is an open annulus of the form

$$A = \{z \in \mathbb{C} : r < |z| < R\},$$

with $0 \leq r < R$. The integral of an analytic function on $A$ around a closed path in $A$ may fail to be zero. In fact, the integral of $1/z$ around any positively oriented circle in $A$, centered at 0, is $2\pi i$, not 0, even though $1/z$ is analytic on $A$. Thus, Cauchy's Integral Theorem, as stated in Theorem 2.6.2, is not valid if we simply drop the condition that $U$ be convex. The problem here is that there is a hole in the domain $A$ on which $f$ is analytic (the missing disc of radius $r$) and the path goes around this hole. In our final version of Cauchy's Theorem we will avoid this problem, not with a hypothesis on the shape of $U$, but rather with a hypothesis that ensures that the path $\gamma$ does not go around any holes in $U$. We do this by adding the hypothesis that the path $\gamma$ has index 0 around every point in the complement of $U$.

We begin by introducing a compact and efficient notation for expressing an otherwise complicated sum of integrals over a number of different paths. This is the language of chains and cycles, as used in the study of homology in algebraic topology. Using this language will allow for relatively simple statements and proofs of theorems that otherwise could be, at least notationally, quite complicated.

## 4.1. Chains and Cycles

In many calculations involving path integrals of an analytic function, we may integrate the same function over several paths and add the results, or add some of the results and subtract others. We may integrate over the same path a number of times. We may break a single path up into a collection of several paths, integrate over each and then add the results. We may replace a path by one which simply reverses the direction of travel, resulting in an integral which is the negative of the original. We have already seen examples of some of these things (Examples 2.6.4 and 2.3.4 come to mind). The language of chains and cycles, introduced below, will provide us with a compact and efficient way to deal with these kinds of integrals.

In this section, all paths will have parameter interval $[0, 1]$.

**1-Chains.** An *abelian group* is a set $X$ with a single operation $(x, y) \to x + y : X \times X \to X$ which is associative and commutative, has an identity (denoted by 0) and in which every element $x$ has an inverse (denoted by $-x$). So an abelian group satisfies the addition axioms of a vector space, but it is not assumed that there is also a scalar multiplication operation.

Let $U$ be an open subset of $\mathbb{C}$. We will construct an abelian group called *the group of 1-chains* in $U$ from the set of all paths $\gamma : [0, 1] \to U$ as follows: we define a *1-chain* in $U$ to be a finite formal linear combination, with integral coefficients,

$$\Gamma = \sum_{j=1}^{p} m_j \gamma_j,$$

of distinct paths $\gamma_j$ in $U$. We agree that a chain remains the same if we either throw out or add in some summands $m_j \gamma_j$ with $m_j = 0$.

The operation in our group of chains is addition. We add two chains in the obvious way: if $\Gamma$ and $\Lambda$ are chains and $\gamma_1, \gamma_2, \cdots, \gamma_n$ are the paths that occur in summands of either $\Gamma$ or $\Lambda$, then

$$\Gamma = \sum_{j=1}^{p} n_j \gamma_j \quad \text{and} \quad \Lambda = \sum_{j=1}^{p} m_j \gamma_j$$

for coefficients $\{n_j\}$ and $\{m_j\}$ some of which may be zero. Then we define

$$\Gamma + \Lambda = \sum_{j=1}^{p} (n_j + m_j) \gamma_j.$$

The operation of addition, so defined, is clearly associative and commutative and so the set of 1-chains forms an abelian group.

In group theory, the group of chains, as defined above, is known as the *free abelian group* generated by the set of paths in $U$.

A path itself is a 1-chain (a 1-chain where there is only one summand and its coefficient is 1). However, often it is useful to break up a path into pieces that form the summands of a 1-chain. The original path and the 1-chain formed by its pieces are equivalent in a sense that will be discussed later in the section.

**Example 4.1.1.** Show how to use a path in $U$, in the sense of the previous chapter, to produce a 1-chain in $U$ which is a linear combination of smooth paths.

**Solution:** Suppose $\gamma : [a,b] \to U$ is a closed path. Let

$$a = t_0 < t_1 < \cdots < t_n = b$$

be a partition such that $\gamma$ is smooth on each subinterval. Let $z_j = \gamma(t_j)$ for $j = 1, \cdots, n$. For $j = 1, \cdots, n$ let $\gamma_j$ be a reparameterization of the restriction of $\gamma$ to $[t_{j-1}, t_j]$ which changes the parameter interval to $[0,1]$. If we define a chain $\Gamma$ to be the formal sum

$$\Gamma = \sum_{j=1}^{n} \gamma_j,$$

then this is a chain consisting of smooth paths.

**Integration Over Chains.** We introduced 1-chains in order to integrate over them. The definition of the integral is as follows:

**Definition 4.1.2.** Let

$$\Gamma = \sum_{j=1}^{p} m_j \gamma_j$$

be a 1-chain. With $I = [0,1]$, we set

$$\Gamma(I) = \bigcup_{m_j \neq 0} \gamma_j(I).$$

If $f$ is a continuous function on a set $E \subset \mathbb{C}$ with $\Gamma(I) \subset E$, then we define the integral of $f$ over $\Gamma$ by

$$\int_\Gamma f(z)\,dz = \sum_{j=1}^{p} m_j \int_{\gamma_j} f(z)\,dz. \tag{4.1.1}$$

It is an immediate consequence of the definition of the integral over a 1-chain, that if $\Gamma$ and $\Lambda$ are two 1-chains and $f$ is a function defined and continuous on a set containing both $\Gamma(I)$ and $\Lambda(I)$, then

$$\int_{\Gamma+\Lambda} f(z)\,dz = \int_\Gamma f(z)\,dz + \int_\Lambda f(z)\,dz. \tag{4.1.2}$$

**Definition 4.1.3.** Suppose $\Gamma$ and $\Lambda$ are chains with $\Gamma(I)$ and $\Lambda(I)$ both subsets of a set $E \subset \mathbb{C}$. We will say that $\Gamma$ and $\Lambda$ are *E-equivalent* if

$$\int_\Gamma f(z)\,dz = \int_\Lambda f(z)\,dz$$

for every continuous function $f$ on $E$. If $\Gamma(I) = \Lambda(I) = E$ and $\Gamma$ and $\Lambda$ are $E$-equivalent, then will will simply say that $\Gamma$ and $\Lambda$ are *equivalent*.

**Example 4.1.4.** Let $\Delta$ be the triangle with vertices $\{a,b,c\}$ listed in counterclockwise order. Recall that $\partial\Delta$ denotes the path which traverses the boundary of $\Delta$ once in the counterclockwise direction. Show that $\partial\Delta$ and $\Gamma = [a,b] + [b,c] + [c,a]$ are equivalent, as are $\partial\Delta$ and $\Lambda = [a,b] + [b,c] - [a,c]$.

**Solution:** Clearly $\partial\Delta(I) = \Gamma(I) = \Lambda(I)$ – each of them is just the boundary of the triangle $\Delta$. Also, it follows from Theorem 2.4.6, Parts (b) and (c), that $\partial\Delta, \Gamma$, and $\Lambda$ all determine the same integral for continuous functions on the boundary of $\Delta$. Thus, by definition, the three chains are equivalent.

**0-Chains.** Just as the group of 1-chains is the free abelian group generated by paths, the group of *0-chains* in $U$ is the free abelian group generated by the singleton subsets of $\mathbb{C}$ (sets $\{z\}$ containing a single point $z \in \mathbb{C}$) – that is, it is the group of formal linear combinations,

$$\sum_{j=1}^{p} m_j\{z_j\},$$

with integral coefficients, of distinct singleton subsets of $U$. Again, linear combinations which differ only by summands with 0's as coefficients are all identified. Addition is defined as before, and the result is another abelian group.

A path $\gamma$ determines both a 1-chain (the chain with $\gamma$ as its only summand and a 0-chain $\{\gamma(1)\} - \{\gamma(0)\}$. This 0-chain is called the *boundary* of $\gamma$. Notice that it is the 0 element of the group of 0-chains if and only if $\{\gamma(1)\} = \{\gamma(0)\}$ – that is, if and only if $\gamma$ is a closed path. Both the boundary of a path in this sense and the concept of a closed path generalize to the context of chains. The result is as follows:

### The Boundary Map and Cycles.

**Definition 4.1.5.** If $U$ is an open subset of $\mathbb{C}$, we define the *boundary map* $\partial$ from 1-chains in $U$ to 0-chains in $U$ by

$$\partial\left(\sum_{j=1}^{p} m_j\gamma_j\right) = \sum_{j=1}^{p}(m_j\{\gamma_j(1)\} - m_j\{\gamma_j(0)\}).$$

This may not result in a linear combination of distinct singleton sets, but we fix this, combining terms involving the same singleton subset by adding their coefficients. In order for $\partial$ to determine a well defined map from 1-chains to 0-chains, one must show that it preserves the identifications made in our definition of 1-chains. This simply requires that any summand in a 1-chain that has zero as coefficient is sent by $\partial$ to a 0-chain with all zeroes as coefficients, and this is obviously true.

Once we know that $\partial$ is well defined, it is quite evident from the definition that it is a group homomorphism $\partial$ from 1-chains to 0-chains (meaning $\partial(\Gamma+\Lambda) = \partial\Gamma+\partial\Lambda$). This ensures that $\ker\partial = \{\Gamma : \partial\Gamma = 0\}$ is closed under addition and taking additive inverses and, hence, is a subgroup of the set of 1-chains in $U$. Elements of this

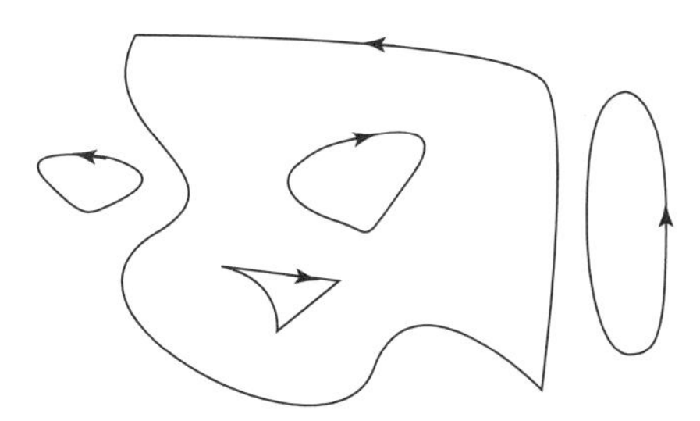

**Figure 4.1.1.** A Cycle which is a Sum of Closed Paths.

subgroup are called *cycles* in $U$ – that is:

**Definition 4.1.6.** A 1-chain $\Gamma$ in $U$ is called a *1-cycle* if $\partial\Gamma = 0$.

As we shall see below, a cycle is a chain whose summands can be fit together to form a series of loops (closed paths).

Given a chain $\Gamma$, there are always many other chains which are equivalent to it. This is due to the fact that, as in the previous example, paths in a chain may be partitioned into sums of subpaths without changing the integral, and several paths may be joined to form a single path if their endpoints correspond in the right way. This is what is going on in the following theorem.

**Theorem 4.1.7.** *If $\Gamma$ is a 1-cycle, then there is a 1-cycle $\Lambda$ which is equivalent to $\Gamma$ and which is a sum of closed paths.*

**Proof.** We will make a succession of changes to $\Gamma$, none of which will change $\Gamma(I)$ or the integral (4.1.1).

We first express $\Gamma$ as a linear combination of paths in which the coefficients are all 1 or $-1$. That is, each term $m_j\gamma_j$ is replaced by a sum of $m_j$ copies of $\gamma_j$ if $m_j$ is positive and by a sum of $-m_j$ copies of $(-1)\gamma_j$ if $m_j$ is negative. This does not change the chain $\Gamma$.

The next step is to replace each summand of the form $(-1)\gamma_j$ with the path $-\gamma_j$, where we recall that $-\gamma_j(t) = \gamma_j(1-t)$, so that $-\gamma_j$ is $\gamma_j$ traversed in the reverse direction. Theorem 2.4.6, Part (c), implies that this change results in a cycle $\tilde{\Gamma}$ which is equivalent to $\Gamma$.

At this point we have replaced $\Gamma$ with a cycle $\tilde{\Gamma}$ which is a simple sum of some number of paths. Say the number is $n$. We next show that if these paths are not all closed paths, then we can replace $\tilde{\Gamma}$ with a sum of fewer than $n$ paths without changing the integral.

Suppose $\gamma_j$ is a path in $\tilde{\Gamma}$ which is not closed. Then $\gamma_j(0) \neq \gamma_j(1)$. Since $\partial\tilde{\Gamma} = 0$, $\gamma_j(1)$ must be equal to $\gamma_k(0)$ for some path $\gamma_k$ in $\tilde{\Gamma}$; otherwise there would be no term to cancel with $\{\gamma_j(1)\}$ in the expression for $\partial\Gamma$. We may join $\gamma_k$ and $\gamma_j$, as in (2.4.2), to form a new path which begins at $\gamma_j(0)$ and ends at $\gamma_k(1)$. By Theorem 2.4.6, Part (b), replacing $\gamma_j$ and $\gamma_k$ in $\tilde{\Gamma}$ by this new path does not change the integral. Thus, if its summands are not all closed paths, we may replace $\tilde{\Gamma}$ with a path with fewer summands, which determines the same integral. This forms the basis for an induction argument, which is carried out below.

If a cycle consists of a single path $\gamma$, then the condition

$$\partial\gamma = \{\gamma(1)\} - \{\gamma(0)\} = 0$$

implies $\gamma$ is a closed path. Suppose we know that any cycle which can be written as a sum of fewer than $n$ paths may be replaced by an equivalent cycle which is a sum of closed paths. Then the argument of the previous paragraph shows that any cycle which can be written as a sum of $n$ paths is equivalent to some cycle which is a sum of closed paths. By induction, every cycle is equivalent to a sum of closed paths. □

**Index of a Cycle.** The notion of index of a path around a point can be extended to cycles as follows.

**Definition 4.1.8.** If $\Gamma$ is a 1-cycle in $\mathbb{C}$ and $z \in C$ is a point which does not lie on $\Gamma(I)$, then we set

$$\mathrm{Ind}_\Gamma(z) = \frac{1}{2\pi i}\int_\Gamma \frac{1}{w-z}\,dw.$$

The number $\mathrm{Ind}_\Gamma(z)$ is called the *index* of $\Gamma$ around $z$.

By Theorem 4.1.7 the cycle $\Gamma$ may be replaced by one which is a sum of closed paths without changing $\Gamma(I)$ or the integral defining $\mathrm{Ind}_\Gamma(z)$. Thus, $\mathrm{Ind}_\Gamma(z)$ is a finite sum of numbers of the form $\mathrm{Ind}_\gamma(z)$, where $\gamma$ is a closed path. This, together with Theorems 2.6.6 and 2.7.5 proves the first three parts of the following theorem. The last part follows immediately from (4.1.2).

**Theorem 4.1.9.** *Let $\Gamma$ be a 1-cycle in $\mathbb{C}$. Then*

(a) $\mathrm{Ind}_\Gamma$ *is an integer-valued function defined on the complement of* $\Gamma(I)$;
(b) $\mathrm{Ind}_\Gamma$ *is constant on each component of* $\mathbb{C}\setminus\Gamma(I)$;
(c) $\mathrm{Ind}_\Gamma$ *is* 0 *on the unbounded component of* $\mathbb{C}\setminus\Gamma(I)$;
(d) *if* $\Lambda$ *is also a cycle, and* $z \notin \Gamma(I)\cup\Lambda(I)$, *then*

$$\mathrm{Ind}_{\Gamma+\Lambda}(z) = \mathrm{Ind}_\Gamma(z) + \mathrm{Ind}_\Lambda(z).$$

**Example 4.1.10.** For each integer $j$, let

$$\gamma_j(t) = j + \frac{1}{4}\,\mathrm{e}^{2\pi i t}, \quad t \in [0,1].$$

For a positive integer $N$ set

$$\gamma(t) = (N+1/2)\,\mathrm{e}^{2\pi i t}, \quad t \in [0,1]$$

and

$$\Gamma_1 = \sum_{j=-N}^{N} \gamma_j, \quad \Gamma = \gamma - \Gamma_1.$$

Find $\mathrm{Ind}_\gamma(z), \mathrm{Ind}_{\Gamma_1}(z)$, and $\mathrm{Ind}_\Gamma(z)$ for each value of $z$ which is an integer.

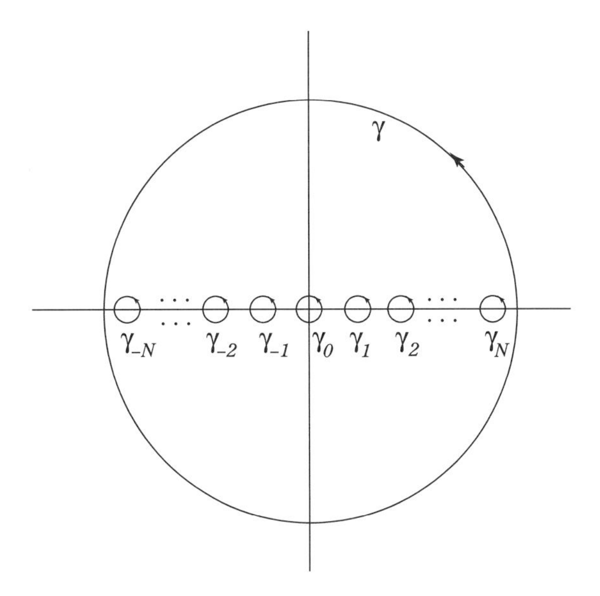


**Figure 4.1.2.** The Paths Making up the Cycle Γ of Example 4.1.10.

**Solution:** The paths $\gamma$ and $\gamma_j$ are circles traversed once in the counterclockwise direction. For such a path we know the index is 1 inside the circle and 0 outside the circle. Thus, if $k$ is an integer,

$$\text{Ind}_\gamma(k) = \begin{cases} 1, & \text{if } |k| \le N; \\ 0, & \text{if } |k| > N; \end{cases}$$

and

$$\text{Ind}_{\gamma_j}(k) = \begin{cases} 1, & \text{if } k = j; \\ 0 & \text{if } k \neq j. \end{cases}$$

Thus,

$$\text{Ind}_{\Gamma_1}(k) = \sum_{j=-k}^{k} \text{Ind}_{\gamma_j}(k) = \begin{cases} 1, & \text{if } |k| \le N; \\ 0, & \text{if } |k| > N; \end{cases}$$

and so,

$$\text{Ind}_\Gamma(k) = \text{Ind}_\gamma(k) - \text{Ind}_{\Gamma_1}(k) = 0$$

for every integer $k$.

**Homologous Cycles.** In the general Cauchy Theorem of the next section, the condition that $U$ be convex is replaced by a condition on the path or cycle over which the integration takes place. The condition is that the path or cycle be *homologous* to 0 in the sense of the following definition.

**Definition 4.1.11.** Let $U$ be an open subset of $\mathbb{C}$ and let $\Gamma$ and $\Lambda$ be 1-cycles in $U$. We will say that $\Gamma$ and $\Lambda$ are *homologous* in $U$ if

$$\text{Ind}_\Gamma(z) = \text{Ind}_\Lambda(z)$$

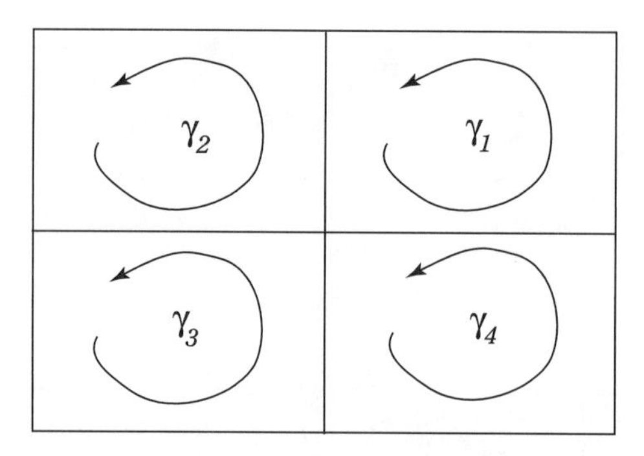


**Figure 4.1.3.** The Paths Making up the Cycle $\Gamma$ of Exercise 4.1.6.

for every $z$ in the complement of $U$. We will say that $\Gamma$ is *homologous* to 0 in $U$ if

$$\mathrm{Ind}_\Gamma(z) = 0$$

for every $z$ in the complement of $U$.

For a closed path $\gamma$, intuitively, the statement that $\gamma$ is homologous to 0 in $U$ means that $\gamma$ does not go around any points of the complement of $U$ – that is, it does not go around any "holes" in $U$. For cycles, the meaning is more complicated. For example, the cycle $\Gamma$ of Example 4.1.10 is homologous to 0 in the complement of the integers in $\mathbb{C}$.

Note that two 1-cycles $\Gamma$ and $\Lambda$ are homologous if and only if $\Gamma - \Lambda$ is homologous to 0.

**Example 4.1.12.** If $\gamma_1(t) = e^{2\pi i t}$ and $\gamma_2(t) = 2\,e^{2\pi i t}$ for $t \in [0,1]$, show that $\Gamma = \gamma_2 - \gamma_1$ is a 1-cycle which is homologous to 0 in $U = \{z \in \mathbb{C} : 1/2 < |z| < 3\}$, but is not homologous to 0 in the set $V = U \setminus \{3/2\}$.

**Solution:** The curve $\gamma_1$ has index 1 about points inside the unit disc $D_1(0)$, and index 0 about points exterior to the closed unit disc, while $\gamma_2$ has index 1 about points in $D_2(0)$ and 0 about points exterior to the closure of this disc. The complement of $U$ is

$$\overline{D}_{1/2}(0) \cup (\mathbb{C} \setminus D_3(0)).$$

Since

$$\mathrm{Ind}_{\gamma_1} = \mathrm{Ind}_{\gamma_2} = 1 \quad \text{on} \quad \overline{D}_{1/2}(0)$$

and

$$\mathrm{Ind}_{\gamma_1} = \mathrm{Ind}_{\gamma_2} = 0 \quad \text{on} \quad \mathbb{C} \setminus D_3(0),$$

we conclude that $\Gamma = \gamma_2 - \gamma_1$ is homologous to 0 in $U$.

However, at the point $z = 3/2$, $\gamma_1$ has index 0 and $\gamma_2$ has index 1. Since 3/2 is in the complement of $V = U \setminus \{3/2\}$, $\Gamma$ is not homologous to 0 in $V$.

## Exercise Set 4.1

1. Let $a, b, c$ be three points in $\mathbb{C}$ and let $\gamma_1$ be a path beginning at $a$ and ending at $b$, $\gamma_2$ a path beginning at $a$ and ending at $c$, and $\gamma_3$ a path beginning at $b$ and ending at $c$. Is $\gamma_1 + \gamma_2 + \gamma_3$ a cycle? How about $\gamma_1 - \gamma_2 + \gamma_3$?
2. Consider the following directed line segments: $\gamma_1$ from $-1$ to $-1 + i$, $\gamma_2$ from $1$ to $1 + i$, $\gamma_3$ from $-1$ to $0$, $\gamma_4$ from $0$ to $1$, $\gamma_5$ from $-1 + i$ to $i$, $\gamma_6$ from $i$ to $i + 1$, and $\gamma_7$ from $0$ to $i$. Show that $\Gamma = \gamma_1 + \gamma_2 - \gamma_3 + \gamma_4 + \gamma_5 - \gamma_6 - 2\gamma_7$ is a cycle.
3. Find a sum of closed paths which is equivalent to the cycle of the previous exercise.
4. For $t \in [0, 1]$, let $\gamma_1(t) = 1 + e^{2\pi i(t+1/2)}$ and $\gamma_2(t) = -1 + e^{2\pi it}$. Find a single closed path $\gamma$ with parameter interval $[0, 1]$, such that $\gamma$ is equivalent to $\gamma_1 - \gamma_2$.
5. Let $f$ be a function which is analytic on the open unit disc $D$ and $\gamma_1$ and $\gamma_2$ two closed paths in $D$ each of which has index 1 about 0. If $\Gamma = \gamma_1 - \gamma_2$, compute $\int_\Gamma f(z)/z\, dz$?
6. Let $R \subset \mathbb{C}$ be a rectangle and let $\gamma$ be a path which traverses the boundary of $R$ once in the positive direction. Partition $R$ into four subrectangles $R_j$, $j = 1, 2, 3, 4$, by joining each pair of opposite sides of $R$ with a perpendicular line. Let $\gamma_j$ be a path which traverses the boundary of $R_j$ once in the positive direction. Let $\Gamma = \sum_{j=1}^{4} \gamma_j$ and $E = \Gamma(I)$. Show that $\gamma$ and $\Gamma$ are $E$-equivalent but not equivalent (see Definition 4.1.3).
7. For $t \in I = [0, 1]$, let

$$\gamma_1(t) = 4\, e^{2\pi it},$$
$$\gamma_2(t) = i + e^{2\pi it},$$
$$\gamma_3(t) = -i + e^{2\pi it}.$$

   If $\Gamma = \gamma_1 - \gamma_2 - \gamma_3$, then find $\mathrm{Ind}_\Gamma(z)$ for $z$ in each of the components of $\mathbb{C} \setminus \Gamma(I)$. Is $\Gamma$ homologous to 0 in $\mathbb{C} \setminus \{i, -i\}$?
8. Find $\mathrm{Ind}_\Gamma(z)$ for $z$ in each component of $\mathbb{C} \setminus \Gamma(I)$ if $\Gamma$ is the cycle of Exercise 4.1.2.
9. Find $\mathrm{Ind}_\Gamma(z)$ for $z$ in each component of $\mathbb{C} \setminus \Gamma(I)$ if $\Gamma = 2\gamma_1 - \gamma_2 - \gamma_3$ with $\gamma_1$, $\gamma_2$, $\gamma_3$ the positively oriented circles of radii 4, 2, and 2, centered at 0, $-1$, and 1, respectively.
10. Let $U$ be an open set. Prove that if $\Gamma$ and $\Gamma_1$ are cycles that are $U$-equivalent, then they are homologous in $U$.
11. Let $U$ be an open subset of $\mathbb{C}$ and $S = \{z_1, \cdots, z_n\}$ a finite subset of $U$. Find a cycle $\Gamma$ which is homologous to 0 in $U$ and which has index 1 about each point of $S$.
12. If $n$ is an integer, let $\gamma(t) = e^{2n\pi it}$ and $\gamma_1(t) = 2\, e^{2\pi it}$ for $t \in [0, 1]$. Show that the cycle $\gamma - n\gamma_1$ is homologous to 0 in $A = \{z \in \mathbb{C} : 0 < |z| < 3\}$.
13. Find a cycle $\Gamma$ which is homologous to 0 in the annulus $\{z : 1 < |z| < 4\}$, but has index 1 about every point in the annulus $\{z : 2 < |z| < 3\}$.

## 4.2. Cauchy's Theorems

We will first prove the general version of Cauchy's Formula and then use it to prove the general version of Cauchy's Theorem. This is the reverse of the order in which we proved the analogous results on convex sets.

**A Continuity Lemma.** Let $f$ be an analytic function on an open set $U$. In the proof of the Cauchy Integral Formula for convex sets (Theorem 2.6.7) a key role was played by the function

$$g(z,w) = \begin{cases} \dfrac{f(w)-f(z)}{w-z} & \text{if } w \neq z; \\ f'(z) & \text{if } w = z. \end{cases} \tag{4.2.1}$$

For each fixed $z$, this function is clearly a continuous function of $w$ everywhere on $U$, and is analytic on $U$ except possibly at $w = z$. This was enough for our purposes in the proof of Theorem 2.6.7. We will use the same function in the proof of our more general Cauchy Integral Theorem, but we need to know that it satisfies a stronger continuity condition. We need that it is continuous as a function of two complex variables on

$$U \times U = \{(z,w) : z \in U, w \in U\}.$$

This is equivalent to the condition that $g$ be continuous on $U \times U$ as a function of four real variables with values in $\mathbb{R}^2 = \mathbb{C}$. It means simply that

$$\lim_{(z,w)\to(z_0,w_0)} g(z,w) = g(z_0,w_0) \tag{4.2.2}$$

for all $(z_0, w_0) \in U \times U$. Here, the limit has the usual $\epsilon$-$\delta$ definition, using the appropriate definition of the distance between $(z, w)$ and $(z_0, w_0)$, which is

$$|(z,w)-(z_0,w_0)| = \sqrt{|z-z_0|^2 + |w-w_0|^2}.$$

This is the same as the Euclidean distance between these two points, considered as points of $\mathbb{R}^4$. It is easy to see that this type of limit satisfies all the usual limit rules regarding sums, products, quotients, etc.

**Lemma 4.2.1.** *If $g(z,w)$ is defined by (4.2.1), then $g$ is a continuous function of two complex variables on $U \times U$.*

**Proof.** We must prove that (4.2.2) holds at every point $(z_0, w_0) \in U \times U$. If $z_0 \neq w_0$, then this follows from the continuity of $f$ and the quotient rule for limits.

In the case where $z_0 = w_0$, we have $g(z_0, z_0) = f'(z_0)$. Thus, in order to prove (4.2.2) we just need to prove

$$\lim_{(z,w)\to(z_0,z_0)} g(z,w) = f'(z_0). \tag{4.2.3}$$

Since $f'$ is continuous on $U$, given $\epsilon > 0$, we may choose $\delta > 0$ such that $\overline{D}_\delta(z_0) \subset U$ and

$$|f'(z) - f'(z_0)| < \epsilon \quad \text{whenever} \quad |z - z_0| < \delta. \tag{4.2.4}$$

If $|(z, w) - (z_0, z_0)| < \delta$, then both $z$ and $w$ are in $D_\delta(z_0)$, as is any point $\lambda$ on the line segment joining them. If $z \neq w$, integrating over this line segment yields

$$f(w) - f(z) = \int_z^w f'(\lambda)\, d\lambda,$$

and so

$$|g(z, w) - f'(z_0)| = \left| \frac{1}{w - z} \int_z^w (f'(\lambda) - f'(z_0))\, d\lambda \right| < \epsilon.$$

If $z = w$, then $|g(z, w) - f'(z_0)| = |f'(z) - f'(z_0)| < \epsilon$. Hence, $g$ is continuous at $(z_0, z_0)$. This completes the proof. □

### Cauchy's Integral Formula.

**Theorem 4.2.2** (Cauchy's Integral Formula). *Let $U$ be any open subset of $\mathbb{C}$, $f$ a function which is analytic on $U$, and $\Gamma$ a 1-cycle in $U$ which is homologous to 0 in $U$. Then*

$$\operatorname{Ind}_\Gamma(z) f(z) = \frac{1}{2\pi i} \int_\Gamma \frac{f(w)}{w - z} dw, \tag{4.2.5}$$

*for every point $z \in U$ which does not lie on $\Gamma(I)$.*

**Proof.** Let $h$ be the function defined by

$$h(z) = \int_\Gamma g(z, w)\, dw,$$

where $g$ is the function of the previous lemma. The fact that $g$ is continuous on $U \times U$ as a function of two complex variables implies that $h$ is continuous on $U$.

If $z \notin \Gamma(I)$, then $z \neq w$ for all $w \in \Gamma(I)$ and so the integral defining $h$ can be broken apart as

$$h(z) = \int_\Gamma \frac{f(w)}{z - w} dw - \int_\Gamma \frac{f(z)}{z - w} dw = \int_\Gamma \frac{f(w)}{z - w} dw - 2\pi i \operatorname{Ind}_\Gamma(z) f(z).$$

Thus, to prove the theorem, we simply need to prove that $h$ is identically zero in $U$. In the proof of Theorem 2.6.7, we proved $h(z) = 0$ using the Cauchy Theorem in a convex set. Here, since $U$ is not convex, and $\Gamma$ may not be a single closed path, we must proceed differently.

We will show that $h(z) = 0$ for all $z \in U$ by showing that it is analytic on $U$ and then extending it to an entire function which is bounded and has limit 0 at infinity. Then Liouville's Theorem will imply that it is identically 0.

If $\Delta$ is a triangle contained in $U$, we have

$$\begin{aligned} \int_{\partial\Delta} h(z) dz &= \int_{\partial\Delta} \int_\Gamma g(z, w)\, dw dz \\ &= \int_\Gamma \int_{\partial\Delta} g(z, w)\, dz dw. \end{aligned} \tag{4.2.6}$$

That the order of integration can be reversed follows from the fact that, ultimately, each iterated integral is a sum of iterated Riemann integrals, over the rectangle $I \times I$ in $\mathbb{R}^2$, of functions continuous on the rectangle. The fact that the order of integration can be reversed in each of these integrals is then Fubini's Theorem from calculus.

For each fixed $z \in U$, $g(z, w)$ is an analytic function of $w \in U$, except at $w = z$, where it is at least continuous. Thus, by Theorem 2.5.8 the last integral in (4.2.6) is 0. Hence, the integral of $h$ around each triangle in $U$ is 0. It follows from Morera's Theorem that $h$ is analytic in $U$.

Now let $V$ be the set of all points $z$ in the complement of $\Gamma(I)$ for which $\mathrm{Ind}_\Gamma(z) = 0$. This is an open set, since it is a union of some of the components of $\mathbb{C} \setminus \Gamma(I)$. It is also a set which contains the complement of $U$, by hypothesis. So $U \cup V = \mathbb{C}$. If $z \in V \cap U$, then

$$\frac{1}{2\pi i} \int_\Gamma \frac{f(z)}{z - w} dw = f(z) \, \mathrm{Ind}_\Gamma(z) = 0.$$

Since $z$ is not on $\Gamma(I)$, we may break the integral defining $h$ apart and write

$$h(z) = \int_\Gamma \frac{f(w)}{w - z} dw - \int_\Gamma \frac{f(z)}{w - z} dw = \int_\Gamma \frac{f(w)}{z - w} dw \tag{4.2.7}$$

for $z \in U \cap V$. This means we can extend the definition of $h$ to all of $\mathbb{C}$ by defining it to be $\displaystyle\int_\Gamma \frac{f(w)}{z - w} dw$ on $V$. That is, the original definition of $h$ on $U$ and this new definition of $h$ on $V$ agree on the overlap $U \cap V$. The resulting function $h$ is analytic on all of $\mathbb{C}$ and, hence, is an entire function. On the unbounded component of $\mathbb{C} \setminus \Gamma(I)$, $h(z)$ is given by (4.2.7) and, from this, it is easy to see that

$$\lim_{z \to \infty} h(z) = 0. \tag{4.2.8}$$

We conclude that $h$ is a bounded entire function and, hence, is a constant, by Liouvilles's Theorem. By (4.2.8) this constant must be zero. Thus, $h(z)$ is identically zero and this immediately implies (4.2.5) for $z \notin \Gamma(I)$. □

**Example 4.2.3.** Find $\displaystyle\int_\Gamma \frac{1}{z^2 - z} dz$ if $\Gamma = \gamma_2 - \gamma_1$, where $\gamma_1 = (1/2)\, e^{2\pi i t}$ and $\gamma_2 = 2\, e^{2\pi i t}$ for $t \in I = [0, 1]$.

**Solution:** The cycle $\Gamma$ has index 0 at all points outside the annulus bounded by the two circles $\gamma_1(I)$ and $\gamma_2(I)$ and index 1 at points inside this annulus. If we set $f(z) = 1/z$, then $f$ is analytic in $U = \mathbb{C} \setminus \{0\}$ and the cycle $\Gamma$ is a cycle homologous to 0 in $U$. Thus, the Cauchy Integral Formula tells us that

$$\int_\Gamma \frac{1}{z^2 - z} dz = \int_\Gamma \frac{f(z)}{z - 1} dz = 2\pi i f(1) = 2\pi i.$$

**The Cauchy Integral Theorem.** The Cauchy Integral Theorem follows easily from Theorem 4.2.2.

**Theorem 4.2.4.** *If $f$ is analytic in the open set $U$ and if $\Gamma$ is a 1-cycle in $U$ which is homologous to 0 in $U$, then*

$$\int_\Gamma f(z)\, dz = 0.$$

**Proof.** Let $z_0$ be any point in $U$ which is not on $\Gamma(I)$ and define

$$g(z) = f(z)(z - z_0).$$

Then $g(z)$ is also analytic in $U$ and $g(z_0) = 0$. Theorem 4.2.2, applied to the function $g$, tells us that

$$\int_\Gamma f(z)\,dz = \int_\Gamma \frac{g(z)}{z - z_0}\,dz = 2\pi i \operatorname{Ind}_\Gamma(z_0) g(z_0) = 0.$$

This completes the proof. □

**Example 4.2.5.** Find $\displaystyle\int_\Gamma \frac{1}{\sin(\pi z)}\,dz$ if $\Gamma$ is the cycle of Example 4.1.10.

**Solution:** By Example 4.1.10, $\Gamma$ has index 0 at every integer $k$ and so it is homologous to 0 in $U = \mathbb{C} \setminus \mathbb{Z}$, where $\mathbb{Z}$ is the set of integers. The function $\sin(\pi z)$ vanishes exactly on the set of integers (Exercise 4.2.1) and so $1/\sin(\pi z)$ is analytic in $U$. It follows from the above theorem that the integral in question is 0.

**Simple Closed Paths.** Although we have stated the general Cauchy theorems in terms of integration over 1-cycles, they also hold if the integration is over an ordinary closed path. After all, a closed path is just a particularly simple 1-cycle. Classically, these theorems are stated and proved for integrals over a particular kind of closed path – a *simple* closed path.

A simple closed curve is a closed curve which does not intersect itself except at the endpoints of the parameter interval $I = [a, b]$. That is,

**Definition 4.2.6.** A closed curve $\gamma$, defined on $[a, b]$, is called a *simple closed curve* if $\gamma(s) \neq \gamma(t)$ for $a \le s < t \le b$ unless $s = a$ and $t = b$.

The *Jordan Curve Theorem* says that if $\gamma$ is a simple closed curve, then the complement of $\gamma(I)$ has two components – one unbounded, and one bounded. Here we will prove a weak version of the Jordan Curve Theorem – one that is sufficient for our purposes.

Recall from Definition 2.7.10 that a simple point of a path $\gamma$ is a point that the path passes through just once and at which the left and right derivatives of $\gamma$ are both non-zero.

**Definition 4.2.7.** A closed path will be called a *simple closed path* if each of its points is a simple point.

Thus, a simple closed path is a simple closed curve which is piecewise smooth with non-zero left and right derivatives at each point. Our weak version of the Jordan Curve Theorem is the following.

**Theorem 4.2.8.** *If $\gamma$ is a simple closed path, then $\mathbb{C} \setminus \gamma(I)$ has exactly two components – a bounded component on which* $\operatorname{Ind}_\gamma(z) = \pm 1$ *and an unbounded component on which* $\operatorname{Ind}_\gamma(z) = 0$.

**Proof.** By Theorems 2.7.8 and 2.7.11, we may choose for each point $z \in \gamma(I)$ an open disc $D_z$, centered at $z$, such that $D_z \setminus (D_z \cap \gamma(I))$ has exactly two components – a left component $L_z$ and a right component $R_z$, and $\operatorname{Ind}_\gamma(z)$ is one unit greater on $L_z$ than on $R_z$ (see Figure 4.2.2).

It is not difficult to see that two sufficiently close points of $\gamma(I)$ have discs with overlapping left components and overlapping right components. It follows that the

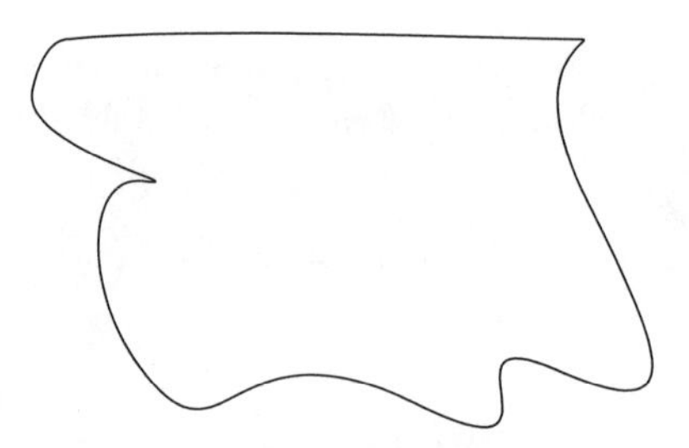

**Figure 4.2.1.** A Simple Closed Path.

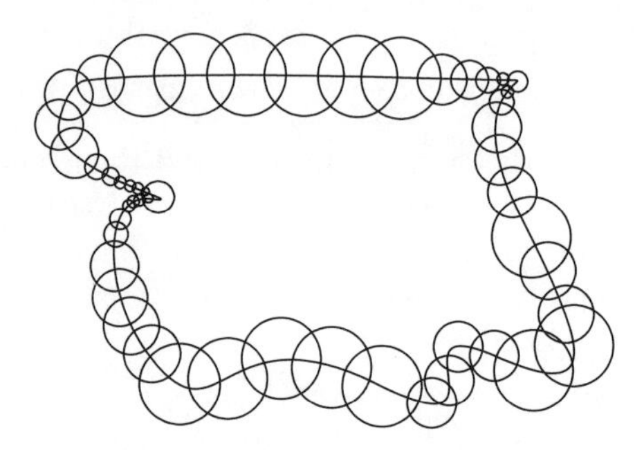

**Figure 4.2.2.** Some of the Discs $D_z$ in the Proof of Theorem 4.2.8.

union of all the left components is a connected open set $L$, while the union of all the right components is a connected open set $R$ (see Exercise 4.2.11). Furthermore, $\text{Ind}_\gamma(z) = \text{Ind}_\gamma(w) + 1$ if $z \in L$ and $w \in R$. It follows that $L$ and $R$ are disjoint. The union of all the discs $D_z$ is an open set $U$ containing $\gamma(I)$ and, in fact, $U$ is the union of the disjoint sets $\gamma(I)$, $L$, and $R$.

Now every component of the complement of $\gamma(I)$ has a subset of $\gamma(I)$ as its boundary (Exercise 4.2.12). This implies that every such component must have non-empty intersection with $U$ and, hence, must meet either $L$ or $R$. But since $L$ and $R$ are connected, a component which meets one of them must, in fact, contain it. It follows that there are only two components of the complement of $\gamma(I)$ – one containing $L$ and one containing $R$. One of these must be the unbounded component and $\text{Ind}_\gamma$ is zero on it. The other one is a bounded component and $\text{Ind}_\gamma$ must be plus or minus one on it, depending on whether it contains $L$ or $R$. This completes the proof of the theorem. □

From the proof of the above theorem, it follows that $\text{Ind}_\gamma(z) = 1$ on the bounded component of $\mathbb{C} \setminus \gamma(I)$ if this component contains $L$. Intuitively, this means that the bounded component is on the left as the path is traversed in the positive direction. In this case, we say that the path has *positive orientation.* The other possibility is that $\text{Ind}_\gamma(z) = -1$ on the bounded component, which implies that this component contains $R$ – i.e., that the bounded component is on the right as the path is traversed in the positive direction. In this case, we say that the path has *negative orientation.*

If $\gamma$ is a simple closed path, we will call the bounded component of $\mathbb{C}\setminus\gamma(I)$ the *inside* of $\gamma$ and the unbounded component the *outside* of $\gamma$.

We may now apply Theorems 4.2.2 and 4.2.4 in the case where the cycle $\Gamma$ is a single simple closed path $\gamma$ to obtain the Cauchy Theorem and Formula in their classical forms.

**Theorem 4.2.9.** *If $\gamma$ is a simple closed path and $f$ is a function analytic in an open set $U$ containing both $\gamma(I)$ and its inside, then*

$$\int_\gamma f(w)\,dw = 0 \tag{4.2.9}$$

*and*

$$f(z) = \frac{1}{2\pi i}\int_\gamma \frac{f(w)}{w-z}\,dw \tag{4.2.10}$$

*for each $z$ on the inside of $\gamma(I)$.*

**Proof.** We know that $\mathrm{Ind}_\gamma(z) = 0$ on the outside of $\gamma$, and this contains the complement of $U$ (since $U$ contains $\gamma(I)$ and its inside). Thus, the hypothesis of Theorem 4.2.2 is satisfied, and so we know that (4.2.9) is true and that

$$\mathrm{Ind}_\gamma(z)f(z) = \frac{1}{2\pi i}\int_\gamma \frac{f(w)}{w-z}\,dw$$

for $z$ on the inside of $\gamma$. Then (4.2.10) follows if we just observe that $\mathrm{Ind}_\gamma(z) = 1$ on the inside of $\gamma$. □

## Exercise Set 4.2

1. Verify the fact, used in Example 4.2.5, that $\sin(\pi z) = 0$ if and only if $z$ is an integer. Hint: We know this if $z$ is real, but how do we know there are no zeroes of $\sin z$ except those on the real line?
2. Let $\Gamma = \gamma_1 - \gamma_2 - \gamma_3$, where $\gamma_1, \gamma_2$, and $\gamma_3$ are positively oriented circles with radii 5, 1, and 1 and centers 0, $-2$, and 3, respectively. Find
$$\int_\Gamma \frac{1}{(z+2)(z-3)}\,dz$$
without calculating the integrals around the individual paths $\gamma_i$.
3. Check your answer to the previous problem by calculating
$$\int_{\gamma_j} \frac{1}{(z+2)(z-3)}\,dz$$
for each of the individual paths $\gamma_j$. Hint: The integral around $\gamma_1$ may be calculated by using a partial fraction decomposition of the integrand.
4. If $\gamma$ is any simple closed path, with $-2$ and 3 inside $\gamma$, find
$$\int_\gamma \frac{1}{(z+2)(z-3)}\,dz.$$

5. For the cycle $\Gamma$ of Exercise 2, find
$$\int_\Gamma \frac{1}{z(z+2)(z-3)}\,dz$$
by applying the Cauchy Integral Formula for the cycle $\Gamma$ and the function $f(z) = \dfrac{1}{(z+2)(z-3)}$.
6. If $\gamma_1$ and $\gamma_2$ are positively oriented circles, centered at 0, with radii 1 and 2, respectively, and if $\Gamma = \gamma_2 - \gamma_1$, find the integrals of $e^z/(z-1/2)$, $e^z/(z-3/2)$, and $e^z/(z-3)$ around $\Gamma$.
7. If $\gamma$ is a simple closed path with $-1$ and $1$ on its inside and $f$ is an entire function, show that
$$\int_\gamma \frac{f(z)}{z^2-1}\,dz = \pi i(f(1) - f(-1)).$$
8. If $\gamma$ is a simple closed path with 0 on its inside and all other integral multiples of $\pi$ on its outside, find
$$\int_\gamma \frac{1}{\sin z}\,dz.$$
9. For positive numbers $r < R$, set $\gamma_1(t) = r\,e^{2\pi i t}$ and $\gamma_2(t) = R\,e^{2\pi i t}$ for $t \in [0,1]$ and set $\Gamma = \gamma_2 - \gamma_1$. Prove that if $f$ is analytic in an open set containing the set $\{z : r \le |z| \le R\}$, then
$$f(z) = \frac{1}{2\pi i}\int_\Gamma \frac{f(w)}{w-z}\,dw \tag{4.2.11}$$
for every $z$ with $r < |z| < R$.
10. For the cycle $\Gamma$ and function $f$ of the previous problem, what is the integral (4.2.11) if $|z| < r$ or $|z| > R$?
11. Let $\gamma$ be a simple closed path and, for each $z \in \gamma(I)$, let $D_z$, $L_z$ and $R_z$ be the sets used in the proof of Theorem 4.2.8. Prove that if $z$ and $w$ are two points of $\gamma(I)$ such that $D_z \cap D_w \cap \gamma(I) \neq \emptyset$, then $L_z \cap L_w \neq \emptyset$ and $R_z \cap R_w \neq \emptyset$. Show that this implies that the union of all the $L_z$ is a connected set as is the union of all the $R_z$.
12. Prove that if $K$ is a closed subset of $\mathbb{C}$, then each component of $\mathbb{C} \setminus K$ has its boundary contained in $K$.

## 4.3. Laurent Series

In this section, we will show that a function which is analytic on an annulus has a special kind of power series expansion. We first make a preliminary observation about functions which have limit 0 at infinity.

Recall that a neighborhood of $\infty$ is any open set which contains the complement of some closed bounded disc. If $f$ is a function which is analytic in a neighborhood of $\infty$, then we say that $f$ *vanishes at infinity* if
$$\lim_{z\to\infty} f(z) = 0.$$

**Lemma 4.3.1.** *If $h$ is a function which is analytic in the exterior of the disc $\overline{D}_r(z_0)$ and vanishes at $\infty$, then the function*

$$q(w) = \begin{cases} h(1/w + z_0), & \text{if } w \neq 0; \\ 0, & \text{if } w = 0; \end{cases}$$

*is analytic in $D_{1/r}(0)$.*

**Proof.** The number $1/w + z_0$ is in $\mathbb{C} \setminus \overline{D}_r(z_0)$ if and only if $1/|w| > r$, that is, if and only if $w \in D_{1/r}(0)$. So the domain of $q$ is $D_{1/r}(0)$. Clearly $q$ is analytic on the complement of $\{0\}$ in this disc.

The fact that $\lim_{z \to \infty} h(z) = 0$ implies that $\lim_{w \to 0} q(w) = 0$. Thus, $q$ is continuous at 0. It follows from the Removable Singularity Theorem (Theorem 3.4.8) that it is analytic on all of $D_{1/r}(0)$. □

**Analytic Functions on an Annulus.** An open *annulus* centered at $z_0$ is a set of the form

$$A = \{z \in \mathbb{C} : r < |z - z_0| < R\}, \tag{4.3.1}$$

where $0 \leq r < R \leq \infty$. The numbers $r$ and $R$ are the inner and outer radii of $A$, respectively.

For the annulus $A$ of (4.3.1), and numbers $s$ and $S$ with

$$r < s < S < R,$$

consider two paths $\gamma_s$ and $\gamma_S$ in $A$, which traverse the circles $|w - z_0| = s$ and $|w - z_0| = S$ once in the positive direction. If we define a cycle $\Gamma$ by

$$\Gamma = \gamma_S - \gamma_s,$$

then

$$\mathrm{Ind}_\Gamma(z) = \begin{cases} 0, & \text{if } |z - z_0| > S; \\ 1, & \text{if } s < |z - z_0| < S; \\ 0, & \text{if } |z - z_0| < s. \end{cases}$$

In particular, this means that $\Gamma$ is homologous to 0 in $A$. From this and the general Cauchy theorems, it follows that if $f$ is a function which is analytic in $A$, then

$$\frac{1}{2\pi i} \int_\Gamma \frac{f(w)}{w - z}\, dw = \begin{cases} 0, & \text{if } |z - z_0| > S; \\ f(z), & \text{if } s < |z - z_0| < S; \\ 0, & \text{if } |z - z_0| < s. \end{cases}$$

In other words,

$$\frac{1}{2\pi i} \int_{\gamma_S} \frac{f(w)}{w - z}\, dw = \frac{1}{2\pi i} \int_{\gamma_s} \frac{f(w)}{w - z}\, dw \tag{4.3.2}$$

if $s$ and $S$ are on the same side of $|z - z_0|$ and

$$f(z) = \frac{1}{2\pi i} \int_{\gamma_S} \frac{f(w)}{w - z}\, dw - \frac{1}{2\pi i} \int_{\gamma_s} \frac{f(w)}{w - z}\, dw \tag{4.3.3}$$

if $s < |z - z_0| < S$. This is the basis for the following result.

**Theorem 4.3.2.** *If $f$ is an analytic function on an annulus $A$ of the form (4.3.1), then there exists a unique way of writing $f$ as*

$$f(z) = g(z) - h(z) \quad \textit{for} \quad z \in A,$$

*where $g$ is an analytic function on $D_R(z_0)$ and $h$ is an analytic function on the complement of $D_r(z_0)$, which vanishes at infinity.*

**Proof.** We define the function $g$ on $D_R(z_0)$ as follows: If $|z - z_0| < R$, we choose $S$ such that $|z - z_0| < S$ and $r < S < R$. We choose a closed path $\gamma_S$ which traces the circle $|w - z_0| = S$, once in the positive direction; and we set

$$g(z) = \frac{1}{2\pi i} \int_{\gamma_S} \frac{f(w)}{w - z}\, dw.$$

It follows from (4.3.2) that this does not depend on the choice of $S$ as long as $|z - z_0| < S < R$.

Similarly, we define the function $h$ on $\mathbb{C} \setminus \overline{D}_r(z_0)$ as follows. If $|z - z_0| > r$, we choose $s$ such that $r < s < R$ and $s < |z - z_0|$; we choose a closed path $\gamma_s$ which traces the circle $|w - z_0| = s$ once in the positive direction; and we set

$$h(z) = \frac{1}{2\pi i} \int_{\gamma_s} \frac{f(w)}{w - z}\, dw. \tag{4.3.4}$$

Again, by (4.3.2), this does not depend on the choice of $s$.

By (4.3.3)

$$f(z) = g(z) - h(z)$$

if $z \in A$. We emphasize that $z$ was an arbitrary point of $A$, and that $g$ and $h$ do not depend on the choices for the circles $\gamma_S$ and $\gamma_s$.

It follows from Morera's Theorem that $g$ is analytic in the open disc $D_R(z_0)$ and $h$ is analytic in the exterior of the closed disc $\overline{D}_r(z_0)$ (see Exercise 3.2.16).

Also, for a fixed $s$ with $r < s < R$, $|z - z_0| > s$, and $\gamma_s$ as above,

$$|z - w| \geq |z| - |z_0| - |w - z_0| = |z| - |z_0| - s \quad \text{if} \quad w \in \gamma_s(I),$$

and so

$$\left| \frac{f(w)}{z - w} \right| \leq \frac{M}{|z| - |z_0| - s} \quad \text{if} \quad w \in \gamma_s(I),$$

where $M = \sup\{|f(w)| : w \in \gamma_s(I)\}$. From this bound on the integrand in (4.3.4), it follows that

$$|h(z)| \leq \frac{Ms}{|z| - |z_0| - s}$$

and so, since $|z_0|$ and $s$ are fixed,

$$\lim_{z \to \infty} h(z) = 0,$$

as required.

The uniqueness of $g$ and $h$ follows from Liouville's Theorem. In fact, suppose we also have

$$f = g_1 - h_1$$

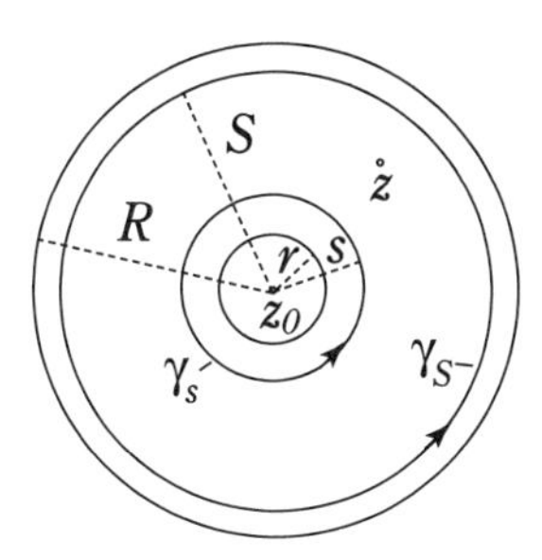


**Figure 4.3.1.** The Paths $\gamma_S$ and $\gamma_s$ in the Proof of Theorem 4.3.2.

with $g_1$ analytic on $D_R(z_0)$, $h_1$ analytic on $\mathbb{C}\setminus\overline{D}_r(z_0)$, and $h_1$ vanishing at infinity. Then

$$g - g_1 = h_1 - h \quad \text{on} \quad A.$$

This means that the analytic function $g - g_1$ on $D_R(z_0)$ and the analytic function $h_1 - h$ on $\mathbb{C}\setminus\overline{D}_r(z_0)$ agree on the intersection of these two open sets. It follows that they define an entire function on the plane which vanishes at infinity. By Liouville's Theorem, the only such function is the identically 0 function. Thus, $g = g_1$ and $h = h_1$. □

**Example 4.3.3.** Find a decomposition like that in the previous theorem if the function $f$ is

$$f(z) = \frac{1}{(z-1)(z-2)}$$

and the annulus $A$ is $A = \{z \in \mathbb{C} : 1 < |z| < 2\}$.

**Solution:** This is just the partial fraction decomposition of $f$. We set

$$g(z) = \frac{1}{z-2} \quad \text{for} \quad |z| < 2,$$

and

$$h(z) = \frac{1}{z-1} \quad \text{for} \quad |z| > 1.$$

Then $g$ and $h$ are analytic, $f = g - h$ on $A$, and $h$ vanishes at infinity.

**Laurent Series Expansion.** A function which is analytic on an annulus has a special kind of series expansion called a *Laurent Series* expansion.

**Theorem 4.3.4.** *Let $f$ be a function which is analytic on the annulus $A$ of* (4.3.1). *Then $f$ has a unique series expansion of the form*

$$f(z) = \sum_{n=-\infty}^{\infty} c_n (z - z_0)^n, \tag{4.3.5}$$

*which converges to $f$ at all points of $A$ and converges uniformly on compact subsets of $A$.*

**Proof.** We write $f = g - h$ with $g$ and $h$ as in the previous theorem. Since $g$ is analytic on $D_R(z_0)$, it has a power series expansion

$$g(z) = \sum_{n=0}^{\infty} c_n(z - z_0)^n,$$

which converges on $D_R(z_0)$ and converges uniformly on compact subsets of this disc.

By Lemma 4.3.1 the function $q(w) = h(1/w + z_0)$ is, if given the value 0 at $w = 0$, an analytic function of $w$ in the disc $D_{1/r}(0)$, and it vanishes at $w = 0$. Therefore, it has a power series expansion

$$q(w) = \sum_{n=1}^{\infty} b_n w^n$$

which converges uniformly on compact subsets of $D_{1/r}(0)$. If we set $c_n = -b_{-n}$ for negative integers $n$ and make the substitution $w = (z - z_0)^{-1}$, then this becomes

$$h(z) = -\sum_{n=-\infty}^{-1} c_n(z - z_0)^n.$$

On combining the series for $g$ and $h$, using the fact that $f = g - h$ in $A$, we obtain (4.3.5). The series converges uniformly on compact subsets of $A$ because the series for $g$ and $h$ both have this property.

The uniqueness of the expansion follows from the uniqueness statements in Theorem 4.3.2 and Corollary 3.2.4. ◻

**Example 4.3.5.** For the function $f$ and annulus $A$ of Example 4.3.3 find the Laurent series expansion for $f$ on $A$.

**Solution:** We have $f = g - h$, where $g(z) = \dfrac{1}{z-2} = -\dfrac{1/2}{1 - z/2}$ has power series expansion

$$g(z) = -\frac{1}{2}\sum_{n=0}^{\infty} \frac{z^n}{2^n}$$

in $D_2(0)$, and $h(z) = \dfrac{1}{z-1} = \dfrac{1/z}{1 - 1/z}$ has an expansion

$$h(z) = \sum_{n=1}^{\infty} \frac{1}{z^n}$$

on $\mathbb{C} \setminus \overline{D}_1(0)$. Thus, the function $f = g - h$ has Laurent expansion

$$f(z) = \sum_{n=-\infty}^{-1} (-1)z^n + \sum_{n=0}^{\infty} \left(-\frac{1}{2^{n+1}}\right) z^n$$

in $A = D_2(0) \cap (\mathbb{C} \setminus \overline{D}_1(0))$.

**Example 4.3.6.** The function $f = \dfrac{1}{(z-1)(z-2)}$ of the previous exercise is also analytic on the annulus

$$B = \{z \in \mathbb{C} : 0 < |z - 1| < 1\},$$

which is centered at 1. Find its Laurent expansion in this annulus.

**Solution:** Again, $f = g - h$, with $g(z) = \dfrac{1}{z-2}$ and $h(z) = \dfrac{1}{z-1}$. Here $g$ is analytic in the disc $D_1(1)$, with power series expansion

$$g(z) = -\sum_{n=0}^{\infty}(z-1)^n,$$

while $h(z)$ is analytic in $\mathbb{C} \setminus \{1\}$ with limit 0 at infinity. Since it is already a power of $(z-1)$, it needs no further expansion. Thus, the Laurent expansion of $f$ in the annulus $B$ is

$$f(z) = -\sum_{n=0}^{\infty}(z-1)^n - (z-1)^{-1} = -\sum_{n=-1}^{\infty}(z-1)^n.$$

**Example 4.3.7.** Find the Laurent expansion of $e^{1/z}$ in the annulus $\mathbb{C} \setminus \{0\}$.

**Solution:** Since $f(z) = e^{1/z}$ is analytic except at 0 and has limit 1 at infinity, in the decomposition $f = g - h$ of Theorem 4.3.2 the function $g$ is 1 and the function $h$ is $1 - e^{1/z}$. The Laurent expansion is

$$e^{1/z} = \sum_{n=-\infty}^{0} \frac{z^n}{|n|!}.$$

The coefficients of the Laurent expansion of a function analytic in an annulus are given by an integral formula similar to the one for the power series expansion of a function analytic in a disc (Theorem 3.2.5).

**Theorem 4.3.8.** *If $A = \{z \in \mathbb{C} : r < |z - z_0| < R\}$, $f$ is analytic in $A$ and $r < s < R$, then the Laurent series (4.3.5) for $f$ in $A$ has coefficients given by*

$$c_k = \frac{1}{2\pi i}\int_{|w-z_0|=s} \frac{f(w)}{(w-z_0)^{k+1}}\,dw. \tag{4.3.6}$$

**Proof.** With $f$ given by (4.3.5), we have

$$\frac{1}{2\pi i}\int_{|w-z_0|=s} \frac{f(w)}{(w-z_0)^{k+1}}\,dw = \sum_{n=-\infty}^{\infty} c_n \int_{|w-z_0|=s} (w-z_0)^{n-k-1}\,dw. \tag{4.3.7}$$

A path which traverses the circle $|w - z_0| = s$ once in the positive direction is given by $\gamma(t) = z_0 + s\,e^{it}$, for $0 \le t \le 2\pi$. Then

$$\gamma'(t) = si\,e^{it} \quad \text{and} \quad \gamma(t) - z_0 = s\,e^{it}.$$

Thus,

$$\int_{|w-z_0|=s} (w-z_0)^{n-k-1}\,dw = \int_0^{2\pi} is^{n-k}\,e^{i(n-k)t}\,dt.$$

This is 0 if $n \neq k$ and $2\pi i$ if $n = k$. Combined with (4.3.7), this proves the theorem. ∎

## Exercise Set 4.3

1. In Examples 4.3.5 and 4.3.6, Laurent expansions for
$$f(z) = \frac{1}{(z-1)(z-2)}$$
are given in two different annuli. Find a third annulus in which this function is analytic and find its Laurent expansion there. Are there other such annuli?
2. Find the Laurent expansion of
$$f(z) = \frac{1}{z - z^2} \quad \text{in the annulus} \quad A = \{z \in \mathbb{C} : 0 < |z| < 1\}.$$
3. Find the Laurent expansion for $f(z) = z^3\, e^{1/z}$ in $\mathbb{C} \setminus \{0\}$.
4. Find the Laurent expansion for $f(z) = z^{-3}\, e^z$ in $\mathbb{C} \setminus \{0\}$.
5. Find the Laurent expansion for
$$f(z) = \frac{\log(1+z)}{z^4} \quad \text{in} \quad \mathbb{C} \setminus \{0\}.$$
6. Find the Laurent expansion for
$$f(z) = \frac{z}{z^2+1} \quad \text{in the annulus} \quad A = \{z \in \mathbb{C} : 0 < |z - i| < 2\}.$$
7. Find the Laurent expansion for $f(z) = \sin(1/z)$ in $\mathbb{C} \setminus \{0\}$.
8. Find the Laurent expansion for $f(z) = e^z + e^{1/z}$ in $\mathbb{C} \setminus \{0\}$.
9. Find the Laurent expansion for $f(z) = e^{z+1/z}$ in $\mathbb{C} \setminus \{0\}$. Hint: You will need the binomial formula.
10. Prove that if $f$ is analytic in a disc $D_r(z_0)$ except at $z_0$, where it has a pole of order $k$, then, in the annulus $\{z : 0 < |z - z_0| < r\}$, the Laurent expansion for $f$ has only finitely many terms with negative exponent and the most negative exponent that appears is $-k$.
11. Suppose $f$ is analytic in the annulus $A = \{z \in \mathbb{C} : R < |z|\}$ and satisfies the inequality $|f(z)| \leq |z|^k$ in this set. Then prove that the Laurent expansion of $f$ in $A$ has no terms with positive exponent greater than $k$.
12. If $c_n$ is the $n$th coefficient in the Laurent series expansion of $\dfrac{1}{\sin z}$ in the annulus $\{z : 0 < |z| < \pi\}$, show that $c_n = 0$ if $n$ is even or if $n < -1$.
13. With $c_n$ as in the previous exercise, find $c_{-1}, c_1$, and $c_3$ Hint: use long division to divide $\sin z$ into 1.
14. For the function $\dfrac{1}{\sin z}$ of the previous two exercises, how do the Laurent coefficients change when we compute the Laurent expansion in the annulus $\{z : \pi < |z| < 2\pi\}$ instead of the annulus $\{z : 0 < |z| < \pi\}$? Hint: Use Cauchy's formula to compute the difference in formula (4.3.6) if the radius $s$ of the circle of integration is changed from one with $0 < s < \pi$ to one with $\pi < s < 2\pi$.

## 4.4. The Residue Theorem

In this section we return to the study of analytic functions with isolated singularities, as discussed in Section 3.4. Recall that a function $f$ has isolated singularities on an open set $U$ if it is analytic on $U$ except on a discrete subset.

In particular, if a function $f$ is analytic on an open set containing the annulus $D_r(z_0) \setminus \{z_0\}$, then it has an isolated singularity at $z_0$. Since $D_r(z_0) \setminus \{z_0\}$ is an annulus centered at $z_0$, $f$ has a Laurent expansion of the form

$$f(z) = \sum_{n=-\infty}^{\infty} c_n (z - z_0)^n. \tag{4.4.1}$$

The coefficient $c_{-1}$ of $(z - z_0)^{-1}$ in (4.4.1) plays a special role in complex analysis.

**Definition 4.4.1.** If $f$ is a function which is analytic on $A = D_r(z_0) \setminus \{z_0\}$, with an isolated singularity at $z_0$, then the coefficient of $(z - z_0)^{-1}$ in the Laurent expansion of $f$ on $A$ is called the *residue* of $f$ at $z_0$ and is denoted by $\mathrm{Res}(f, z_0)$.

The residue of $f$ at $z_0$ is given by a simple integral formula, which follows immediately from Theorem 4.3.8.

**Theorem 4.4.2.** *Let $f$ be an analytic function with an isolated singularity at $z_0 \in \mathbb{C}$. Let $R > r > 0$ be numbers with $f$ analytic on $D_R(z_0) \setminus \{z_0\}$. Then*

$$\mathrm{Res}(f, z_0) = \frac{1}{2\pi i} \int_{|z-z_0|=r} f(z)\, dz.$$

This leads to the following important application of the general Cauchy Theorem.

**Theorem 4.4.3** (Residue Theorem). *Let $f$ be a function which is analytic on $U \setminus E$, where $U$ is an open subset of $\mathbb{C}$ and $E$ is a discrete subset of $U$. If $\gamma$ is a closed path in $U \setminus E$ which is homologous to 0 in $U$, then*

(a) *there are only finitely many points of $E$ at which $\mathrm{Ind}_\gamma$ is non-zero;*
(b) *if these points are $\{z_1, z_2, \cdots, z_n\}$, then*

$$\frac{1}{2\pi i} \int_\gamma f(z)\, dz = \sum_{j=1}^{n} \mathrm{Ind}_\gamma(z_j)\, \mathrm{Res}(f, z_j). \tag{4.4.2}$$

**Proof.** We first prove (a). Recall from the proof of Theorem 2.7.5 that if $r$ is chosen so that $\gamma(I) \subset D_r(0)$, then the bounded components of $\mathbb{C} \setminus \gamma(I)$ are also contained in $D_r(0)$. Also, $\mathrm{Ind}_\gamma(z)$ is non-zero only on certain bounded components of $\mathbb{C} \setminus \gamma(I)$. It follows that the union of $\gamma(I)$ and the components of its complement where $\mathrm{Ind}_\gamma(z)$ is non-zero is a bounded set $K$. The set $K$ is also the complement of the union of the components of $\mathbb{C} \setminus \gamma(I)$ on which $\mathrm{Ind}_\gamma = 0$, and so it is closed. Thus, $K$ is compact.

Since $\gamma$ is homologous to 0 in $U$, every point of the complement of $U$ is a point where $\mathrm{Ind}_\gamma$ is 0. This implies $K \subset U$. Since the singularities of $f$ form a discrete subset $E$ of $U$, we may choose for each point of $U$ an open disc, centered at that point and contained in $U$, which either contains one singularity (the center of the

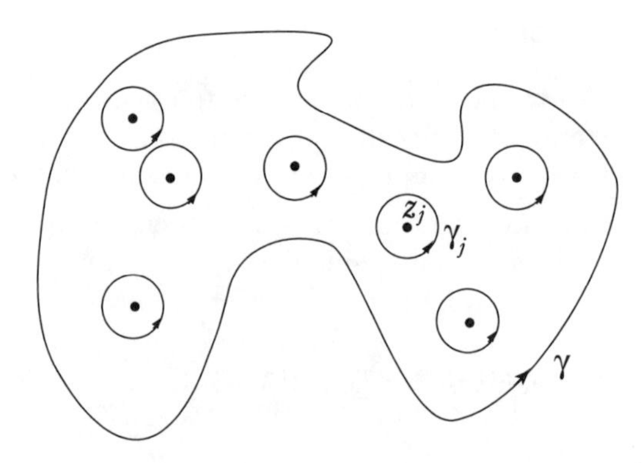


**Figure 4.4.1.** The Paths $\gamma$ and $\gamma_j$ in the Proof of Theorem 4.4.3.

disc) or no singularities. This collection of open discs covers $K$ (since its union is $U$), and so some finite subcollection also covers $K$. But this means there are only finitely many singularities of $f$ in $K$, and, hence, there are only finitely many singularities of $f$ at which $\mathrm{Ind}_\gamma$ is nonzero. This completes the proof of (a).

Let $z_1, z_2, \cdots, z_n$ be the singularities of $f$ at which $\mathrm{Ind}_\gamma$ is nonzero. For each of these points $z_j$, we choose an $r_j > 0$ such that $D_{r_j}$ is contained in the open set $U \setminus \gamma(I)$. We choose $r > 0$ such that $r < \min\{r_1, \cdots, r_n\}$ and the discs $\overline{D}_r(z_j)$ are non-overlapping. We then set

$$m_j = \mathrm{Ind}_\gamma(z_j) \tag{4.4.3}$$

and define a 1-cycle $\Gamma$ by

$$\Gamma = \gamma - \sum_{j=1}^{n} m_j \gamma_j,$$

where $\gamma_j(t) = z_j + r\,e^{2\pi i t}$ for $t \in [0,1]$ (we may assume that $\gamma$ also has $[0,1]$ as its parameter interval).

Now each $z_j$ is a point where $\mathrm{Ind}_\gamma$ is non-zero, and so is every point in the open disc $D_r(z_j)$. Hence the closure $\overline{D}_r(z_j)$ of this disc is contained in $K \subset U$. This means that the complement of $U$ is contained in the complement of $\overline{D}_r(z_j)$. Thus, $\mathrm{Ind}_{\gamma_j}(z) = 0$ on the complement of $U$. Since this is true for every $j$ and is also true of $\gamma$, we have that $\mathrm{Ind}_\Gamma(z) = 0$ on the complement of $U$ – that is, $\Gamma$ is homologous to 0 in $U$.

Note, $f$ is not analytic in $U$ and so the general Cauchy Theorem does not yet apply. However, we also have that

$$\mathrm{Ind}_\gamma(z_j) = m_j = m_j \,\mathrm{Ind}_{\gamma_j}(z_j)$$

while

$$m_j \,\mathrm{Ind}_{\gamma_j}(z_k) = 0 \quad \text{for} \quad k \neq j.$$

It follows that $\Gamma$ is also homologous to 0 in $U \setminus E$. Now $U \setminus E$ is a set on which $f$ is analytic, and so, by the general Cauchy Integral Theorem,

$$0 = \int_{\Gamma} f(z)\,dz = \int_{\gamma} f(z)\,dz - \sum_{j=1}^{n} m_j \int_{\gamma_j} f(z)\,dz.$$

Now (4.4.2) follows from this, Theorem 4.4.2, and (4.4.3). This completes the proof. □

The Residue Theorem has a vast number of applications. We will explore many of these in the remainder of this section and the next chapter. The next three examples give just a taste of how the Residue Theorem is used to calculate integrals.

**Example 4.4.4.** Let $f$ be a function of the form

$$f(z) = \frac{g(z)}{z - z_0},$$

where $g$ is analytic in an open set containing $z_0$. Show that

$$\operatorname{Res}(f, z_0) = g(z_0). \tag{4.4.4}$$

**Solution:** The function $g(z)$ is analytic in some disc centered at $z_0$ and so it has a power series expansion in this disc with constant term $g(z_0)$. Hence the Laurent series expansion of $f(z)$ about $z_0$ has $g(z_0)$ as the coefficient of $(z - z_0)^{-1}$. By definition, this means that (4.4.4) holds.

**Example 4.4.5.** Let $\gamma$ be a simple closed path with 1 and 2 inside $\gamma$. Compute

$$\int_{\gamma} \frac{z+1}{(z-1)(z-2)}\,dz.$$

**Solution:** By the Residue Theorem, this integral is

$$2\pi i(\operatorname{Res}(f, 1) + \operatorname{Res}(f, 2)),$$

where $f(z) = \dfrac{z+1}{(z-1)(z-2)}$. The function $g(z) = \dfrac{z+1}{z-2}$ is analytic in a disc centered at 1 and $f(z) = \dfrac{g(z)}{z-1}$. Hence, by the previous example,

$$\operatorname{Res}(f, 1) = g(1) = -2.$$

By the same reasoning, if $h(z) = \dfrac{z+1}{z-1}$, then

$$\operatorname{Res}(f, 2) = h(2) = 3.$$

We conclude that

$$\int_{\gamma} \frac{z+1}{(z-1)(z-2)}\,dz = 2\pi i(3-2) = 2\pi i.$$

**Example 4.4.6.** If $\gamma$ is a simple closed path with 0 and $\pi$ inside $\gamma$ and all other multiples of $\pi$ outside $\gamma$, find

$$\int_{\gamma} \frac{1}{\sin z}\,dz.$$

**Solution:** The integrand $f(z) = \dfrac{1}{\sin z}$ has isolated singularities inside $\gamma$ at 0 and $\pi$. The function

$$g(z) = \frac{z}{\sin z}$$

has a removable singularity at 0 and the value of the resulting analytic function at 0 is 1. Thus,

$$\text{Res}(f, 0) = 1.$$

Since $\sin(z) = -\sin(z - \pi)$, the function

$$h(z) = \frac{z-\pi}{\sin z} = -\frac{z-\pi}{\sin(z-\pi)}$$

has a removable singularity at $z = \pi$. The value at $\pi$ of the resulting analytic function is $-1$. Thus,

$$\text{Res}(f, \pi) = -1.$$

It follows from the Residue Theorem that the integral of $f$ around $\gamma$ is zero.

**Counting Zeroes and Poles.** Recall that a meromorphic function on an open set $U$ is a function which is analytic on $U$ except on a discrete subset $E$ where it has poles.

A *simple pole* is a pole of order 1. If $f$ is a meromorphic function in an open set $U$, then, as we will show in the proof of the next theorem, the function $f'/f$ has a simple pole at each zero and at each pole of $f$. Furthermore, these simple poles have a special form which leads directly to an integral formula for the sum of the number of zeroes minus the number of poles, counting multiplicity, surrounded by a closed path in $U$.

**Theorem 4.4.7.** *If $f$ is a meromorphic function on $U$ and $z_0 \in U$, then*

$$\text{Res}(f'/f, z_0) = k,$$

*where $k$ is the order of the zero of $f$ at $z_0$, or minus the order of the pole of $f$ at $z_0$, or 0 if $f$ has no zero or pole at $z_0$.*

**Proof.** We may factor $f$ as

$$f(z) = (z - z_0)^k g(z),$$

where $g$ is meromorphic on $U$ and has no zero or pole at $z_0$. The integer $k$ is positive, negative, or zero, depending on whether $f$ has a zero, a pole, or neither at $z_0$. Then

$$f'(z) = k(z - z_0)^{k-1} g(z) + (z - z_0)^k g'(z),$$

and so

$$\frac{f'(z)}{f(z)} = \frac{k}{z - z_0} + \frac{g'(z)}{g(z)}. \tag{4.4.5}$$

If $\overline{D}_r(z_0)$ is a disc contained in $U$, which contains no zero or pole of $f$ except possibly $z_0$, then when we integrate (4.4.5) around the boundary of this disc and divide by $2\pi i$, we get

$$\text{Res}(f'/f, z_0) = k,$$

since $g'/g$ is analytic on an open set containing $\overline{D}_r(z_0)$ and, thus, has integral 0. □

If we combine this with the Residue Theorem, the result is the following theorem.

**Theorem 4.4.8.** *Let $f$ be a meromorphic function defined in a open set $U$ and let $\gamma$ be a closed path in $U$ which is homologous to 0 in $U$. Assume there are no zeroes or poles of $f$ on $\gamma(I)$ and the zeroes and poles of $f$ at which $\operatorname{Ind}_\gamma$ is not zero occur at the points $z_1, z_2, \cdots, z_n$. Then*

$$\sum_{j=1}^{n} m_j k_j = \frac{1}{2\pi i}\int_\gamma \frac{f'(z)}{f(z)}\,dz, \tag{4.4.6}$$

*where, for $j = 1, \cdots, n$, the integer $k_j$ is the order of the zero of $f$ at $z_j$ or minus the order of the pole at $z_j$, and $m_j = \operatorname{Ind}_\gamma(z_j)$.*

**Corollary 4.4.9.** *Let $U$, $f$, $\{z_1, \cdots, z_n\}$, $\{k_1, \cdots k_n\}$, and $\gamma$ be as in Theorem 4.4.8. If we compose $f$ with $\gamma$ to form a new path $f \circ \gamma$, then*

$$\sum_{j=1}^{n} m_j k_j = \operatorname{Ind}_{f\circ\gamma}(0), \tag{4.4.7}$$

*where $m_j = \operatorname{Ind}_\gamma(z_j)$.*

**Proof.** If the parameter interval of $\gamma$ is $[a, b]$, we have

$$\begin{aligned}\operatorname{Ind}_{f\circ\gamma}(0) &= \frac{1}{2\pi i}\int_{f\circ\gamma} \frac{1}{z}\,dz = \frac{1}{2\pi i}\int_a^b \frac{(f\circ\gamma)'(t)}{f\circ\gamma(t)}\,dt \\ &= \frac{1}{2\pi i}\int_a^b \frac{f'(\gamma(t))\gamma'(t)}{f(\gamma(t))}\,dt = \frac{1}{2\pi i}\int_\gamma \frac{f'(z)}{f(z)}\,dz.\end{aligned}$$

By Theorem 4.4.8, this last integral is equal to $\sum_{j=1}^n m_j k_j$. □

If the path $\gamma$ is simple, then the previous theorem and corollary are simpler.

**Corollary 4.4.10.** *Let $f$ be a meromorphic function on the open set $U$. If $\gamma$ is a simple closed path in $U$, with its inside contained in $U$, which does not pass through a zero or pole of $f$, and $\{z_1, \cdots, z_n\}$ is the set of zeroes and poles of $f$ inside $\gamma$, then*

$$\sum_{j=1}^{n} k_j = \frac{1}{2\pi i}\int_\gamma \frac{f'(z)}{f(z)}\,dz = \operatorname{Ind}_{f\circ\gamma}(0), \tag{4.4.8}$$

*where, for $j = 1, \cdots, n$, $k_j$ is the order of the zero or minus the order of the pole at $z_j$.*

**Proof.** Since the inside of $\gamma$ is contained in $U$, $\gamma$ is homologous to 0 in $U$. Thus, Theorem 4.4.8 and the previous corollary apply. Since $\operatorname{Ind}_\gamma$ is 1 on the inside of $\gamma$, it is 1 at each $z_j$. Equation (4.4.8) follows. □

**Example 4.4.11.** Suppose $f$ is a meromorphic function in a convex open set $U$. Suppose that the only zero of $f$ in $U$ occurs at $z_1$ and has order $k$, and the only pole of $f$ in $U$ occurs at $z_2$ and also has order $k$. Prove that there is an analytic logarithm of $f$ defined in $V = U \setminus [z_1, z_2]$. That is, prove that there is an analytic function $g$ on $V$ such that $f(z) = e^{g(z)}$ for every $z \in V$.

**Solution:** If $\gamma$ is any closed path in $V$, then $\gamma$ is homologous to 0 in $U$, since $U$ is convex. Since $z_1$ and $z_2$ are connected by a line segment in the complement of $V$, they lie in the same component of the complement of $\gamma(I)$. This implies $\mathrm{Ind}_\gamma(z_1) = \mathrm{Ind}_\gamma(z_2)$. Then, by Theorem 4.4.8,

$$\frac{1}{2\pi i}\int_\gamma \frac{f'(w)}{f(w)}\,dw = \mathrm{Ind}_\gamma(z_1)k - \mathrm{Ind}_\gamma(z_2)k = 0.$$

It follows from this that, if $z_0$ is a fixed point and $z$ is a variable point of $V$, and $\gamma_z$ is a path in $V$ which begins at $z_0$ and ends at $z$, then

$$h(z) = \int_{\gamma_z} \frac{f'(w)}{f(w)}\,dw$$

is independent of which path $\gamma_z$ is chosen. Furthermore, we know that $h$ is an antiderivative of $\dfrac{f'(w)}{f(w)}$, by Theorem 2.5.6. Thus, $h' = f'/f$. This implies

$$(f\,\mathrm{e}^{-h})' = f'\,\mathrm{e}^{-h} - fh'\,\mathrm{e}^{-h} = 0,$$

and, hence, that $f = C\,\mathrm{e}^h$ for some non-zero constant $C$. Then

$$g(z) = h(z) + \log(C)$$

has the required properties, where log is any branch of the log function.

---

## Exercise Set 4.4

1. If $f(z) = \dfrac{1}{2z^2 - 5z + 2}$, find the residue of $f$ at each of its singularities.
2. If $\gamma$ is a simple closed path with 0 and 3 inside $\gamma$, find $\displaystyle\int_\gamma \frac{1}{z^2 - 3z}\,dz$.
3. Find $\displaystyle\int_{|z|=2} \frac{\mathrm{e}^z}{z^2 - 1}$.
4. If $\gamma$ is a simple closed path with 0 inside, but no other multiples of $2\pi i$ inside, find
$$\int_\gamma \frac{1}{\mathrm{e}^z - 1}\,dz.$$
5. If $\gamma$ is any simple closed path with $0, 1$, and 2 inside, find
$$\int_\gamma \frac{1}{z(z-1)(z-2)}\,dz.$$
6. Find $\displaystyle\int_{|z|=2} \frac{3z^2 - 1}{z^3 - z}\,dz$.
7. Find $\displaystyle\int_{|z|=\pi} \tan z\,dz$.
8. Find $\displaystyle\int_{|z-5|=4} \frac{\log z}{\sin z}\,dz$.
9. Let $f$ be a meromorphic function on $\mathbb{C}$ with a zero of order 1 at $z = 1$ and a pole of order 1 at $z = -1$ and no other zeroes or poles. If $\gamma$ is a simple closed

path which does not pass through 1 or $-1$, what are the possible values for $\int_\gamma \frac{f'(z)}{f(z)}$ and what determines which value is achieved?

10. Use Theorem 4.4.8 to derive a formula for $\int_\gamma \cot z\, dz$, where $\gamma$ is any closed path which does not pass through an integral multiple of $\pi$.
11. Prove that there is an analytic logarithm for $\frac{z+1}{z-1}$ defined in the open set $U = \mathbb{C} \setminus [-1, 1]$.
12. Example 4.4.11 is only one of many possible theorems concerning the existence of analytic logarithms that can be proved using Theorem 4.4.8. Invent and prove another one.

## 4.5. Rouché's Theorem and Inverse Functions

Corollary 4.4.10 leads to a proof of Rouché's Theorem, which is a very useful tool for analyzing the zeroes of an analytic function.

**Theorem 4.5.1** (Rouché's Theorem). *Let $f$ and $g$ be analytic in an open set $U$ and let $\gamma$ be a simple closed path in $U$, with its inside contained in $U$, and with parameter interval $I$. If $f$ has no zero on $\gamma(I)$, and*

$$|f(z) - g(z)| \le |g(z)| \quad \text{on} \quad \gamma(I), \tag{4.5.1}$$

*then $f$ and $g$ have the same number of zeroes, counting order, inside $\gamma$.*

**Proof.** If $f$ has no zeroes on $\gamma(I)$, then (4.5.1) implies that $g$ also has no zeroes on $\gamma(I)$. If we set $h(z) = f(z)/g(z)$, then $h$ is meromorphic in $U$ and has no zeroes or poles on $\gamma(I)$. The inequality (4.5.1) implies that the curve $h \circ \gamma$ satisfies

$$|h \circ \gamma(t) - 1| \le 1$$

for all $t \in I$. It follows from this that 0 is in the unbounded component of $\mathbb{C} \setminus h \circ \gamma(I)$. Hence,

$$\mathrm{Ind}_{h\circ\gamma}(0) = 0.$$

The path $\gamma$ is homologous to 0 in $U$ and so we may apply Corollary 4.4.9. It tells us that the number of zeroes minus the number of poles of $h$ inside $\gamma$, counting order, is 0. However, this is the number of zeroes of $f$ minus the number of zeroes of $g$ inside $\gamma$, counting order. Hence, these two numbers are the same. □

**Example 4.5.2.** How many zeroes, counting order, does the polynomial

$$4z^5 - z^3 + z^2 - 2$$

have inside the unit circle?

**Solution:** We apply Rouché's Theorem with $f(z) = 4z^5 - z^3 + z^2 - 2$ and $g(x) = 4z^5$. On the unit circle $|z| = 1$, we have

$$|f(z) - g(z)| = |-z^3 + z^2 - 2| \le |z|^3 + |z|^2 + 2 = 4 = |g(z)|.$$

By Rouché's Theorem, $f$ and $g$ have the same number of zeroes inside the unit circle. Since $g(z) = 4z^5$ has 5 zeroes, counting order, inside this circle, so does $f$.

**Example 4.5.3.** Let $\{f_n\}$ be a sequence of non-vanishing analytic functions on a connected open set $U$. If $\{f_n\}$ converges uniformly to $f$ on each compact subset of $U$, prove that $f$ is either also non-vanishing on $U$ or is identically 0.

**Solution:** Assume $f$ is not identically 0. Let $z_0$ be any point of $U$. We will prove that $f(z_0) \neq 0$.

We may choose a closed disc $\overline{D}_r(z_0)$ on which $f$ has no zeroes except possibly at $z_0$. This is possible because $f$ is analytic and not identically zero and, hence, has a discrete set of zeroes in $U$. Since $f$ is non-vanishing on the boundary of $\overline{D}_r(z_0)$, it has a minimum modulus $m > 0$ on $\partial\overline{D}_r(z_0)$. Since $f_n \to f$ uniformly on $\overline{D}_r(z_0)$, there is an $N$ such that

$$n \geq N \quad \text{implies} \quad |f(z) - f_n(z)| < m \quad \forall\, z \in \overline{D}_r(z_0).$$

Since $m \leq |f(z)|$ for all $z \in \partial\overline{D}_r(z_0)$, Rouché's Theorem implies that $f$ and $f_n$ have the same number of zeroes in $D_r(z_0)$ if $n \geq N$. Since $f_n$ has no zeroes, $f$ has no zeroes in $D_r(z_0)$. In particular, $f(z_0) \neq 0$.

**The Inverse Mapping Theorem.** An application which demonstrates the utility of Rouché's Theorem is the following proof of the Inverse Function Theorem for analytic functions.

A function from a set $U$ to a set $V$ is said to be *one-to-one* if $f(z) \neq f(w)$ for every pair of distinct points $z, w \in U$. It is said to be *onto* if $f(U) = V$ – that is, if every point of $V$ is the image of some point of $U$. A function $f : U \to V$ is one-to-one and onto if and only if it has a well-defined inverse function $f^{-1}$. This is defined by the condition that $f(z) = w$ for $z \in U$ if and only if $f^{-1}(w) = z$. This is the same as saying that

$$\begin{aligned} f^{-1} \circ f(z) &= z \quad \text{for all} \quad z \in U, \quad \text{and} \\ f \circ f^{-1}(w) &= w \quad \text{for all} \quad w \in V. \end{aligned}$$

The inverse function for a continuous function $f$ may or may not be continuous; if it is, we say that $f$ has a continuous inverse function on $U$. Similarly, if $U$ and $V$ are open, $f : U \to V$ is analytic, and $f^{-1}$ is also analytic, then we say that $f$ has an analytic inverse function on $U$.

Typically, analytic functions are not one-to-one and, hence, do not have inverse functions on their full domains. However, they quite often have local analytic inverse functions in the sense of the following definition.

**Definition 4.5.4.** If $f$ is analytic in $U$ and $z_0 \in U$, then we say that $f$ has a *local analytic inverse* at $z_0$ if there are neighborhoods $V$ of $z_0$ and $W$ of $w_0 = f(z_0)$ such that $f$ is a one-to-one map of $V$ onto $W$ and its inverse function $f^{-1} : W \to V$ is analytic.

**Theorem 4.5.5** (Inverse Mapping Theorem). *If $f$ is analytic on $U$, $z_0 \in U$, and $f'(z_0) \neq 0$, then $f$ has a local analytic inverse $f^{-1}$ at $z_0$. Furthermore, the derivative of $f^{-1}$ is* $(f^{-1})'(w) = \dfrac{1}{f'(f^{-1}(w))}$.

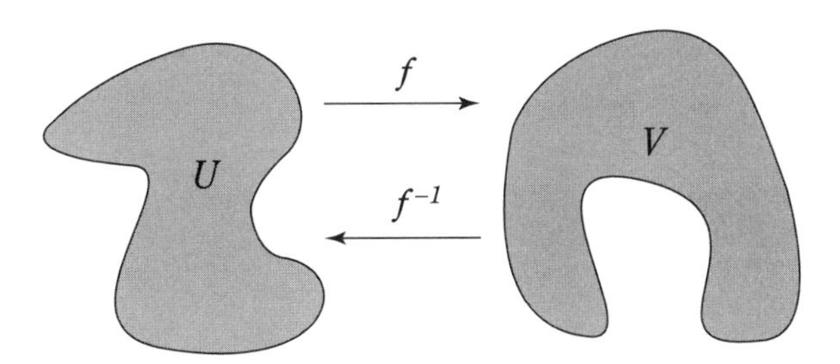


**Figure 4.5.1.** A Function $f$ and its Inverse Function $f^{-1}$.

**Proof.** We set $w_0 = f(z_0)$. If $f'(z_0) \neq 0$, then the function $f(z) - w_0$ also has non-vanishing derivative at $z_0$ and, hence, it has a zero of order 1 at $z_0$. Furthermore, by Theorem 3.4.2 there is an $r > 0$ such that, in the disc $D_r(z_0)$, this is the only zero of $f(z) - w_0$.

We choose $\delta$ such that $0 < \delta < r$ and $f'(z) \neq 0$ for all $z \in D_\delta(z_0)$. This is possible because $f'$ is continuous and so $f^{-1}(\mathbb{C} \setminus \{0\})$ is an open set containing $z_0$. We next let $\gamma$ be the circle

$$\gamma(t) = z_0 + \delta\, e^{2\pi i t}, \quad t \in [0, 1].$$

Then $f(z) - w_0$ has no zero on the compact set $\gamma(I)$. This means that the minimum value $\epsilon$ of $|f(z) - w_0|$ in $\gamma(I)$ is positive. Then if $|w - w_0| < \epsilon$ and $z \in \gamma(I)$, we have

$$|(f(z) - w) - (f(z) - w_0)| = |w - w_0| < \epsilon \leq |f(z) - w_0|.$$

By Rouché's Theorem, the two functions of $z$, $f(z) - w$ and $f(z) - w_0$, have the same number of zeros, counting order, in $D_\delta(z_0)$. Since, $f(z) - w_0$ has one zero in this disc, so does $f(z) - w$, and this is true for each $w \in D_\epsilon(w_0)$.

We conclude from the above that, for each $w \in D_\epsilon(w_0)$, there is exactly one $z \in D_\delta(z_0)$ such that $f(z) = w$. In other words, if we set

$$W = D_\epsilon(w_0) \quad \text{and} \quad V = f^{-1}(W),$$

then $f : V \to W$ is one-to-one and onto, and, hence, has a well-defined inverse function $f^{-1}$.

We claim that $f^{-1}$ is continuous on $W$. By Theorem 2.1.13, to show this, we only need to show that its inverse function, $f$, takes open sets to open sets. However, we just showed that an analytic function $f$ on an open set $U$, which has a non-zero derivative at a point $z_0$, takes $U$ to a set which contains a neighborhood of $f(z_0)$. This result applies equally well to any open set containing any point $z$ of $V$, since $f'(z) \neq 0$ for $z \in V$. Thus, $f$ takes any open subset of $V$ to an open subset of $W$ and this implies that $f^{-1} : W \to V$ is continuous.

We next show that the inverse function is analytic and has the indicated derivative. If $w$ and $w_1$ are two points of $W$, with $z = f^{-1}(w)$ and $z_1 = f^{-1}(w_1)$ the corresponding points of $V$, then $w = f(z)$, $w_1 = f(z_1)$, and

$$\lim_{w \to w_1} \frac{f^{-1}(w) - f^{-1}(w_1)}{w - w_1} = \lim_{z \to z_1} \left( \frac{f(z) - f(z_1)}{z - z_1} \right)^{-1} = \frac{1}{f'(z_1)} = \frac{1}{f'(f^{-1}(w_1))}.$$

This shows that $(f^{-1})'(w_1)$ exists and equals $\dfrac{1}{f'(f^{-1}(w_1))}$. Of course, the fact that $z \to z_1$ as $w \to w_1$ follows from the continuity of $f^{-1}$. □

The hypothesis that $f'(z_0) \neq 0$ in the above theorem is necessary. In fact, we have the following theorem, the proof of which is left as an exercise (Exercise 4.5.9).

**Theorem 4.5.6.** *If $f$ is analytic in $U$, $z_0 \in U$, and $f'(z_0) = 0$, then $f$ has no local analytic inverse at $z_0$. In fact, if $f'(z_0) = 0$, then $f$ is not one-to-one in any neighborhood of $z_0$.*

**Example 4.5.7.** At which points $z_0$ is it true that there is an analytic inverse function for $\sin z$? What is the derivative of the inverse function at $w = \sin(z)$ for each such point $z$?

**Solution:** By Theorems 4.5.5 and 4.5.6, sin has an analytic inverse exactly at those points $z_0$ for which $\sin' z_0 \neq 0$. Since, $\sin' z = \cos z$, the points we are looking for are those which are not odd multiples of $\pi i/2$. The derivative of the inverse function $\sin^{-1}$ at $w = \sin z$ for any such point $z$ is

$$\frac{1}{\sin'(\sin^{-1} w)} = \frac{1}{\cos(\sin^{-1} w)} = \frac{1}{\cos z}.$$

Since $\cos^2 z = 1 - \sin^2 z = 1 - w^2$, $\cos z$ will be some square root of $1 - w^2$. Thus, we may write

$$(\sin^{-1})'(w) = \frac{1}{\sqrt{1 - w^2}}$$

with the understanding that we are using some branch of the square root function. Which branch is being used depends on the two points $z_0$ and $w_0 = \sin z_0$ and the neighborhood on which the inverse function is defined. For example, for $z_0 = w_0 = 0$, we must use a branch of the square root function for which $\sqrt{1} = 1$, since $\cos 0 = 1$. On the other hand, if $z_0 = \pi$ and $w_0 = 0$, then we must use a branch of the square root function for which $\sqrt{1} = -1$, since $\cos \pi = -1$.

**The Open Mapping Theorem.** If $U$ is an open subset of $\mathbb{C}$, then an open mapping from $U$ to $\mathbb{C}$ is a function $f : U \to \mathbb{C}$ such that $f(V)$ is open for every open subset $V \subset U$.

**Theorem 4.5.8** (Open Mapping Theorem). *A non-constant analytic function on a connected open set $U$ is an open mapping of $U$ to $\mathbb{C}$.*

**Proof.** Let $f$ be a non-constant analytic function defined on a connected open set $U$. If $V$ is any open subset of $U$, we need to show that $f(V)$ is open. We will do this by showing that for each $z_0 \in V$ there is a neighborhood of $f(z_0)$ contained in $f(V)$.

Let $w_0 = f(z_0)$ and let $k$ be the order of the zero of $f(z) - w_0$ at $z_0$. By Exercise 3.4.8, there is a neighborhood $V_1$ of $z_0$ (which we may assume is contained in $V$) and an analytic function $g$ on $V_1$ such that, on $V_1$,

$$f(z) - w_0 = g^k(z) \quad \text{and} \quad g'(z_0) \neq 0.$$

The previous theorem implies that there is a neighborhood $V_2$ of $z_0$, with $V_2 \subset V_1$ such that $g(V_2)$ is open. In fact, we may assume $g(V_2)$ is an open disc $D_\epsilon(0)$ centered at $g(z_0) = 0$. Then the image of $V_2$ under $g^k$ is the open disc $D_{\epsilon^k}(0)$, and so the image of $V_2$ under $f$ is $f(z_0) + D_{\epsilon^k}(0)$. Since this is a neighborhood of $f(z_0)$ which is contained in $f(V)$, the proof is complete. □

## Exercise Set 4.5

1. How many zeroes, counting order, does the polynomial $3z^7 - z^3 + 1$ have in the open unit disc?
2. How many zeroes, counting order, does the polynomial $z^5 - 4z^3 + z - 1$ have in the open unit disc?
3. Prove that if $0 \leq C \leq 1/\mathrm{e}$, then the equation $z = C\,\mathrm{e}^z$ has exactly one solution in the unit disc.
4. Prove that if $h$ and $g$ are analytic functions defined in a neighborhood of the closed unit disc and if $h$ has $k$ zeroes in the open unit disc and no zeroes on the unit circle, then there is a $\delta > 0$ such that $h + \lambda g$ also has $k$ zeroes in the open unit disc for all positive numbers $\lambda < \delta$.
5. Prove that if $\{f_n\}$ is a sequence of one-to-one analytic functions on a connected open set $U$, and if $f_n \to f$ uniformly on compact subsets of $U$, then $f$ is either also one-to-one or is a constant.
6. Prove that if $f$ is analytic in an open set $U$ and has a zero of order $k$ at a point $z_0$ in $U$, then there is a neighborhood $V$ of $z_0$ and a neighborhood $W$ of 0, such that the equation $f(z) = w$ has exactly $k$ solutions in $V$ for each $w \in W \setminus \{0\}$.
7. At which points in the plane does the function $f(z) = 2z^3 + 3z^2 - 12z + 6$ have a local analytic inverse?
8. Without using any knowledge of log functions, use Theorem 4.5.5 to decide at which points $z_0 \in \mathbb{C}$ it is true that the function $f(z) = \mathrm{e}^z$ has a local analytic inverse. Also, use Theorem 4.5.5 to compute the derivative of this inverse function.
9. Prove Theorem 4.5.6.
10. Prove that if $f$ is an analytic function defined on an open set $U$ and if $f$ is one-to-one on $U$, then $f$ has an analytic inverse function $f^{-1} : W \to U$, where $W$ is the open set $f(U)$.
11. Prove that if $f$ is an entire function with the property that $f^{-1}(B)$ is bounded whenever $B$ is bounded, then $f$ is onto (that is, $f(\mathbb{C}) = \mathbb{C}$). Hint: Show that $f(\mathbb{C})$ is both open and closed.

## 4.6. Homotopy

This section is devoted to developing tools for calculating the index of a closed path (or more generally a cycle) about a point. This is crucial, since both the hypotheses and conclusions of the Cauchy theorems involve the index. The tools we will develop involve ideas from algebraic topology applied in the plane.

We begin by extending the index function to closed curves which are not necessarily piecewise smooth. The key to doing this is to show that all closed paths sufficiently near a given closed curve have the same index, and then to show that every closed curve can be approximated arbitrarily closely by closed paths. The common value of the index of all closed paths near a given closed curve $\gamma$ is then defined to be the index of $\gamma$. The key to the first part of this program is the following theorem, which is closely related to Rouché's Theorem and has a similar proof.

**Theorem 4.6.1.** *Suppose $z \in \mathbb{C}$ and $\gamma_1$ and $\gamma_2$ are two closed paths which do not pass through $z$ and which satisfy*

$$|\gamma_1(t) - \gamma_2(t)| \le |\gamma_2(t) - z|$$

*for all $t \in I$. Then $\text{Ind}_{\gamma_1}(z) = \text{Ind}_{\gamma_2}(z)$.*

**Proof.** By translating $\gamma_1$ and $\gamma_1$ by $z$ and using the fact, obvious from the definition of Ind, that $\text{Ind}_\gamma(z) = \text{Ind}_{\gamma - z}(0)$, we may assume that $z = 0$. Then our hypothesis is that $\gamma_1$ and $\gamma_2$ are two paths which do not pass through 0 and which satisfy

$$|\gamma_1(t) - \gamma_2(t)| \le |\gamma_2(t)|.$$

We define a new path $\gamma$ by

$$\gamma(t) = \frac{\gamma_1(t)}{\gamma_2(t)}.$$

This path does not pass through 0, and it satisfies

$$|\gamma(t) - 1| \le 1.$$

In other words, $\gamma(I)$ lies in the closed unit disc of radius 1 centered at 1 and does not contain 0. It follows from this that 0 is in the unbounded component of $\mathbb{C} \setminus \gamma(I)$. Hence,

$$\text{Ind}_\gamma(0) = 0.$$

To complete the proof, we will show that

$$\text{Ind}_\gamma(0) = \text{Ind}_{\gamma_1}(0) - \text{Ind}_{\gamma_2}(0). \tag{4.6.1}$$

In fact,

$$\text{Ind}_\gamma(0) = \frac{1}{2\pi i} \int_\gamma \frac{1}{z}\, dz = \frac{1}{2\pi i} \int_a^b \frac{\gamma'(t)}{\gamma(t)}\, dt,$$

and so (4.6.1) follows from the calculation

$$\frac{\gamma'}{\gamma} = \frac{\gamma_1'\gamma_2 - \gamma_1\gamma_2'}{\gamma_1\gamma_2} = \frac{\gamma_1'}{\gamma_1} - \frac{\gamma_2'}{\gamma_2}.$$

This completes the proof. □

**Index for Closed Curves.** Recall that a curve is just a continuous function $\gamma : I \to \mathbb{C}$, where $I$ is a closed bounded interval – no differentiability is assumed. For simplicity in the following discussion, we will just work with curves for which the parameter interval $I$ is $[0, 1]$.

There is a notion of distance between two curves parameterized on $I = [0, 1]$.

**Definition 4.6.2.** The distance between curves $\gamma_1$ and $\gamma_2$ on $I$ is defined to be

$$||\gamma_1 - \gamma_2|| = \sup_{t \in I} |\gamma_1(t) - \gamma_2(t)|.$$

With this notion of distance, we will show that every curve can be approximated arbitrarily closely by curves which are piecewise smooth, in fact, piecewise linear. Here, by a linear curve, we mean a curve of the form

$$\gamma(t) = u + tv$$

for fixed complex numbers $u$ and $v$ and $t$ ranging over some parameter interval. Such a curve traces a straight line segment. The linear curve which traces the line segment from $w$ to $z$ and has parameter interval $[a, b]$ is given by

$$\gamma(t) = \frac{b-t}{b-a}w + \frac{t-a}{b-a}z \tag{4.6.2}$$

(see Exercise 4.6.1).

A curve on $I$ is piecewise linear if there is a partition $0 = t_0 < t_1 < \cdots < t_n = 1$ of $I$ such that $\gamma$ is linear on each subinterval $[t_{j-1}, t_j]$. Obviously, piecewise linear curves are piecewise smooth and, hence, are paths.

**Theorem 4.6.3.** *If $\gamma : I \to \mathbb{C}$ is a curve, then for each $\epsilon > 0$ there is a piecewise linear curve $\tilde{\gamma}$ such that $||\tilde{\gamma} - \gamma|| < \epsilon$.*

**Proof.** Since $\gamma$ is continuous and $I$ is a closed bounded interval, $\gamma$ is uniformly coninuous on $I$. This means that, given $\epsilon > 0$, there is a $\delta > 0$ such that

$$|f(s) - f(t)| < \epsilon \quad \text{whenever} \quad s, t \in I, \ |s - t| < \delta.$$

We choose a partition $0 = t_0 < t_1 < \cdots < t_n = 1$ of $I$ such that $t_j - t_{j-1} < \delta$. By (4.6.2) the linear curve $\gamma_j$, with parameter interval $[t_{j-1}, t_j]$, which traces the line segment from $\gamma(t_{j-1})$ to $\gamma(t_j)$ is given by

$$\gamma_j(t) = \frac{t_j - t}{t_j - t_{j-1}}\gamma(t_{j-1}) + \frac{t - t_{j-1}}{t_j - t_{j-1}}\gamma(t_j) \quad \text{for} \quad t \in [t_{j-1}, t_j]$$

Note that, for $j = 1, \cdots, n$,

$$\gamma_j(t_j) = \gamma(t_j) = \gamma_{j+1}(t_j).$$

This implies that the linear curves $\gamma_j$, defined on the subintervals $[t_{j-1}, t_j]$, fit together to form a continuous piecewise linear curve $\tilde{\gamma}$ on $I = [0, 1]$. Also, for each $j$,

$$\begin{aligned}
|\gamma(t) - \gamma_j(t)| &= \left| \frac{t_j - t}{t_j - t_{j-1}}(\gamma(t) - \gamma(t_{j-1})) + \frac{t - t_{j-1}}{t_j - t_{j-1}}(\gamma(t) - \gamma(t_j)) \right| \\
&\leq \frac{t_j - t}{t_j - t_{j-1}}|\gamma(t) - \gamma(t_{j-1})| + \frac{t - t_{j-1}}{t_j - t_{j-1}}|\gamma(t) - \gamma(t_j)| \\
&\leq \frac{t_j - t}{t_j - t_{j-1}}\epsilon + \frac{t - t_{j-1}}{t_j - t_{j-1}}\epsilon = \epsilon
\end{aligned}$$

if $t \in [t_{j-1}, t_j]$. Since this is true for every $j$ and since $\tilde{\gamma}(t) = \gamma_j(t)$ for $t \in [t_{j-1}, t_j]$, we conclude that $||\gamma - \tilde{\gamma}|| < \epsilon$. This completes the proof. □

The next theorem tells us that, given a closed curve $\gamma$ and a point $z$ not on $\gamma(I)$, all closed paths sufficiently near $\gamma$ have the same index about $z$.

**Theorem 4.6.4.** *If $\gamma$ is a closed curve and $z \in \mathbb{C}$, a point not on $\gamma(I)$, then there is a $\delta > 0$ such that if $\gamma_1$ and $\gamma_2$ are paths with $||\gamma - \gamma_j|| < \delta$ for $j = 1, 2$, then*

$$\mathrm{Ind}_{\gamma_1}(z) = \mathrm{Ind}_{\gamma_2}(z).$$

**Proof.** This is an application of Theorem 4.6.1. We choose

$$\delta = \frac{1}{3} \inf_{t \in I} |\gamma(t) - z|,$$

so that

$$|\gamma(t) - z| \geq 3\delta \quad \text{for all} \quad t \in I.$$

If $||\gamma - \gamma_j|| < \delta$ for $j = 1, 2$, then

$$|\gamma(t) - \gamma_j(t)| < \delta \quad \text{for all} \quad t \in I,\ j = 1, 2,$$

and so

$$|\gamma_j(t) - z| \geq |\gamma(t) - z| - |\gamma(t) - \gamma_j(t)| \geq 2\delta \quad \text{for all} \quad t \in I,\ j = 1, 2.$$

It follows that

$$|\gamma_1(t) - \gamma_2(t)| \leq |\gamma(t) - \gamma_1(t)| + |\gamma(t) - \gamma_2(t)| < 2\delta \leq |\gamma_1(t) - z|$$

for all $t \in I$. By Theorem 4.6.1, $\mathrm{Ind}_{\gamma_1}(z) = \mathrm{Ind}_{\gamma_2}(z)$. □

This leads to the following definition.

**Definition 4.6.5.** Let $\gamma$ be a closed curve in $\mathbb{C}$ and $z \in \mathbb{C}$ a point not on $\gamma(I)$. We define $\mathrm{Ind}_\gamma(z)$ in the following way: we choose a $\delta$ as in Theorem 4.6.4. Then all closed paths within a distance $\delta$ of $\gamma$ have the same index about $z$. By Theorem 4.6.3, there are closed paths within a distance $\delta$ of $\gamma$. We define $\mathrm{Ind}_\gamma(z)$ to be the common value of $\mathrm{Ind}_{\gamma_1}(z)$ for all such paths $\gamma_1$.

With this definition, Theorem 4.6.4 implies that $\mathrm{Ind}_\gamma(z)$ is a locally constant function of the closed curve $\gamma$. That is, the following theorem holds.

**Theorem 4.6.6.** *If $z \in \mathbb{C}$ and $\gamma$ is a closed curve in $\mathbb{C}$ such that $z \notin \Gamma(I)$, then there is a $\delta > 0$ such that if $\gamma_1$ is any other closed curve in $\mathbb{C}$ and $||\gamma - \gamma_1|| < \delta$, then*

$$\mathrm{Ind}_\gamma(z) = \mathrm{Ind}_{\gamma_1}(z).$$

**Proof.** We choose $\delta$ as in Theorem 4.6.4. Then all closed paths within a distance $\delta$ of $\gamma$ have the same index around $z$, and this is the index of $\gamma$ around $z$, by definition. If $\gamma_1$ is a closed curve with $||\gamma - \gamma_1|| < \delta$, set

$$\delta_1 = \delta - ||\gamma - \gamma_1||.$$

Then all closed paths within a distance $\delta_1$ of $\gamma_1$ are within a distance $\delta$ of $\gamma$, and so they all have the same index about $z$ as $\gamma$ does. By definition, this is also the index of $\gamma_1$ about $z$. □

**Homotopy.** Let $U$ be an open set in $\mathbb{C}$ and let $\gamma_0$ and $\gamma_1$ be two closed curves in $U$ with parameter interval $[0,1]$. The curves are said to be *homotopic* if it is possible to continuously deform $\gamma_0$ to $\gamma_1$ while remaining in $U$. This is made precise in the following definition.

**Definition 4.6.7.** Two closed curves $\gamma_0$ and $\gamma_1$ in $U$, with parameter interval $[0,1]$, are said to be *homotopic* in $U$ if there is a continuous function $h$, defined on the square $[0,1]\times[0,1]$ in $\mathbb{R}^2$, with values in $U$, such that

(a) $h(0,t)=\gamma_0(t)$ for all $t\in[0,1]$;
(b) $h(1,t)=\gamma_1(t)$ for all $t\in[0,1]$; and
(c) $h(s,0)=h(s,1)$ for all $s\in[0,1]$.

In the above definition, for each fixed $s$, the function $\gamma_s(t)=h(s,t)$ defines a curve $\gamma_s$ in $U$. Parts (a) and (b) say that when $s=0$ and $1$, these curves are the original curves $\gamma_0$ and $\gamma_1$. Part (c) says that each of the curves $\gamma_s$ is closed.

The family of curves $\{\gamma_s\}$ determined by a homotopy $h$, as in the above definition, is a continuous one-parameter family of curves in the sense that the following theorem is true.

**Theorem 4.6.8.** *If $\{\gamma_s\}$ is the family of curves determined by a homotopy, as in Definition 4.6.7, then for each $s_0\in I$ and each $\epsilon>0$, there is a $\delta>0$ such that*

$$||\gamma_s-\gamma_{s_0}||<\epsilon \quad \textit{whenever} \quad |s-s_0|<\delta.$$

**Proof.** This proof uses the fact that a continuous function on a compact set such as the square $I\times I$, is uniformly continuous. For the continuous function $h$, this means that given $\epsilon>0$ there is a $\delta>0$, such that

$$|h(s,t)-h(s_0,t_0)|<\epsilon \quad \text{whenever} \quad |(s,t)-(s_0,t_0)|<\delta,$$

for any pair of points $(s,t),(s_0,t_0)\in I\times I$. In particular, in the case where $t=t_0$ this says that if $|s-s_0|<\delta$, then

$$|\gamma_s(t)-\gamma_{s_0}(t)|<\epsilon \quad \text{for all} \quad t\in I,$$

which implies $||\gamma_s-\gamma_{s_0}||<\epsilon$, and this completes the proof. □

We are now in a position to prove the homotopy theorem for the index function.

**Theorem 4.6.9.** *If $\gamma_0$ and $\gamma_1$ are two closed curves in $U$ which are homotopic in $U$, then*

$$\mathrm{Ind}_{\gamma_0}(z)=\mathrm{Ind}_{\gamma_1}(z)$$

*for every $z\in\mathbb{C}\setminus U$.*

**Proof.** Let $\{\gamma_s\}$, $s\in I$ be the family of curves determined by a homotopy $h$ joining $\gamma_0$ to $\gamma_1$ in $U$. Since $z\in\mathbb{C}\backslash U$, none of these curves passes through $z$ and so $\mathrm{Ind}_{\gamma_s}(z)$ is defined for every $s\in I$.

If $s_0\in I$, then Theorem 4.6.6 implies that there is an $\epsilon>0$ such that

$$\mathrm{Ind}_{\gamma_s}(z)=\mathrm{Ind}_{\gamma_{s_0}}(z) \quad \text{whenever} \quad ||\gamma_s-\gamma_{s_0}||<\epsilon.$$

Then Theorem 4.6.8 implies that there is a $\delta>0$ such that

$$||\gamma_s-\gamma_{s_0}||<\epsilon \quad \text{whenever} \quad |s-s_0|<\delta.$$

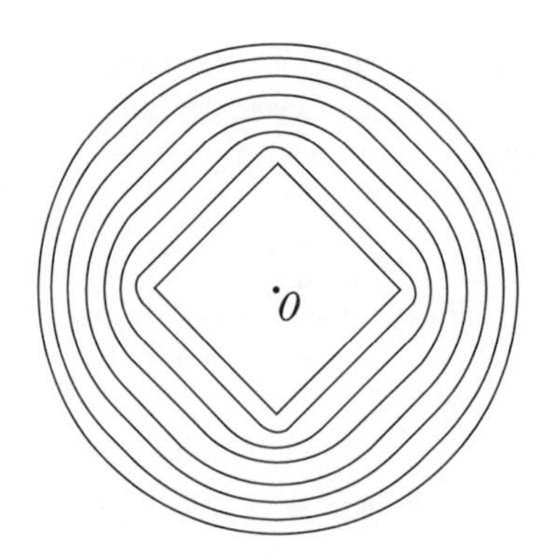


**Figure 4.6.1.** Some Paths $\gamma_s$ for a Homotopy of a Circle to a Square.

We conclude that $\mathrm{Ind}_{\gamma_s}(z)$ is constant in the open interval $(s_0-\delta, s_0+\delta)$. Since $s_0$ is an arbitrary point of $I$, we conclude that the set on which $\mathrm{Ind}_{\gamma_s}(z)$ takes on any given value is an open set. Since $I$ is connected, and the union of these open sets is $I$, there can be only one of them that is non-empty. Thus, $\mathrm{Ind}_{\gamma_s}(z)$ is constant on $I$ and, in particular, $\mathrm{Ind}_{\gamma_1}(z) = \mathrm{Ind}_{\gamma_0}(z)$. □

If $\gamma_0$ and $\gamma_1$ are homotopic closed paths in an open set $U$, then the above theorem implies that the cycle $\Gamma = \gamma_1 - \gamma_0$ has index 0 about every point of the complement of $U$. Hence, $\Gamma$ is homologous to 0. The general Cauchy Theorem then says that

$$0 = \int_\Gamma f(z)\,dz = \int_{\gamma_1} f(z)\,dz - \int_{\gamma_0} f(z)\,dz$$

for every function $f$ which is analytic on $U$. This proves the homotopy version of Cauchy's Theorem.

**Theorem 4.6.10.** *If $U$ is an open subset of $\mathbb{C}$, $\gamma_0$ and $\gamma_1$ are homotopic closed paths in $U$, and $f$ is an analytic function on $U$, then*

$$\int_{\gamma_1} f(z)\,dz = \int_{\gamma_0} f(z)\,dz.$$

**Example 4.6.11.** Prove that the circle of radius 2 centered at 0, with its standard parameterization, is homotopic in $\mathbb{C}\setminus\{0\}$ to the piecewise linear curve which traces once in the positive direction around the square with vertices at $1, i, -1, -i$ (see Figure 4.6.1).

**Solution:** We parameterize the circle by $\gamma_0(t) = 2\,e^{2\pi i t}$, $t \in [0,1]$ and the square by

$$\gamma_1(t) = \begin{cases} (1-4t)+4ti, & \text{if } 0 \le t \le 1/4; \\ (4t-1)+(2-4t)i, & \text{if } 1/4 \le t \le 1/2; \\ (4t-3)+(2-4t)i, & \text{if } 1/2 \le t \le 3/4; \\ (4t-3)+(4t-4)i, & \text{if } 3/4 \le t \le 1. \end{cases} \tag{4.6.3}$$

We define a homotopy $h$ between the two by simply letting $h(s,t)$ be the point

$$h(s,t) = (1-s)\gamma_0(t) + s\gamma_1(t)$$

on the line segment joining $\gamma_0(t)$ to $\gamma_1(t)$. The resulting point $h(s,t)$ is always on or outside the boundary of the square and so it is never 0. For $0 \leq t \leq 1/4$, the formula for $h$ is

$$h(s,t) = 2(1-s)\cos t + s(1-4t) + [2(1-s)\sin t + 4st]i.$$

The formula for $h$ if $t$ is in one of the other subintervals of the partition $0 < 1/4 < 1/2 < 3/4 < 1$ is also easily calculated using (4.6.3).

**Example 4.6.12.** Show why the curves $\gamma_0$ and $\gamma_1$ of the previous exercise cannot be homotopic in $V = \mathbb{C} \setminus \{3/2\}$.

**Solution:** The point $3/2$ is in the complement of $V$, and the curve $\gamma_0$ has index 1 about $3/2$. However, the curve $\gamma_1$ has index 0 about $3/2$, since $3/2$ is in the unbounded component of $\mathbb{C} \setminus \gamma_1(I)$. By Theorem 4.6.9, the two curves cannot be homotopic in $V$.

### Simply Connected Sets.

**Definition 4.6.13.** A connected open set $U$ is said to be *simply connected* if every closed curve in $U$ is homotopic to a point (that is, a constant curve).

**Example 4.6.14.** Prove that a convex open set $U$ is simply connected.

**Solution:** Fix a point $z_0 \in U$. If $\gamma : I \to U$ is a closed curve in $U$, we define a homotopy between $\gamma$ and the constant curve $z_0$ as follows:

$$h(s,t) = (1-s)\gamma(t) + sz_0.$$

Note that, for each $t \in I$, $h(x,t)$ is on the line segment joining $\gamma(t)$ to $z_0$ and, hence, is in $U$. Clearly, $h$ is continuous on $I \times I$, $h(0,t) = \gamma(t)$, $h(1,t) = z_0$, and $h(s,0) = h(s,1)$, since $\gamma(0) = \gamma(1)$. This establishes that $h$ is a homotopy between $\gamma$ and the constant curve $z_0$. Since this can be done for every closed curve $\gamma$ in $U$, the set $U$ is simply connected.

We have talked about an open set in $\mathbb{C}$ possibly having a "hole". What do we mean by a hole in an open set? This is usually understood to mean a bounded component of the closed set $\mathbb{C} \setminus U$. However, we have not explored the ideas of connectedness and connected components for closed sets so far in this text. A few words concerning these concepts now will make the statement and proof of the next theorem much easier to understand.

Recall that a set $S$ is *separated* by a pair of open sets $U$, $V$ if $U \cap S$ and $V \cap S$ are both non-empty, have empty intersection, and have union equal to $S$. If there is no pair of open sets which separate $S$, then $S$ is connected. A *component* of $S$ is a maximal connected subset of $S$.

If the pair $U$, $V$ separates $S$ and $S$ happens to be closed, then $A = S \setminus (\mathbb{C} \setminus V)$ is closed and is equal to $U \cap S$. Similarly, $B = S \setminus (\mathbb{C} \setminus U)$ is closed and equal to $V \cap S$. Thus, a closed set $S$ is separated by a pair of open sets $U, V$ if and only if it is separated by a pair of its closed subsets $A, B$ in the sense that $A, B$ is a disjoint pair of closed, non-empty subsets of $S$ with union equal to $S$. In this case, $A = U \cap S$ and $B = V \cap S$.

In the next theorem we will need the following lemma concerning components and separation of a close subset of $\mathbb{C}$. The proof is left to the exercises.

**Lemma 4.6.15.** *Let $S$ be a closed subset of $\mathbb{C}$. If $C$ is a component of $S$, then $C$ is bounded if and only if $C$ is contained in a closed bounded subset $A$ of $S$ such that $B = S \setminus A$ is also closed.*

Note that the set $B$ of the above lemma could be empty. If it is empty, then $S$ itself is bounded. If it is not empty, then the pair $A, B$ separates $S$.

The next theorem shows that simply connected open sets are connected open sets with no holes and that they are exactly the sets on which all the things we would like to be true in complex analysis are, in fact, true.

**Theorem 4.6.16.** *Let $U$ be a connected open subset of $\mathbb{C}$. Then the following statements are equivalent.*

(a) *$U$ is simply connected;*
(b) *every cycle in $U$ is homologous to $0$;*
(c) *$U$ has no holes, that is, $\mathbb{C} \setminus U$ has no bounded components;*
(d) *$\int_\Gamma f(z)\,dz = 0$ for every cycle $\Gamma$ in $U$ and every analytic function $f$ on $U$;*
(e) *every analytic function on $U$ has an antiderivative;*
(f) *every harmonic function on $U$ has a harmonic conjugate;*
(g) *every non-vanishing analytic function $f$ on $U$ has an analytic logarithm;*
(h) *every non-vanishing analytic function $f$ on $U$ has an analytic square root.*

**Proof.** We can only give an incomplete proof at this point. We will prove that each of the statements (a) through (h) implies the next statement on the list. However, the proof that (h) implies (a) will have to wait until we prove the Riemann Mapping Theorem in Chapter 6.

A constant path $\gamma(t) \equiv c$ in $U$ has $\gamma'(t) \equiv 0$, and so its index about any point other than $c$ is 0. In particular, it is homologous to 0 in $U$. Since each cycle is equivalent to a sum of closed paths, it follows from Theorem 4.6.9 that (a) implies (b).

We next prove that (b) implies (c). Suppose (c) fails, that is, suppose there is a bounded component $C$ of $S = \mathbb{C} \setminus U$. Then, by the previous lemma, $C$ is contained in a closed bounded subset $A$ of $S$ such that $B = S \setminus A$ is also closed.

The set $A$ is compact. It is contained in some bounded open rectangle $R$ as well as in the open set $\mathbb{C} \setminus B$. Thus, $A$ is contained in the bounded open set $V = R \setminus (\mathbb{C} \setminus B)$. We fix a point $z_0 \in A$. Let $\delta > 0$ be chosen so that $\sqrt{2}\,\delta$ is smaller than the distance from $A$ to the complement of $V$. Suppose the open rectangle $R$ is defined by inequalities

$$a < \mathrm{Re}(z) < b, \quad c < \mathrm{Im}(z) < d.$$

Let $a = s_0 < s_1 < \cdots < s_n = b$ and $c = t_0 < t_1 < \cdots < t_n = d$ be partitions of $[a, b]$ and $[c, d]$ into subintervals of length at most $\delta$. We are careful to choose these so that $\mathrm{Re}(z_0)$ is not one of the $s_j$'s and $\mathrm{Im}(z_0)$ is not one of the $t_k$'s.

Now each of the rectangles

$$R_{jk} = \{z : s_{j-1} \le \mathrm{Re}(z) \le s_j, t_{k-1} \le \mathrm{Im}(z) \le t_k\}$$

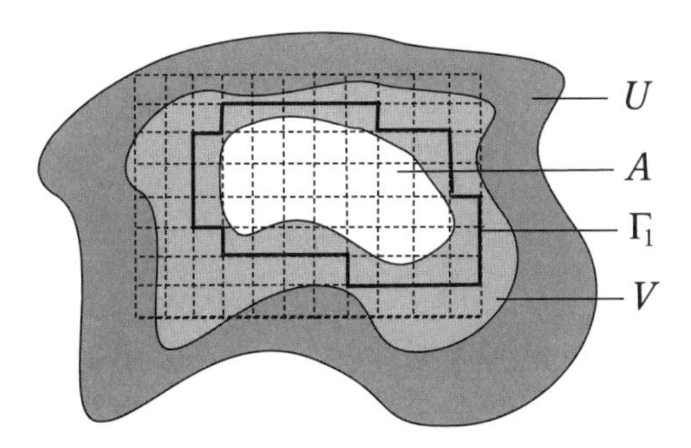


**Figure 4.6.2.** Creating a Cylce $\Gamma_1$ Surrounding a Hole.

which meets $A$ also lies in $V$. Let $\Gamma$ be the cycle which is the sum of all the paths $\partial R_{jk}$ for which $R_{jk} \cap A \neq \emptyset$. Now if an edge of one of these rectangles meets $A$, then this edge belongs to two of the rectangles whose boundaries make $\Gamma$. The edge occurs with opposite orientation in the two paths and, hence, their contributions will cancel out in any integral around $\Gamma$ as well as in the expression for $\partial\Gamma$. This means that if $\Gamma_1$ is the chain which is left after all edges which meet $A$ are thrown out of $\Gamma$, then $\Gamma_1$ is still a cycle, and if $f$ is a function continuous on $\Gamma(I)$, then

$$\int_\Gamma f(z)\,dz = \int_{\Gamma_1} f(z)\,dz. \tag{4.6.4}$$

Note that $\Gamma_1(I)$ is contained in $V \setminus A$ which is contained in $U$ (see Figure 4.6.2).

Since $z_0$ is in a rectangle $R_{pq}$ whose boundary contributes to $\Gamma$, but $z_0$ is not on this boundary, we may calculate the index of $\Gamma$ around $z_0$. By the Cauchy Integral Theorem, the contributions of the paths $\partial R_{jk}$ different from $\partial R_{pq}$ are all 0 and so

$$\operatorname{Ind}_\Gamma(z_0) = \frac{1}{2\pi i}\int_\Gamma \frac{1}{z-z_0}\,dz = \frac{1}{2\pi i}\int_{\partial R_{pq}} \frac{1}{z-z_0}\,dz = 1.$$

Thus, in view of (4.6.4), we have

$$\operatorname{Ind}_{\Gamma_1}(z_0) = 1.$$

Since $\Gamma_1$ is a cycle in $U$, we conclude that (b) fails. This completes the proof that (b) implies (c).

To show that (c) implies (d), suppose $\gamma$ is a closed path in $U$. If $V$ is a bounded component of $\mathbb{C}\setminus\gamma(I)$, then $A = V\cap(\mathbb{C}\setminus U)$ is a closed set since it is the intersection of $\mathbb{C} \setminus U$ with the complement of the union of the other components of $\mathbb{C} \setminus \gamma(I)$. The set $B = (\mathbb{C} \setminus U) \setminus A$ is also closed since it is equal to $\mathbb{C} \setminus (U \cup V)$. By Lemma 4.6.15, if $A$ is non-empty, then it contains a bounded component of $\mathbb{C} \setminus U$. Thus, if (c) holds, then $A$ must be empty – that is, $V \cap (\mathbb{C} \setminus U) = \emptyset$ for every bounded component $V$ of $\mathbb{C}\setminus\gamma(I)$. Then $z \in \mathbb{C}\setminus U$ implies $z$ is in the unbounded component of $\mathbb{C}\setminus\gamma(I)$ and so $\operatorname{Ind}_\gamma(z) = 0$. The hypotheses of the Cauchy Theorem are satisfied and so (d) holds. Thus, (c) implies (d).

If (d) holds, then we may define an antiderivative $g$ for the analytic function $f$ on $U$ by fixing a point $z_0 \in U$ and setting

$$g(z) = \int_{\gamma_z} f(w)\, dw,$$

where $\gamma_z$ is any path in $U$ beginning at $z_0$ and ending at $z$. That this is independent of the path chosen follows from (d). That it yields an antiderivative for $f$ follows from Theorem 2.6.1. Thus, (d) implies (e).

That (e) implies (f) follows if we notice that, in the proof of Theorem 3.5.7, the only use of the hypotheses that $U$ was convex was to ensure that an analytic function has an antiderivative.

If $f$ is analytic and non-vanishing on $U$, then $\log|f|$ is a harmonic function on $U$. Thus, if (f) is true, then $\log|f|$ is the real part of an analytic function $g$ on $U$. Then $e^g$ and $f$ are two analytic functions on $U$ with the same modulus $|f|$. This means $f\,e^{-g}$ has constant modulus and, hence, is a constant $a$, by the Maximum Modulus Theorem. If we choose a logarithm $b$ for the non-zero number $a$, then $a = e^b$ and

$$f = e^{g+b}.$$

In other words, $g + b$ is an analytic logarithm for $f$. This proves that (f) implies (g).

If $f$ has an analytic logarithm $h$, then $e^{h/2}$ is an analytic square root for $f$. Thus, (g) implies (h).

We have now proved that each of the statements (a) through (g) implies the next statement on the list. To complete the proof we must show that (h) implies (a). This is done in Corollary 6.4.6 of Chapter 6. □

**Example 4.6.17.** Prove that $U = \mathbb{C} \setminus D_1(0)$ is not simply connected.

**Solution:** The point 0 is in the complement of $U$ and $\gamma(t) = 2\,e^{2\pi i t}$ defines a curve in $U$ which has index 1 about 0. Thus, $\gamma$ is a closed curve in $U$ which is not homologous to 0. Since (c) of the preceding theorem fails to hold for $U$, we conclude that $U$ is not simply connected.

**Homotopy for Non-Closed Curves.** There is a type of homotopy for non-closed curves that have a fixed starting point $z_0$ and a fixed ending point $z_1$.

**Definition 4.6.18.** If $\gamma_0$ and $\gamma_1$ are curves in $U$, parameterized on $I = [0,1]$, and if $\gamma_0(0) = \gamma_1(0) = z_0$ and $\gamma_0(1) = \gamma_1(1) = z_1$, then $\gamma_0$ and $\gamma_0$ are said to be *homotopic* curves joining $z_0$ to $z_1$ in $U$ if there exists a continuous function

$$h : I \times I \to U$$

such that

$$\begin{aligned} h(0,t) &= \gamma_0(t), \quad \text{for all} \quad t \in I, \\ h(1,t) &= \gamma_1(t), \quad \text{for all} \quad t \in I, \\ h(s,0) &= z_0, \quad \text{for all} \quad s \in I, \quad \text{and} \\ h(s,1) &= z_1, \quad \text{for all} \quad s \in I. \end{aligned}$$

As with a homotopy for closed curves, a homotopy of the above type determines a continuous one-parameter family of curves $\{\gamma_s\}$ by

$$\gamma_s(t) = h(s,t).$$

In this case, the curves in the family all begin at $z_0$ and end at $z_1$. For this type of homotopy, we have the following results, which follow easily from the preceding material, and whose proofs are left as exercises.

**Theorem 4.6.19.** *If $\gamma_0$ and $\gamma_1$ are homotopic paths in $U$ connecting $z_0$ to $z_1$, then*

$$\int_{\gamma_0} f(z)\,dz = \int_{\gamma_1} f(z)\,dz$$

*for every function $f$ which is analytic in $U$.*

**Theorem 4.6.20.** *Let $U$ be a connected open set and let $z_0, z_2$ be points of $U$. Then $U$ is simply connected if and only if any two curves in $U$ connecting $z_0$ to $z_1$ are homotopic in $U$.*

## Exercise Set 4.6

1. Show that the path defined by (4.6.2) is, as claimed in the text, a linear curve, with parameter interval $[a, b]$, which traces the line segment from $w$ to $z$.
2. Find $||\gamma - \gamma_1||$ if $\gamma(t) = e^{2\pi it}$, and $\gamma_1(t) = 2\cos(2\pi t) + 3i\sin(2\pi t)$, for $t \in [0,1]$.
3. If $\gamma(t) = e^{2\pi it}$ for $t \in [0,1]$, describe a piecewise linear path $\gamma_1$ such that $||\gamma - \gamma_1|| < \pi/8$.
4. Show that the curves $\gamma$ and $\gamma_1$ of Exercise 2 are homotopic in $\mathbb{C}\setminus\{0\}$ by finding an explicit homotopy in $\mathbb{C}\setminus\{0\}$ between them.
5. Explain why the curves $\gamma$ and $\gamma_1$ of the previous exercise cannot be homotopic in the set $\mathbb{C}\setminus\{2i\}$.
6. Which of the following open sets in $\mathbb{C}$ are simply connected:

   (a) $\mathbb{C}\setminus\{0\}$, (b) $\mathbb{C}\setminus[-\infty,0]$,
   (c) $D_2(0)\setminus[-1,1]$, (d) $D_1(0)\setminus[0,1)$,
   (e) $\mathbb{C}\setminus(\{0\}\cup\{1\})$, (f) $D_2(0)\setminus\overline{D}_1(0)$?
7. For the set $\mathbb{C}\setminus\{0\}$, show by example that (b), (c), and (d) of Theorem 4.6.16 all fail.
8. Also show by example that (e), (f), and (g) of Theorem 4.6.16 fail for the set $\mathbb{C}\setminus\{0\}$.
9. Prove Theorem 4.6.19.
10. Prove Theorem 4.6.20.
11. Prove that if $U$ and $V$ are homeomorphic open subsets of $\mathbb{C}$, and if $U$ is simply connected, then so is $V$. Here, $U$ and $V$ are homeomorphic if there is a one-to-one continuous function $f$ of $U$ onto $V$ which has a continuous inverse function.
12. Prove Lemma 4.6.15 – that is, prove that if $S$ is a closed subset of $\mathbb{C}$ and $C$ is a component of $S$, then $C$ is bounded if and only if it contains a bounded closed subset $A$ for which $S\setminus A$ is also closed.

Chapter 5

# Residue Theory

The Residue Theorem (Theorem 4.4.3) has a wide range of applications. This chapter is devoted to exploring some of them. We begin with a section on techniques for computing residues.

## 5.1. Computing Residues

Recall the discussion of residues in Section 4.4. If $f$ is analytic in a neighborhood of $z_0$, except at $z_0$ itself, then $f$ has an isolated singularity at $z_0$. In this case, it has a Laurent expansion

$$f(z) = \sum_{n=-\infty}^{\infty} a_n(z - z_0)^n$$

in $D \setminus \{z_0\}$ for some disc $D$, centered at $z_0$. The *residue* of $f$ at $z_0$, denoted $\mathrm{Res}(f, z_0)$, is then the coefficient $a_{-1}$ of $(z - z_0)^{-1}$ in this expansion. Computing the residue can be easy. It can also be hard.

If the singularity of $f$ at $z_0$ is a pole of order $k$, then there is a formula for $\mathrm{Res}(f, z_0)$ which makes it, in principle, computable. If $f$ has a pole of order $k$ at $z_0$, then it has the form

$$f(z) = \frac{g(z)}{(z - z_0)^k},$$

where $g$ is analytic and non-vanishing in a neighborhood of $z_0$.

**Theorem 5.1.1.** *If* $f(z) = \dfrac{g(z)}{(z - z_0)^k}$, *where* $g$ *is analytic in a neighborhood of* $z_0$, *then* $\mathrm{Res}(f, z_0)$ *is the coefficient of degree* $k - 1$ *in the power series expansion of* $g$ *about* $z_0$. *That is,*

$$\mathrm{Res}(f, z_0) = \frac{g^{(k-1)}(z_0)}{(k - 1)!}.$$

**Proof.** Since $g$ is analytic in a neighborhood of $z_0$, Theorem 3.2.5 implies that it has a power series expansion

$$g(z) = \sum_{n=0}^{\infty} b_n (z - z_0)^n,$$

where, by Theorem 3.2.1,

$$b_n = \frac{g^{(n)}(z_0)}{n!}.$$

The Laurent expansion of $f$ is then given by

$$f(z) = b_0(z - z_0)^{-k} + \cdots + b_{k-1}(z - z_0)^{-1} + \sum_{n=k}^{\infty} b_n (z - z_0)^{n-k}.$$

Thus, the residue of $f$ at $z_0$ is

$$\text{Res}(f, z_0) = b_{k-1} = \frac{g^{k-1}(z_0)}{(k-1)!},$$

as claimed. □

The first couple of cases of the preceding theorem are worth highlighting. We do this in the following corollary, which follows immediately from the theorem.

**Corollary 5.1.2.** *Given a function $g$, analytic in a neighborhood of $z_0$,*

(a) *if* $f(z) = \dfrac{g(z)}{z - z_0}$, *then* $\text{Res}(f, z_0) = g(z_0)$;

(b) *if* $f(z) = \dfrac{g(z)}{(z - z_0)^2}$, *then* $\text{Res}(f, z_0) = g'(z_0)$.

**Example 5.1.3.** Find $\text{Res}(f, 1)$ if $f(z) = \dfrac{z^2 + 1}{(z-1)(z^2 - 2z + 5)}$.

**Solution:** The function $f$ is of the form

$$f(z) = \frac{g(z)}{z - 1},$$

where

$$g(z) = \frac{z^2 + 1}{z^2 - 2z + 5}.$$

Since $g$ is analytic in a neighborhood of 1, Corollary 5.1.2(a) implies that

$$\text{Res}(f, 1) = g(1) = \frac{1}{2}.$$

**Example 5.1.4.** Find $\text{Res}(f, 0)$ if $f(z) = \dfrac{\sin z}{z^2}$.

**Solution:** This is clearly a situation where Corollary 5.1.2(b) applies, with $g(z) = \sin z$. Thus,

$$\text{Res}(f, 0) = g'(0) = \cos 0 = 1.$$

We could also have done this one directly, by writing out the power series expansion for $\sin z$ about 0, and dividing by $z^2$ to obtain the Laurent series expansion of $f$.

**Long Division.** Often the function $f$, whose residue we wish to compute at some $z_0$, is of the form

$$f(z) = \frac{p(z)}{h(z)},$$

where $p$ and $h$ are analytic functions in a neighborhood of $z_0$ with known power series expansions about $z_0$, and $h$ has a zero of order $k$ at $z_0$. In this situation, we can always derive a formula for $\mathrm{Res}(f, z_0)$ using Theorem 5.1.1. We simply write $h$ as $h(z) = (z - z_0)^k q(z)$, where $q$ is analytic in a neighborhood of $z_0$ and $q(z_0) \neq 0$. Then

$$(5.1.1) \qquad f(z) = \frac{g(z)}{(z - z_0)^k}, \quad \text{where} \quad g(z) = \frac{p(z)}{q(z)}.$$

We then apply Theorem 5.1.1. Of course, to apply this theorem, one has to compute $g^{(k-1)}(z_0)$ (or, equivalently, the coefficient of $(z - z_0)^{k-1}$ in the power series expansion of $g$ about $z_0$), and this may be quite difficult. Rather than doing this directly, through repeated differentiation, it is possible to compute the power series coefficients of $g$ using power series methods – specifically, the method of long division of power series.

Note that $p$ and $q$ are both analytic functions in a neighborhood of $z_0$ and, since the power series expansions of $p$ and $h$ about $z_0$ are known, the same thing is true of the power series expansion of $q$ about $z_0$. Since $q(z_0) \neq 0$, $g = p/q$ is also analytic in a neighborhood of $z_0$ and, hence, has a power series expansion in some disc centered at $z_0$. The method of long division is an algorithm for finding the power series coefficents of $p/q$ in terms of those of $p$ and $q$.

For simplicity, in our discussion of long division, we will work with power series in $z$ – that is, power series centered at 0. Power series centered at some point $z_0$ other than 0 can always be put in this form with a change of variables. The coefficient formulas we derive below are not affected by such a change.

Suppose that $p$ has a zero of order at least $m$ at 0 ($m$ might be 0 and it might be less than the actual order of the zero of $p$ at 0). Also suppose that $q(0) \neq 0$. Then $p$ and $q$ have power series expansions

$$(5.1.2) \qquad \begin{aligned} p(z) &= a_m z^m + a_{m+1} z^{m+1} + a_{m+2} z^{m+2} + \cdots + a_n z^n + \cdots, \\ q(z) &= b_0 + b_1 z + b_2 z^2 + \cdots + b_n z^n + \cdots, \end{aligned}$$

which converge in some disc $D_r(0)$, with $r > 0$. Since $q(0) = b_0 \neq 0$, $q(z)$ is nonvanishing in some disc $D_\delta(0)$, with $0 < \delta < r$. Then the quotient $p/q$ is analytic in $D_\delta(0)$, with a zero of order at least $m$ at 0, and has a convergent power series expansion

$$(5.1.3) \qquad \frac{p(z)}{q(z)} = c_m z^m + c_{m+1} z^{m+1} + c_{m+2} z^{m_2} + \cdots + c_n z^n + \cdots.$$

How can we determine the coefficients $c_n$ if we know the coefficients $a_n$ and $b_n$? A method often taught in calculus for doing this is the method of long division of power series. This is a recursive method that determines the coefficients $a_n$ one after another. It is described as follows: We have

$$(5.1.4) \qquad \frac{p(z)}{q(z)} = \frac{a_m}{b_0} z^m + \frac{p(z) - (a_0/b_0) z^m q(z)}{q(z)},$$

where the fraction on the right has the same denominator as the original fraction, but the numerator has changed and is now a power series with lowest order term of degree at least $m+1$, whereas the original numerator had lowest order term of degree at least $m$. If we write out explicitly the polynomials involved, then 5.1.4 becomes

$$\begin{aligned}\frac{p(z)}{q(z)} &= \frac{a_m z^m + a_{m+1}z^{m+1} + a_{m+2}z^{m+2} + \cdots}{b_0 + b_1 z + b_2 z^2 + \cdots} \\ &= \frac{a_m}{b_0} z^m + \frac{(a_{m+1} - b_1 a_m/b_0)z^{m+1} + (a_{m+2} - b_2 a_m/b_0)z^{m+2} + \cdots}{b_0 + b_1 z + b_2 z^2 + \cdots}.\end{aligned} \tag{5.1.5}$$

This tells us that the first coefficient in (5.1.3) is

$$c_m = a_m/b_0. \tag{5.1.6}$$

The next coefficient $c_{m+1}$ is obtained by repeating the procedure on the new fraction that appears on the right in the above equation. This procedure may be repeated as often as is needed to obtain a given coefficient $c_n$. For example, in Exercises 5.1.16 and 5.1.17 you are asked to use this procedure to show that

$$c_{m+1} = \frac{a_{m+1}}{b_0} - \frac{a_m b_1}{b_0^2} \tag{5.1.7}$$

and

$$c_{m+2} = \frac{a_{m+2}}{b_0} - \frac{a_{m+1}b_1}{b_0^2} + \frac{a_m b_1^2}{b_0^3} - \frac{a_m b_2}{b_0^2}. \tag{5.1.8}$$

**Residue of a Quotient.** We now return to the problem of finding the residue at $z_0$ of a quotient, $f = p/h$, where $h$ has a zero of order $k$ at $z_0$. We define $q$ and $g = p/q$ as in (5.1.1). It follows from Theorem 5.1.1 that:

**Theorem 5.1.5.** *Suppose $f(z) = \dfrac{p(z)}{h(z)}$, where $p(z)$ and $h(z)$ are analytic in a neighborhood of $z_0$ and $h$ has a zero of order $k$ at $z_0$. If we write $h(z) = (z-z_0)^k q(z)$ where $q$ is analytic in a neighborhood of $z_0$ and $q(0) \neq 0$, then*

$$\operatorname{Res}(f, z_0) = c_{k-1},$$

*where $c_{k-1}$ is the coefficient of the degree $k-1$ term in the power series expansion of $g = p/q$ about $z_0$. The numbers $c_{k-1}$, may be computed using long division on the quotient $p/q$; that is, they may be computed through repeated use of* (5.1.5).

For small $k$, the expression for $c_{k-1}$, given by carrying out the method described in the above theorem, is reasonably simple. However, the complexity increases rapidly with increasing $k$.

When $k = 1$, the above theorem and (5.1.6), with $m = 0$, say that if

$$f(z) = \frac{p(z)}{h(z)} = \frac{p(z)}{(z-z_0)q(z)},$$

with $q(z_0) \neq 0$, then

$$\operatorname{Res}(f, z_0) = c_0 = \frac{a_0}{b_0} = \frac{p(z_0)}{q(z_0)}.$$

This is just Corollary 5.1.2(a). We can restate it in terms of the original functions $p$ and $h$ as follows: since $h(z) = (z - z_0)q(z)$, we have

$$q(z_0) = h'(z_0)$$

and so

**Corollary 5.1.6.** *Let $p$ and $h$ be analytic in a neighborhood of $z_0$, where $h$ has a zero of order 1 at $z_0$. Then*

$$\operatorname{Res}(p/h, z_0) = \frac{p(z_0)}{h'(z_0)}.$$

**Example 5.1.7.** Find $\operatorname{Res}(f, 0)$ if

$$f(z) = \frac{\mathrm{e}^z}{\sin z}.$$

**Solution:** Since $\sin z$ has a zero of order 1 at $z = 0$, we may apply Corollary 5.1.6 with $p(z) = \mathrm{e}^z$ and $h(z) = \sin z$. Then

$$\operatorname{Res}(f, 0) = \frac{\mathrm{e}^0}{\cos 0} = 1.$$

**Example 5.1.8.** Find $\operatorname{Res}(f, 0)$ if

$$f(z) = \frac{1}{\mathrm{e}^z - 1 - z}.$$

**Solution:** Here the denominator has a zero of order 2. We set $p(z) = 1$ and

$$h(z) = \mathrm{e}^z - 1 - z = \frac{z^2}{2} + \frac{z^3}{3!} + \cdots .$$

Then $h(z) = z^2 q(z)$ where

$$q(z) = \frac{1}{2} + \frac{z}{3!} + \cdots .$$

By Theorem 5.1.5, the residue we seek is the coefficient of the degree 1 term in the power series expansion of $1/q$, and this may be obtained by long division. In fact, we have

$$\frac{1}{1/2 + z/3! + \cdots} = 2 - \frac{z/3 + \cdots}{1/2 + z/3! + \cdots} = 2 - (2/3)z + \cdots ,$$

and so the residue is $-2/3$.

**Example 5.1.9.** Find the residue at 0 of $f(z) = \dfrac{\cot z}{z^2}$.

**Solution:** We proceed as in Theorem 5.1.5. The function $f$ has a pole of order 3 at 0. We set $p(z) = \cos z$, $h(z) = z^2 \sin z$, and

$$q(z) = \frac{h(z)}{z^3} = \frac{\sin z}{z} = 1 - \frac{z^2}{6} + \frac{z^4}{120} - \cdots .$$

We then use long division to find the first three power series coefficients, $c_0$, $c_1$ and $c_2$ of $p/q$. By Theorem 5.1.5, the residue we seek will then be the coefficient $c_2$. We have,

$$\begin{aligned}\frac{p(z)}{q(z)} = \frac{\cos z}{z^{-1}\sin z} &= \frac{1 - z^2/2 + \cdots}{1 - z^2/6 + \cdots} \\ &= 1 + \frac{(-1/2 + 1/6)z^2 + \cdots}{1 - z^2/6 + \cdots} \\ &= 1 + \frac{-(1/3)z^2 + \cdots}{1 - z^2/6 + \cdots} \\ &= 1 - (1/3)z^2 + \cdots .\end{aligned}$$

Thus, $c_0 = 1$, $c_1 = 0$, and $c_2 = -1/3$. We conclude that $\operatorname{Res}(f, 0) = -1/3$.

---

## Exercise Set 5.1

1. Find $\operatorname{Res}(f,0)$ and $\operatorname{Res}(f,1)$ if $f(z) = \dfrac{z+2}{z(z-1)(z+4)}$.
2. Find $\operatorname{Res}(f,0)$ and $\operatorname{Res}(f,2)$ if $f(z) = \dfrac{\cos z}{2z - z^2}$.
3. Find $\operatorname{Res}(f,0)$ if $f(z) = e^{1/z}$.
4. Find $\operatorname{Res}(f,0)$ if $f(z) = \cot z$.
5. Find $\operatorname{Res}(f,\pi)$ if $f(z) = \cot z$.
6. Find $\operatorname{Res}(f,0)$ if $f(z) = \cot^2 z$.
7. Find $\operatorname{Res}(f,0)$ if $f(z) = \dfrac{e^z}{\cos z - 1}$.
8. Find $\operatorname{Res}(f,0)$ if $f(z) = \dfrac{1}{z^2 \sin z}$.
9. Find $\operatorname{Res}(f,0)$ if $f(z) = \dfrac{1}{z - \log(1+z)}$.
10. Find $\operatorname{Res}(f,0)$ if $f(z) = \dfrac{\sinh z}{z^4}$.
11. Find $\operatorname{Res}(f,i)$ if $f(z) = \dfrac{(\log z)^3}{z^2+1}$.
12. Use long division of power series to find the power series expansion about 0 of $\tan z = \dfrac{\sin z}{\cos z}$ through terms of degree five.
13. Prove that if $f$ has a simple pole at $z_0$ and $g$ is analytic in a neighborhood of $z_0$, then $\operatorname{Res}(gf, z_0) = g(z_0)\operatorname{Res}(f, z_0)$.
14. Show by example that it is not true that $\operatorname{Res}(gf, z_0) = g(z_0)\operatorname{Res}(f, z_0)$ if $g$ is analytic in a neighborhood of $z_0$ and $f$ has a pole of degree greater than 1 at $z_0$.
15. Prove that a meromorphic function which is even and has an isolated singularity at 0 has residue 0 at 0.
16. Derive formula (5.1.7).
17. Derive formula (5.1.8).

---

## 5.2. Evaluating Integrals Using Residues

The Residue Theorem (Theorem 4.4.3) tells us that if $f$ is a function which is analytic in a set $U$, except at a discrete set of isolated singularities, then we can evaluate its integral around a closed path $\gamma$ in $U$, which is homologous to 0 in $U$ and does not hit any of these singularities, provided we can find the residues of $f$ at singularities where $\gamma$ has non-zero index. As we shall see, this turns out to be a very practical method for evaluating a wide variety of integrals.

**Sines and Cosines.** Because of the identities

$$\cos\theta = \frac{e^{i\theta} + e^{-i\theta}}{2}, \quad \sin\theta = \frac{e^{i\theta} - e^{-i\theta}}{2i},$$

the cosine and sine functions may be regarded as the restrictions to the unit circle (parameterized by $z = e^{i\theta}, 0 \le \theta \le 2\pi$) of the meromorphic functions

$$\frac{z + z^{-1}}{2} \quad \text{and} \quad \frac{z - z^{-1}}{2i}.$$

This means that an integral of the form

$$\int_0^{2\pi} f(\theta)\, d\theta, \tag{5.2.1}$$

where $f$ is a rational expression in $\cos\theta$ and $\sin\theta$, may be reformulated as a path integral around the unit circle of a certain meromorphic function of $z$. We simply let $g(z)$ be the function of $z$ obtained by replacing the functions $\cos\theta$ and $\sin\theta$, which make up $f$, with the functions $(z + z^{-1})/2$ and $(z - z^{-1})/2i$. It will then be true that

$$g(e^{i\theta}) = f(\theta).$$

With $\gamma(\theta) = e^{i\theta}$, we then have

$$\int_\gamma \frac{g(z)}{iz}\, dz = \int_0^{2\pi} f(\theta)\, d\theta. \tag{5.2.2}$$

The integral on the left is $2\pi i$ times the sum of the residues of $g(z)/iz$ inside the unit circle, provided $g(z)/iz$ has no poles on the unit circle itself.

**Example 5.2.1.** Find $\displaystyle\int_0^{2\pi} \frac{d\theta}{2 + \sin\theta}$.

**Solution:** If we replace $\sin\theta$ in the integrand by $(z - z^{-1})/2i$, the resulting function is

$$\frac{1}{2 + (z - z^{-1})/2i} = \frac{2iz}{4iz + z^2 - 1}.$$

This is the $g$ of equation (5.2.2). We divide this by $iz$ and integrate around the unit circle to get our answer. Thus, we need to evaluate the integral

$$\int_{|z|=1} \frac{2}{z^2 + 4iz - 1}\, dz = \int_{|z|=1} \frac{2}{(z + (2 - \sqrt{3})i)(z + (2 + \sqrt{3})i)}\, dz.$$

Here we have factored $z^2 + 4iz - 1$ by finding its roots using the Quadratic Formula. The only pole of the integrand that is inside the unit circle is the one at the point

$(-2+\sqrt{3})i$. The residue at this pole is found using Corollary 5.1.2(a). That is, we simply evaluate the function

$$\frac{2}{z+(2+\sqrt{3})i}$$

at the pole $z=(-2+\sqrt{3})i$. The resulting residue is

$$\frac{2}{2\sqrt{3}\,i}=-\frac{\sqrt{3}}{3}i.$$

By the Residue Theorem, the integral we seek is this number times $2\pi i$. Thus,

$$\int_0^{2\pi}\frac{d\theta}{2+\sin\theta}=\frac{2\sqrt{3}}{3}\pi.$$

This technique may be used to evaluate any integral of the form (5.2.1), where $f(\theta)$ is a rational function of $\sin\theta$ and $\cos\theta$ with a denominator which does not vanish for $\theta\in[0,2\pi]$.

**Improper Integrals.** Improper integrals of the form

$$\int_{-\infty}^{\infty} f(t)\,dt \tag{5.2.3}$$

can often be evaluated using residues. Here we are assuming that $f$ is a function which is Riemann integrable on each finite subinterval $[a,b]$ of $\mathbb{R}$.

There are two senses in which such an integral may converge. If the expression

$$\int_{-y}^{x} f(t)\,dt$$

has a limit as $x$ and $y$ approach $+\infty$ independently, then the improper integral is said to converge. This means, given $\epsilon>0$, there is an $M$ such that the integral is within $\epsilon$ of the limit whenever $x\ge M$ and $y\ge M$. This is equivalent to the convergence of the two one-sided improper integrals

$$\int_0^{\infty} f(t)\,dt \quad\text{and}\quad \int_{-\infty}^{0} f(t)\,dt.$$

On the other hand, if the limit of the symmetric integral,

$$\lim_{x\to\infty}\int_{-x}^{x} f(t)\,dt,$$

exists, then the integral (5.2.3) is said to converge in the *principal value* sense and the above limit is called the *principal value* of the integral. This is weaker than ordinary convergence of the integral – that is, (5.2.3) may converge in the principal value sense and not converge in the ordinary sense.

An improper integral (5.2.3) is said to converge absolutely if the integral of $|f|$ converges. This implies the convergence of (5.2.3) and, in fact, we have the following integral analogue of the comparison test for convergence of series.

**Theorem 5.2.2.** *If $f$ and $g$ are continuous functions on the line, with $g(t) \geq 0$ and $|f(t)| \leq g(t)$ for all $t \in \mathbb{R}$, then the integral of $f$ on $\mathbb{R}$ converges and*

$$\left| \int_{-\infty}^{\infty} f(t)\, dt \right| \leq \int_{-\infty}^{\infty} g(t)\, dt.$$

*provided the integral of $g$ on $\mathbb{R}$ converges.*

**Proof.** Assuming the integral of $f$ on $\mathbb{R}$ converges, we will prove the convergence of the improper integral $\int_0^\infty f(t)\, dt$. The convergence of $\int_{-\infty}^0 f(t)\, dt$ follows from the same argument. We will use the fact that, if a function $h$ on $(0, \infty)$ has the property that $\lim_{n\to\infty} h(x_n)$ exists for every increasing sequence $\{x_n\}$ converging to infinity, then the limits of all such sequences are the same number $L$ and $\lim_{x\to\infty} h(x) = L$ (Exercise 5.2.11).

Let $\{x_k\}_{k=1}^\infty$ be an increasing sequence of positive numbers converging to infinity. We set $x_0 = 0$ and, for $k \geq 1$, set

$$a_k = \int_{x_{k-1}}^{x_k} f(t)\, dt, \quad b_k = \int_{x_{k-1}}^{x_k} g(t)\, dt.$$

Then the hypothesis that $|f(t)| \leq g(t)$ implies that $|a_k| \leq b_k$ for all $n$. Furthermore, for each positive integer $n$,

$$\int_0^{x_n} f(t)\, dt = \sum_{k=1}^{n} a_k \quad \text{and} \quad \int_0^{x_n} g(t)\, dt = \sum_{k=1}^{n} b_k.$$

Thus, the convergence of the improper integral of $g(t)$ implies the convergence of the series of positive terms $\sum_{k=1}^\infty b_k$. This, in turn, implies the absolute convergence of the series $\sum_{k=1}^\infty a_k$ and the inequality

$$\left| \sum_{k=1}^{\infty} a_k \right| \leq \sum_{k=1}^{\infty} b_k. \tag{5.2.4}$$

This means that, if $h(x) = \int_0^x f(t)\, dt$, then the sequence $\{h(x_n)\}$ has a limit for every increasing sequence $\{x_n\}$ converging to infinity. By the remark in the first paragraph, $\lim_{x\to\infty} h(x)$ exists and is the common limit of all the sequences $\{h(x_n)\}$. Hence the improper integral of $f$ exists and is, by definition, this number $L$.

It follows from (5.2.4) that

$$\left| \int_0^{\infty} f(t)\, dt \right| = |L| \leq \sum_{k=1}^{\infty} b_k = \int_0^{\infty} g(t)\, dt.$$

The same argument can be used to show that $\int_{-\infty}^0 f(t)\, dt$ exists and satisfies the inequality

$$\left| \int_{-\infty}^{0} f(t)\, dt \right| \leq \int_{-\infty}^{0} g(t)\, dt.$$

The theorem follows on combining these results. □

If a function $f$ is Riemann integrable on each finite subinterval of the real line, and its improper integral over $(-\infty, \infty)$ converges absolutely, then we will say that $f$ is *absolutely integrable* on $(-\infty, \infty)$.

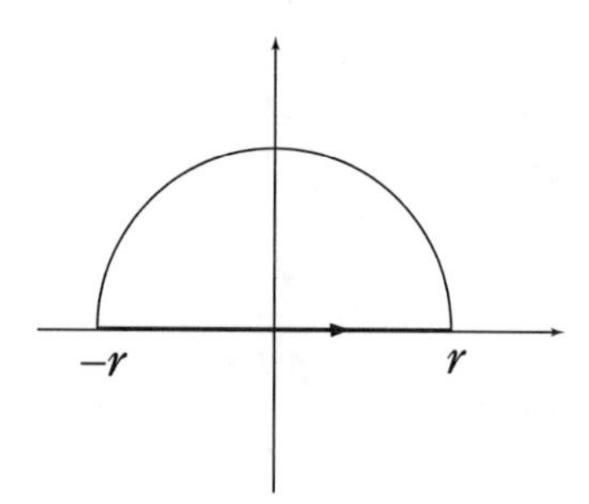


**Figure 5.2.1.** The Path $\gamma_r$.

**Evaluating Integrals on the Line.** Often the integrand $f$ of an integral of the form (5.2.3) is the restriction to the real line of a function which is analytic, except for a discrete set of singularities, on an open set containing either the closed upper half-plane $\{z : \text{Im}(z) \geq 0\}$ or the closed lower half-plane $\{z : \text{Im}(z) \leq 0\}$. If this function has no singularities on the real line and it decreases rapidly enough as $z \to \infty$, then the integral can be approximated arbitrarily closely by the integral of $f$ around a semicirclular closed path in the appropriate half-plane. Such an integral can be evaluated using residues.

Specifically, suppose $f$ is analytic in $U$, except on a discrete subset of $U$, where $U$ is an open set containing the closed upper half-plane. Suppose also that $p$, $R$ and $C$ are positive numbers, with $p > 1$, such that, for $\text{Im}(z) \geq 0$,

$$|f(z)| \leq C|z|^{-p} \quad \text{when} \quad |z| > R. \tag{5.2.5}$$

This ensures that the improper integral (5.2.3) converges (Exercise 5.2.12).

Let $\gamma_r$ be the path which begins at $-r$ on the real line, traverses the interval $[-r, r]$ and then returns to $-r$ along the semicircle which is the upper half of the circle $|z| = r$ (see Figure 5.2.1).

We have

$$\int_{\gamma_r} f(z)\, dz = \int_{-r}^{r} f(x)\, dx + \int_0^{\pi} f(r\,e^{it}) ir\,e^{it}\, dt. \tag{5.2.6}$$

If $r > R$, then (5.2.5) implies that $|f(r\,e^{it}) ir\,e^{it}\,| \leq Cr^{1-p}$ and so

$$\left| \int_0^{\pi} f(r\,e^{it}) ir\,e^{it}\, dt \right| \leq \pi C r^{1-p}.$$

Since $p > 1$, the right side of this inequality has limit 0 as $r \to \infty$. We conclude from this and (5.2.6) that

$$\lim_{r\to\infty} \int_{\gamma_r} f(z)\, dz = \lim_{r\to\infty} \int_{-r}^{r} f(x)\, dx = \int_{-\infty}^{\infty} f(x)\, dx.$$

Now, by the Residue Theorem, $\int_{\gamma_r} f(z)\, dz$ is $2\pi i$ times the sum of the residues of $f$ at singularities inside $\gamma_r$. The condition (5.2.5) ensures that there are no singularities of $f$ in the upper half-plane outside the circle $|z| = R$. This implies there are only finitely many singularieties in the upper half-plane. It also implies that the integral is independent of $r$, for $r > R$, and is equal to $2\pi i$ times the sum of the residues of $f$ at this finite set of singularities.

If $f$ is analytic in an open set containing the lower half-plane, then all of the above still holds with the following changes: the lower half-plane replaces the upper half-plane, the path $\gamma_r$ is replaced by its reflection through the $x$-axis, and, since the new path has negative orientation, the resulting integral is $-2\pi i$ times the sum of the residues of $f$ in the lower half-plane.

This proves the following theorem:

**Theorem 5.2.3.** *Let $H$ be either the closed upper or lower half-plane. If $f$ is analytic, except at a discrete set of singularities, in an open set containing $H$, has no singularities on $\mathbb{R}$, and there are positive numbers $R$, $C$, and $p > 1$ such that (5.2.5) is satisfied for $z \in H$, then*

$$\int_{-\infty}^{\infty} f(x)\,dx = \sigma(H)2\pi i \sum_{j=1}^{m} \mathrm{Res}(f, z_j),$$

*where $z_1, z_2, \cdots, z_m$ are the singularities of $f$ in the half-plane $H$, and $\sigma(H) = 1$ if $H$ is the upper half-plane and $-1$ if $H$ is the lower half-plane.*

**Example 5.2.4.** Find $\displaystyle\int_{-\infty}^{\infty} \frac{x^2}{1+x^4}\,dx$.

**Solution:** We use the previous theorem with

$$f(z) = \frac{z^2}{1+z^4} = \frac{z^{-2}}{z^{-4}+1}.$$

If we fix an $R > 1$, we have

$$|f(z)| \le \frac{|z|^{-2}}{1-R^{-4}} \quad \text{if} \quad |z| \ge R,$$

and so condition (5.2.5) is satisfied with $p = 2$ and $C = (1 - R^{-4})^{-1}$.

The poles of $f$ occur at the 4th roots of $-1$, and these are

$$z_1 = \frac{\sqrt{2}}{2}(1+i), \quad z_2 = \frac{\sqrt{2}}{2}(-1+i),$$
$$z_3 = \frac{\sqrt{2}}{2}(-1-i), \quad z_4 = \frac{\sqrt{2}}{2}(1-i).$$

Thus

$$f(z) = \frac{z^2}{(z-z_1)(z-z_2)(z-z_3)(z-z_4)}.$$

Only $z_1$ and $z_2$ are in the upper half-plane. We evaluate the residue of $f$ at each of these points using Corollary 5.1.2(a). This means, for $j = 1, 2$, we evaluate at $z_j$ what is left of $f$ when $z - z_j$ is removed from the denominator. The results are

$$\mathrm{Res}(f, z_1) = \sqrt{2}\frac{(1+i)^2}{(2)(2+2i)(2i)} = -\frac{\sqrt{2}}{8}(1+i),$$
$$\mathrm{Res}(f, z_2) = \sqrt{2}\frac{(-1-i)^2}{(-2)(2i)(-2+2i)} = \frac{\sqrt{2}}{8}(1-i).$$

The sum of these is $-\frac{\sqrt{2}}{4}i$ and, when multiplied by $2\pi i$, this yields

$$\int_{-\infty}^{\infty} \frac{x^2}{1+x^4}\,dx = \frac{\sqrt{2}}{2}\pi.$$

**Symmetries.** It can be useful to exploit symmetries of the integrand in computing integrals using residues.

**Example 5.2.5.** Find $\displaystyle\int_0^{\infty} \frac{1}{(1+x^2)^2}\,dx$.

**Solution:** Since the integrand is an even function, its integral over the entire real line is twice its integral from 0 to $\infty$. Thus,

$$\int_0^{\infty} \frac{1}{(1+x^2)^2}\,dx = \frac{1}{2}\int_{-\infty}^{\infty} \frac{1}{(1+x^2)^2}\,dx.$$

We evaluate the latter integral using Theorem 5.2.3. We set

$$f(z) = \frac{1}{(1+z^2)^2} = \frac{1}{(z-i)^2(z+i)^2}.$$

The only pole in the upper half-plane is the one at $i$, and this is a pole of order 2. Corollary 5.1.2(b) applies, and it tells us that the residue of $f$ at this pole is obtained by differentiating

$$g(z) = \frac{1}{(z+i)^2},$$

and evaluating at $z = i$. The result is

$$g'(i) = \frac{-2}{(2i)^3} = -\frac{i}{4}.$$

Thus,

$$\int_0^{\infty} \frac{1}{(1+x^2)^2}\,dx = \frac{1}{2}\cdot 2\pi i \cdot \frac{-i}{4} = \frac{\pi}{4}.$$

In the next example, we exploit the fact that the integrand is unchanged if we rotate coordinates by an angle $\pi/n$.

**Example 5.2.6.** Find $\displaystyle\int_0^{\infty} \frac{1}{1+x^{2n}}\,dx$.

**Solution:** The integrand is even, so we could just proceed as in the preceding example, replacing this integral by one over all of $\mathbb{R}$ and then applying Theorem 5.2.3. However, this would mean evaluating residues at each of the $n$ poles in the upper half-plane. There is a better way. We set

$$f(z) = \frac{1}{1+z^{2n}}$$

and note that

$$f(e^{\pi i/n} z) = f(z) \tag{5.2.7}$$

for each $z \in \mathbb{C}$. This means $f$ has the same values along the ray $\{e^{\pi i/n}x : 0 \le x\}$ as it does along the ray $[0, +\infty)$. To exploit this, we choose $r > 1$ and use the path $\gamma_r$ which traverses the boundary of the sector indicated in Figure 5.2.2. That is,

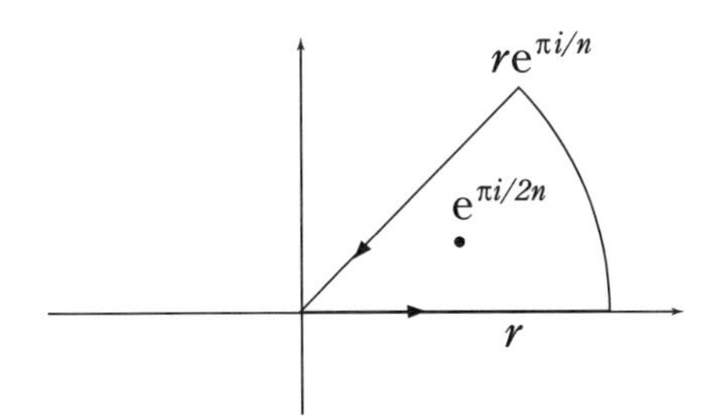


**Figure 5.2.2.** Path for Example 5.2.6.

the path $\gamma_r$ traverses the interval $[0, r]$, then the arc $r\,e^{it}, 0 \le t \le \pi/n$, and then returns to 0 along the interval $[r\,e^{\pi i/n}, 0]$.

Then

$$\begin{aligned}\int_{\gamma_r} f(z)\,dz &= \int_0^r f(x)\,dx + \int_0^{\pi/n} f(r\,e^{it})ri\,e^{it}\,dt - \int_0^r f(e^{\pi i/n}\,x)\,e^{\pi i/n}\,dx\\ &= (1 - e^{\pi i/n}) \int_0^r f(x)\,dx + \int_0^{\pi/n} f(r\,e^{it})ri\,e^{it}\,dt.\end{aligned}$$

The second integral on the right clearly goes to 0 as $r \to \infty$ and the integral on the left is $2\pi\,\mathrm{Res}(f, z_1)$, where $z_1 = e^{\pi i/2n}$ is the only pole of $f$ inside $\gamma_r$. Thus,

$$\int_0^\infty f(x)\,dx = 2\pi i \frac{\mathrm{Res}(f, z_1)}{1 - e^{\pi i/n}}.$$

All that remains is to evaluate this residue.

Since $z_1^{2n} = -1$, we have

$$z^{2n} + 1 = z^{2n} - z_1^{2n} = (z - z_1)(z^{2n-1} + z^{2n-2}z_1 + \cdots + z_1^{2n-1}).$$

Corollary 5.1.2(a) says that, to find the residue of $f$ at $z_1$, we simply evaluate the inverse of the second factor on the right at $z = z_1$. The result is

$$\mathrm{Res}(f, z_1) = \frac{1}{2nz_1^{2n-1}} = -\frac{z_1}{2n} = -\frac{e^{\pi i/2n}}{2n},$$

and this implies

$$\int_0^\infty \frac{1}{1 + x^{2n}}\,dx = -\frac{2\pi i}{2n}\frac{e^{\pi i/2n}}{1 - e^{\pi i/n}} = \frac{\pi/2n}{\sin(\pi/2n)}.$$

## Exercise Set 5.2

1. Find $\displaystyle\int_0^{2\pi} \frac{d\theta}{5 - 4\cos\theta}$.

2. Find $\displaystyle\int_0^{2\pi} \frac{d\theta}{10+6\sin\theta}$.
3. Find $\displaystyle\int_{-\infty}^{\infty} \frac{1}{x^2+2x+2}\, dx$.
4. Find $\displaystyle\int_0^{\infty} \frac{x^2}{(1+x^2)^2}\, dx$.
5. Find $\displaystyle\int_{-\infty}^{\infty} \frac{x}{(x^2+2x+2)^2}\, dx$
6. Find $\displaystyle\int_0^{\infty} \frac{1}{1+x^6}\, dx$ (see Example 5.2.6).
7. Find $\displaystyle\int_0^{\infty} \frac{1}{1+x^3}\, dx$ using contour integration over the boundary of the sector defined in polar coordinates by $\{r\,e^{i\theta} : 0 \le r \le R,\ 0 \le \theta \le 2\pi/3\}$ with $R > 1$ (see Example 5.2.6).
8. Find $\displaystyle\int_{-\infty}^{\infty} \frac{e^{ix}}{1+x^2}\, dx$.
9. Find $\displaystyle\int_{-\infty}^{\infty} \frac{\cos(x)}{(1+x^2)^2}\, dx$ by first finding $\displaystyle\int_{-\infty}^{\infty} \frac{e^{ix}}{(1+x^2)^2}\, dx$ and and then taking the real part of the result.
10. Find $\displaystyle\int_0^{\infty} \frac{\cos x}{1+x^2}\, dx$.
11. Prove the fact, used in the proof of Theorem 5.2.2, that if $h$ is a function on $(0,\infty)$ and $\lim_{k\to\infty} h(x_k)$ exists for every increasing sequence $\{x_k\}$ of positive numbers converging to $\infty$, then these limits are all the same number $L$ and $\lim_{x\to\infty} h(x) = L$.
12. Prove that an improper integral of the form (5.2.3) converges if $f$ is a continuous function on $\mathbb{R}$ which satisfies an inequality $|f(x)| \le C|x|^{-p}$ for $|x| > R$, where $C$ and $R$ are positive constants, and $p > 1$.
13. Suppose $f$ is a continuous function on $(0,\infty)$ (which may have a singularity at 0). Show how to use the substitution $t = e^s$ to convert the improper integral $\int_0^\infty f(t)\, dt$ into an improper integral on $(-\infty,\infty)$. Use this to prove a theorem analogous to Theorem 5.2.2 for improper integrals on $(0,\infty)$.
14. Find $\displaystyle\int_0^{\infty} \frac{dt}{t+t\log^2 t}$. Hint: Use the substitution of the previous exercise.

## 5.3. Fourier Transforms

The technique of evaluating integrals using residue theory is particularly useful in the study of integral transforms such as the Fourier, Laplace, and Mellin transforms. Often the integrals involved cannot be evaluated using standard integration techniques from calculus.

**The Fourier Transform.** The Fourier transform is widely used in Differential Equations, Physics, and Engineering. Its generalizations are the subject of a major area of mathematical research, with applications that impact every branch of mathematics.

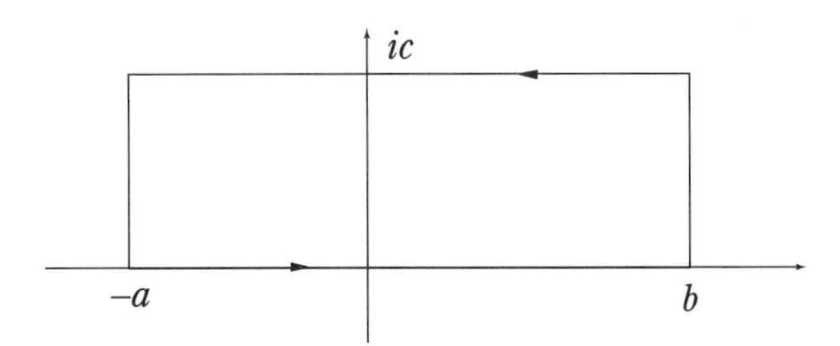


**Figure 5.3.1.** Path $\gamma$ of Theorem 5.3.2.

**Definition 5.3.1.** If $f$ is a function defined on the real line and $t \in \mathbb{R}$, we set

$$f\hat{}(t) = \frac{1}{\sqrt{2\pi}} \int_{-\infty}^{\infty} f(x)\, e^{-itx}\, dx,$$

whenever this improper integral exists. The resulting function of $t$ is called the *Fourier transform* of $f$.

Note that if $f(x)$ is an absolutely integrable function of $x$ on $(-\infty, \infty)$, then so is $e^{-itx} f(x)$ for each $t$, since these two functions have the same absolute value. Thus, the Fourier transform $f\hat{}(t)$, for such a function, exists at every $t \in (-\infty, \infty)$.

We will show how to use residue theory to evaluate Fourier transforms of functions that are analytic on $\mathbb{C}$, except at a discrete set of singularities, and that vanish at infinity. The key to this technique is the fact that $|\, e^{-itz}\, | = e^{ty}$ vanishes rapidly as $y = \mathrm{Im}(z) \to \infty$ if $t < 0$, while it vanishes rapidly as $y \to -\infty$ if $t > 0$. In the first case, we are able to express the integral defining $f\hat{}$ as a sum of residues in the upper half-plane, and in the second case we express it as a sum of residues in the lower half-plane. The final result is the following theorem.

**Theorem 5.3.2.** *Let $f$ be a function analytic on $\mathbb{C}$, except at a discrete set of singularities, with no singularities on the real line and with $\lim_{z\to\infty} f(z) = 0$. Then $f\hat{}(t)$ exists if $t \neq 0$. It is equal to $\sqrt{2\pi}\, i$ times the sum of the residues of $f(z)\, e^{-izt}$ in the upper half-plane if $t < 0$, and is equal to $-\sqrt{2\pi}\, i$ times the sum of the residues of $f(z)\, e^{-izt}$ in the lower half-plane if $t > 0$.*

**Proof.** We give the proof in case $t < 0$. The case $t > 0$ is the same except that the lower half-plane is used instead of the upper half-plane.

Since $\lim_{z\to\infty} f(z) = 0$, $f$ has only finitely many singularities. For positive numbers $a, b$, and $c$ let $\gamma$ be the path which traverses the boundary of the rectangle with vertices $-a, b, b + ic, -a + ic$ in the positive direction (see Figure 5.3.1).

If $a, b$, and $c$ are large enough, all singularities of $f$ in the upper half-plane will be inside $\gamma$. In particular, there will be no singularities on $\gamma(I)$. Then,

$$\begin{aligned} \int_\gamma f(z)\, e^{-izt}\, dz = &\int_{-a}^{b} f(x)\, e^{-itx}\, dx + i \int_0^c f(b+iy)\, e^{-it(b+iy)}\, dy \\ &- \int_{-a}^{b} f(x+ic)\, e^{-it(x+ic)}\, dx - i \int_0^c f(-a+iy)\, e^{-it(-a+iy)}\, dy. \end{aligned} \tag{5.3.1}$$

We pass to the limit, first as $c \to \infty$, and then as $a$ and $b$ approach $\infty$ independently. We will show that each of the last three integrals on the right in (5.3.1) vanishes when we do this.

Since $|f(z)|$ is a continuous function on the complement of its set of singularities, and it has limit 0 at $\infty$, it has a maximum value on each line which does not hit one of the singularities. Let $M_1$ denote its maximum on the vertical line $\text{Re}(z) = b$, $M_2$ its maximum on the horizontal line $\text{Im}(z) = c$, and $M_3$ its maximum on the vertical line $\text{Re}(z) = -a$. These numbers are functions of $b, c$, and $a$, respectively, and they each have limit 0 as $a, b$, and $c$ approach $\infty$, because $\lim_{z\to\infty} f(z) = 0$.

For the third integral on the right in (5.3.1), we have

$$\left|\int_{-a}^{b} f(x+ic)\,\mathrm{e}^{-it(x+ic)}\,dx\right| \le M_2 \int_{-a}^{b} \mathrm{e}^{tc}\,dx = M_2\,\mathrm{e}^{tc}(b+a).$$

Since $t < 0$, this has limit 0 as $c \to \infty$ for fixed values of $a$ and $b$.

For the second integral on the right in (5.3.1) we have

$$\left|\int_{0}^{c} f(b+iy)\,\mathrm{e}^{-it(b+iy)}\,dy\right| \le M_1 \int_{0}^{c} \mathrm{e}^{ty}\,dy = \frac{M_1}{t}(\mathrm{e}^{tc} - 1).$$

This has limit $-M_1/t$ as $c \to \infty$. Then, since $M_1$ has limit 0 as $a, b$ approach $\infty$, the same is true of this integral. The argument that the fourth integral on the right in (5.3.1) has limit 0 as first $c$, then $a, b$ approach $\infty$ is the same.

Of course, the integral on the left in (5.3.1) does not change once $a, b$, and $c$ are large enough that the rectangle they determine contains all the singularities of $f$ in the upper half-plane. Once this is true, its value is $2\pi i$ times the sum of the residues at these singularities. Thus, after passing to the limit as first $c$, then $a$ and $b$ approach $\infty$, (5.3.1) becomes

$$\begin{aligned} 2\pi i \sum_{j=1}^{n} \text{Res}(f, z_j) &= \lim_{(a,b)\to(\infty,\infty)} \int_{-a}^{b} f(x)\,\mathrm{e}^{-itx}\,dx \\ &= \int_{-\infty}^{\infty} f(x)\,\mathrm{e}^{-itx}\,dx = \sqrt{2\pi}\,f\hat{}(t), \end{aligned} \tag{5.3.2}$$

where $z_1, \cdots, z_n$ are the singularities of $f$ in the upper half-plane. On dividing by $\sqrt{2\pi}$, the proof of the theorem is complete in the case $t < 0$.

Note that the existence of the limit in (5.3.2) is exactly what is meant by the convergence of the improper integral defining $f\hat{}(t)$.

The proof in the case $t < 0$ is almost identical. The difference is: the number $c$ is negative. Then the path $\gamma$ lies in the lower half-plane with negative rather that positive orientation. This results in the left side of (5.3.1) being $-2\pi i$ times the sum of the residues of $f\,\mathrm{e}^{-itz}$ in the lower half-plane for large enough $a, b$, and $-c$. □

**Example 5.3.3.** Find the Fourier transform of $f(x) = \dfrac{1}{1+x^2}$.

**Solution:** We set

$$f(z) = \frac{1}{1+z^2}.$$

If $t < 0$, the previous theorem applies, and it tells us that $f\hat{}(t)$ is $\sqrt{2\pi}\, i$ times the sum of the residues of $\mathrm{e}^{-itz} f(z)$ in the upper half-plane. Now $f$ has only one pole in the upper half-plane and that is at $i$. The residue of $\mathrm{e}^{-itz} f(z)$ at this point can be computed using Corollary 5.1.2(a), since

$$f(z) = \frac{1}{(z+i)(z-i)}.$$

The residue is obtained by evaluating $\mathrm{e}^{-itz}/(z+i)$ at $z = i$. The result is

$$\frac{\mathrm{e}^t}{2i},$$

and so

$$f\hat{}(t) = \sqrt{\frac{\pi}{2}}\, \mathrm{e}^t \quad \text{if} \quad t < 0.$$

For $t > 0$, we compute the residue of $\mathrm{e}^{-itz} f(z)$ at $-i$ and then multiply by $-\sqrt{2\pi}\, i$. The result is

$$f\hat{}(t) = \sqrt{\frac{\pi}{2}}\, \mathrm{e}^{-t} \quad \text{if} \quad t > 0.$$

At $t = 0$ we have

$$f\hat{}(0) = \frac{1}{\sqrt{2\pi}} \int_{-\infty}^{\infty} \frac{1}{1+x^2}\, dx = \sqrt{\frac{\pi}{2}},$$

by the case $n = 1$ of Example 5.2.6 (or simply by noting that arctan is an antiderivative for the integrand).

Putting the cases $t < 0$, $t > 0$, $t = 0$ together, we conclude that

$$f\hat{}(t) = \sqrt{\frac{\pi}{2}}\, \mathrm{e}^{-|t|} \quad \text{for all} \quad t \in \mathbb{R}.$$

**Example 5.3.4.** Show that $\displaystyle\int_{-\infty}^{\infty} \frac{\cos x}{1+x^2}\, dx = \pi/\mathrm{e}$.

**Solution:** The integral we seek is the real part of the integral

$$\int_{-\infty}^{\infty} \frac{\mathrm{e}^{-ix}}{1+x^2}\, dx,$$

which is $\sqrt{2\pi}$ times the Fourier transform of $\dfrac{1}{1+x^2}$ evaluated at $t = 1$. By the previous example, this is $\pi/e$.

**Example 5.3.5.** Find the Fourier transform of $f(x) = \dfrac{1}{x+i}$.

Solution: We set

$$f(z) = \frac{1}{z+i}.$$

This function is meromorphic, with no poles on the real axis, and it vanishes at infinity. Thus, Theorem 5.3.2 applies.

The only pole of $f$ occurs at $z = -i$. Since there are no poles in the upper half-plane, Theorem 5.3.2 tells us that $f\hat{}(t) = 0$ if $t < 0$.

For $t > 0$, $f\hat{}(t)$ is $-\sqrt{2\pi}\,i$ times the residue of $f(z)\,\mathrm{e}^{-itz}$ at $z = -i$. This residue is $\mathrm{e}^{-t}$ and so

$$f\hat{}(t) = \begin{cases} 0, & \text{if } t < 0; \\ -\sqrt{2\pi}\,i\,\mathrm{e}^{-t}, & \text{if } t > 0. \end{cases}$$

Note the jump discontinuity at $t = 0$.

What about the value of $f\hat{}$ at $t = 0$? We have

$$f\hat{}(0) = \frac{1}{\sqrt{2\pi}} \int_{-\infty}^{\infty} f(x)\,dx.$$

We can attempt to evaluate this using the fact that $f(z)$ has $\log(z + i)$ as an antiderivative, given any branch of the log function. If we use the principal branch of log, then the line $\mathrm{Im}(z) = i$ is contained in its domain of definition, and so the real line is contained in the domain on which $\log(z + i)$ is defined and has $f(z)$ as its derivative. We conclude that

$$\begin{aligned} \int_{-a}^{b} f(x)\,dx &= \log(b + i) - \log(-a + i) \\ &= \frac{1}{2}(\log(b^2 + 1) - \log(a^2 + 1)) + i(\arg(b + i) - \arg(-a + i)). \end{aligned}$$

The limit as $(a, b) \to (\infty, \infty)$ of this expression does not exist, since its real part approaches $+\infty$ as $b \to \infty$ with $a$ fixed and approaches $-\infty$ as $a \to \infty$ with $b$ fixed. Thus, the improper integral defining $f\hat{}(0)$ does not converge. However, the symmetric integral

$$\int_{-a}^{a} f(x)\,dx = i(\arg(a + i) - \arg(-a + i))$$

has limit $-\pi i$ as $a \to \infty$. In this case, we say that the integral has *principal value* $-\pi i$, even though it is not a convergent improper integral. If we define $f\hat{}(0)$ using this value for the integral, then we have

$$f\hat{}(0) = -\sqrt{\frac{\pi}{2}}\,i,$$

which is exactly half way between $\lim_{t\to 0^+} f\hat{}(t)$ and $\lim_{t\to 0^-} f\hat{}(t)$.

The next example involves a different technique – no residues – but the Cauchy theorems are still involved.

**Example 5.3.6.** Find the Fourier transform of the normal distribution function

$$h(x) = \frac{1}{\sqrt{2\pi}}\,\mathrm{e}^{-x^2/2}.$$

**Solution:** We have

$$h\hat{}(t) = \frac{1}{2\pi} \int_{-\infty}^{\infty} \mathrm{e}^{-(x^2+2itx)/2}\,dx.$$

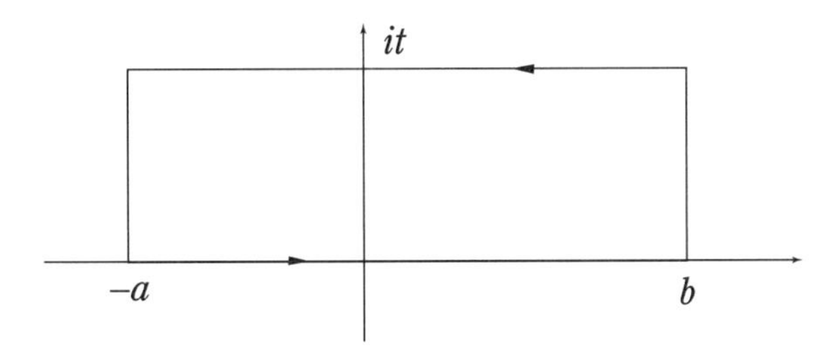


**Figure 5.3.2.** Path for the Fourier Transform in Example 5.3.6.

By completing the square in the exponent, we can write this as

$$h\hat{}(t) = \frac{1}{2\pi}\int_{-\infty}^{\infty} e^{-(x+it)^2/2}\, e^{-t^2/2}\, dx$$
(5.3.3)
$$= \frac{e^{-t^2/2}}{2\pi}\int_{-\infty}^{\infty} e^{-(x+it)^2/2}\, dx.$$

We will use Cauchy's Integral Theorem to show that the last integral above is independent of $t$.

Fix $a, b > 0$ and consider the path $\gamma$ that traces once in the positive direction around the boundary of the rectangle with vertices at $-a, b, b+ti, -a+ti$ (see Figure 5.3.2).

The integral of $e^{-z^2/2}$ over this path is 0 by Cauchy's Theorem. Hence,

$$0 = \int_{-a}^{b} e^{-x^2/2}\, dx + \int_0^t e^{-(b+iy)^2/2}\, i\, dy$$
(5.3.4)
$$- \int_{-a}^{b} e^{-(x+it)^2/2}\, dx - \int_0^t e^{-(-a+iy)^2/2}\, i\, dy.$$

The second of these integrals satisfies the following estimate:

$$\left|\int_0^t e^{-(b+iy)^2/2}\, i\, dy\right| \le |t|\, e^{y^2-b^2},$$

which implies that it has limit 0 as $b \to \infty$. A similar estimate on the fourth integral in (5.3.4) shows that it has limit 0 as $a \to 0$. It follows that

$$\int_{-\infty}^{\infty} e^{-(x+it)^2/2}\, dx = \int_{-\infty}^{\infty} e^{-x^2/2}\, dx,$$

which means that the integral on the left is independent of $t$ and can be evaluated by evaluating the integral on the right. The integral on the right is a standard calculus problem. Its value is $\sqrt{2\pi}$, and it is obtained by expressing the square of the integral as a double integral over the plane and then evaluating this using polar coordinates. We leave it as an exercise (Exercise 5.3.1).

In view of (5.3.3) and the above calculation, we have

$$h\hat{}(t) = \frac{e^{-t^2/2}}{2\pi}\sqrt{2\pi} = \frac{e^{-t^2/2}}{\sqrt{2\pi}} = h(t).$$

In other words, the normal distribution function is its own Fourier transform.

**The Fourier Inversion Formula.** We define the *inverse Fourier transform* $f^{\vee}$ of a function $f$ on $\mathbb{R}$ to be the Fourier transform of $f$ followed by reflection through the origin – that is:

$$f^{\vee}(x) = f^{\wedge}(-x) = \frac{1}{\sqrt{2\pi}} \int_{-\infty}^{\infty} e^{itx} f(t)\, dt. \tag{5.3.5}$$

Under reasonable conditions, at points $x$ where $f$ is continuous, we can recover $f(x)$ from the Fourier transform $f^{\wedge}$ by applying the inverse Fourier transform to $f^{\wedge}$. That is, at points of continuity for $f$,

$$f(x) = f^{\wedge\vee}(x) = \frac{1}{\sqrt{2\pi}} \int_{-\infty}^{\infty} e^{itx} f^{\wedge}(t)\, dt. \tag{5.3.6}$$

There are several versions of this result, depending on the hypotheses that are assumed satisfied by $f$ and $f^{\wedge}$. They all have rather technical proofs that are complicated by the fact that we are working with improper integrals. One fairly standard version assumes that both $f$ and $f^{\wedge}$ are absolutely integrable. A proof of this version can be found in [**8**]. We will not attempt to prove any of the versions of this result here. Instead, we will just illustrate the formula in a couple of examples.

**Example 5.3.7.** By Example 5.3.3 the function $f(x) = 1/(1+x^2)$ has $f^{\wedge}(t) = \sqrt{\pi/2}\, e^{-|t|}$ as Fourier transform. Compute the inverse Fourier transform of $f^{\wedge}$ and verify that the Fourier Inversion Formula holds in this case.

**Solution:** We have

$$\begin{aligned} f^{\wedge\vee}(x) &= \frac{1}{\sqrt{2\pi}} \int_{-\infty}^{\infty} e^{ixt} \sqrt{\pi/2}\, e^{-|t|}\, dt \\ &= \frac{1}{2}\left[\int_{-\infty}^{0} e^{(ix+1)t}\, dt + \int_{0}^{\infty} e^{(ix-1)t}\, dt\right] \\ &= \frac{1}{2}\left[\frac{1}{ix+1} - \frac{1}{ix-1}\right] = \frac{1}{x^2+1} = f(x). \end{aligned}$$

**Indented Contours.** The next example involves a technique in which we integrate only part of the way around a circle centered at a pole of the integrand. The theorem which allows us to compute such integrals is the following:

**Theorem 5.3.8.** *If $f$ is analytic in $D_R(0) \setminus \{0\}$, has a simple pole at $0$, and $\gamma$ is the path $\gamma(t) = r\, e^{it}$, $\theta_1 \le t \le \theta_2$, $r < R$, then*

$$\lim_{r \to 0} \int_{\gamma} f(z)\, dz = i(\theta_1 - \theta_2)\, \mathrm{Res}(f, 0).$$

**Proof.** Since $f$ has a simple pole at 0, its Laurent series expansion in $D_R(0)$ has the form

$$f(x) = \frac{a_{-1}}{z} + a_0 + a_1 z + \cdots + a_n z^n + \cdots,$$

where $a_{-1} = \operatorname{Res}(f, 0)$. The integral of the first term of this expansion over $\gamma$ is

$$\int_{\theta_1}^{\theta_2} ia_{-1}\, d\theta = i(\theta_1 - \theta_2) \operatorname{Res}(f, 0),$$

while the integral of the $n$th term for $n \geq 0$ is

$$\int_{\theta_1}^{\theta_2} ia_n r^{n+1}\, e^{i(n+1)\theta}\, d\theta,$$

which has absolute value less than or equal to $|a_n| r^{n+1}(\theta_2 - \theta_1)$. Since this has limit 0 as $r \to 0$, the proof is complete. □

**Example 5.3.9.** Given real numbers $a < b$, find the Fourier transform of the function $f$ which is equal to 1 for $a \leq x \leq b$ and is equal to 0 otherwise. Verify that the Fourier Inversion Formula holds in this case.

**Solution:** By the definition of $f$,

$$f\hat{}(t) = \frac{1}{\sqrt{2\pi}} \int_a^b e^{-itx}\, dx = \frac{i}{\sqrt{2\pi}} \frac{e^{-ibt} - e^{-iat}}{t}.$$

Then $f\hat{}\check{}$ is given by the improper integral

$$f\hat{}\check{}(x) = \frac{i}{2\pi} \int_{-\infty}^{\infty} \frac{e^{it(x-b)} - e^{it(x-a)}}{t}\, dt.$$

We may write this first as

$$\lim_{r \to 0} \frac{i}{2\pi} \left[ \int_{-\infty}^{-r} \frac{e^{it(x-b)} - e^{it(x-a)}}{t}\, dt + \int_r^{\infty} \frac{e^{it(x-b)} - e^{it(x-a)}}{t}\, dt \right],$$

and then as

$$(5.3.7) \qquad \begin{aligned} &\lim_{r \to 0} \frac{i}{2\pi} \left[ \int_{-\infty}^{-r} \frac{e^{it(x-b)}}{t}\, dt + \int_r^{\infty} \frac{e^{it(x-b)}}{t}\, dt \right], \\ -&\lim_{r \to 0} \frac{i}{2\pi} \left[ \int_{-\infty}^{-r} \frac{e^{it(x-a)}}{t}\, dt + \int_r^{\infty} \frac{e^{it(x-a)}}{t}\, dt \right], \end{aligned}$$

provided these limits exist.

We evaluate the two limits in (5.3.7) using an indented contour. We use the contour described in Figure 5.3.3 in the case where $x - b > 0$ in the first limit and where $x - a > 0$ in the second limit. Since the integrands are analytic except at 0, the integral around this contour is 0. As in the proof of Theorem 5.3.2, the integrals along the vertical lines and the horizontal line at $z = ic$ have limit 0 as $L, R$, and $c$ approach infinity. By the previous theorem, the integral around the semicircle has limit $i\pi$ as $r \to 0$. It follows that the first of the two limits in (5.3.7) is $1/2$ if $x > b$, while the second limit is $1/2$ if $x > a$.

For the cases $x < b$ and $x < a$ we use the reflection of the contour in Figure 5.3.3 through the $x$-axis. This has the result of changing the sign of the integral around the semicircle. Thus, the limits are $-1/2$ for the case $x < b$ and the case $x < a$.

Of course, when $x = b$, the first limit is $\lim_{r \to 0}(\ln r - \ln r) = 0$. The same result holds for the second limit when $x = a$.

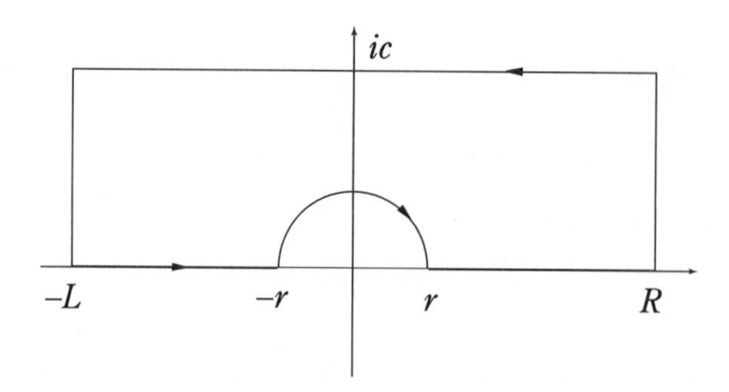


**Figure 5.3.3.** Indented Path Used in Example 5.3.9.

We conclude that $f^{\hat{}\check{}}(x) = 1$ if $a < x < b$ and is 0 otherwise. Note that this is equal to $f(x)$ except at the points $a$ and $b$ where $f$ is not continuous.

## Exercise Set 5.3

1. Prove that
$$\int_{-\infty}^{\infty} \mathrm{e}^{-x^2/2}\, dx = \sqrt{2\pi}$$
by expressing the square of the integral as a double integral over $\mathbb{R}^2$, which can then be evaluated using polar coordinates.
2. Find the Fourier transform of $f(x) = \dfrac{x}{1+x^2}$.
3. Find the Fourier transform of $f(x) = \dfrac{1}{1+x^4}$.
4. Find the Fourier transform of $f$ if $f(x) = 0$ for $x < 0$ and $f(x) = \sqrt{\dfrac{\pi}{2}}\, \mathrm{e}^{-x}$ for $x \geq 0$.
5. Find the Fourier transform of $f(x) = \dfrac{1}{x^2 + 4x + 5}$.
6. Find the Fourier transform of $f(x) = \dfrac{\sin x}{x}$ using the method of indented contours as in Example 5.3.9.
7. If $f$ and $f'$ are both continuous and have Fourier transforms which exist at $t$, and if $f$ has limit 0 at both $\infty$ and $-\infty$, prove that the Fourier transform of $f'$ at $t$ is $itf\hat{}(t)$. Hint: Use integration by parts.
8. Find $\displaystyle\int_{-\infty}^{\infty} \frac{x \sin x}{1+x^2}\, dx$.
9. If $h$ is the normal distribution function, $\lambda > 0$, and $h_\lambda(x) = h(x/\lambda)$, then find $(h_\lambda)\hat{}(t)$.
10. Verify that the Fourier Inversion Formula holds for the function $h_\lambda$ of the previous exercise.

## 5.4. The Laplace and Mellin Transforms

Two other integral transforms of great importance in applications are the Laplace and Mellin transforms. Both are strongly related to the Fourier transform, and residue theory plays a similar role in computations involving these transforms.

**The Laplace Transform.** The *Laplace transform* of a function $f$ on $[0, \infty)$ is defined to be

$$F(\lambda) = \int_0^\infty e^{-x\lambda} f(x)\, dx \tag{5.4.1}$$

on the set of complex numbers $\lambda$ for which this integral exists. If $u$ is a real number such that $e^{-xu} f(x)$ is an absolutely integrable function of $x$, then the function $e^{-x\lambda} f(x)$ will also be integrable for all $\lambda$ with $\text{Re}(\lambda) \geq u$ and the integral $F(\lambda)$ will be an analytic function of $\lambda$ on the open half-plane $\{\lambda \in \mathbb{C} : \text{Re}(\lambda) > u\}$ with limit 0 as $\lambda \to 0$ in this half-plane. In fact, since

$$F(s + it) = \int_0^\infty e^{-ixt} e^{-xs} f(x)\, dx,$$

the Laplace transform of $f$ at $s + it$ is just $\sqrt{2\pi}$ times the Fourier transform of the function $g(x)$ which is $e^{-xs} f(x)$ for $x \geq 0$ and is 0 for $x < 0$. If the Fourier Inversion Formula applies to $g$ and $g\hat{}$, then it results in a formula for recovering $f$ from its Laplace transform $F$. Specifically, with $\lambda = s + it$,

$$f(x) = \frac{e^{xs}}{2\pi} \int_{-\infty}^\infty F(s + it)\, e^{ixt}\, dt = \frac{1}{2\pi i} \int_{s-i\infty}^{s+i\infty} F(\lambda)\, e^{x\lambda}\, d\lambda. \tag{5.4.2}$$

We will call this the *inverse Laplace transform* of $F$.

On the open half-plane where $F$ is defined and analytic, the integral in (5.4.2) is independent of $s$ and, if $x < 0$, it has limit 0 as $s \to \infty$. If $F$ can be extended to a function which is analytic on $\mathbb{C}$ except at finitely many singularities, all to the left of $\text{Re}(\lambda) = u$, then we may calculate the integral in (5.4.2) using residues. That is, we compute

$$\frac{1}{2\pi i} \int_\gamma F(\lambda)\, e^{ix\lambda}\, d\lambda,$$

where $\gamma$ traverses the boundary of a rectangle with interior containing the singularities of $F$ and with right edge along the vertical line $\text{Re}(\lambda) = u$. As in the proof of Theorem 5.3.2, this integral is independent of the rectangle chosen and, on the one hand, is equal to the integral on the right in (5.4.2) and, on the other hand, is equal to the sum of the residues of $F(\lambda)\, e^{x\lambda}$. Thus, if $f$ is continuous on $(0, \infty)$ and has a Laplace transform $F$ which is analytic except at finitely many singularities, then

$$f(x) = \sum \text{Res}(F(\lambda)\, e^{x\lambda}, \lambda_i),$$

where $\{\lambda_i\}$ is the set of singularities of $F$.

**Example 5.4.1.** Find the inverse Laplace transform of $F(\lambda) = \dfrac{\lambda}{(\lambda - 2)(\lambda + 1)^2}$.

**Solution:** The function $F(\lambda)\,e^{x\lambda}$ has a simple pole at $\lambda = 2$ and a pole of order 2 at $\lambda = -1$. By Theorem 5.1.2

$$\operatorname{Res}(F(\lambda)\,e^{x\lambda}, 2) = \frac{2\,e^{2x}}{(2+1)^2} = \frac{2}{9}\,e^{2x},$$

while

$$\operatorname{Res}(F(\lambda)\,e^{x\lambda}, -1) = \frac{d}{d\lambda}\left(\frac{\lambda\,e^{\lambda x}}{\lambda - 2}\right)\Big|_{\lambda=-1} = \frac{3x-2}{9}\,e^{-x}.$$

Thus, $f(x) = \frac{1}{9}[(3x-2)\,e^{-x} + 2\,e^{2x}]$ for $x \geq 0$.

**The Mellin Transform.** If $f$ is a function defined on the positive real numbers, then the function of $t$ defined by

$$\int_0^\infty f(x)x^{t-1}\,dx, \tag{5.4.3}$$

with $t > 0$, is called the *Mellin transform* of $f$. Of course, this improper integral may converge for only certain values of $t$ and possibly for no values of $t$. Using residue theory, we will derive some conditions on $f$ and $t$ which ensure this integral does converge, and give a formula for computing it.

We assume that $f$ is analytic on $\mathbb{C}$ except at a finite set of singularities, none of which lie on the positive real axis. We define a function $z^{t-1}$, analytic on $\mathbb{C}\setminus[0,\infty)$, in the following way: We cut the plane along the positive real axis and denote by $\log(z)$ for $z \neq 0$ the branch of the log function determined by restricting $\arg(z)$ to lie in the interval $[0, 2\pi]$. Note that, since we allow both $\arg(z) = 0$ and $\arg(z) = 2\pi$, this results in there being two values of log at each point on the positive real axis. Think of the cut along $(0,\infty)$ as resulting in two copies of the positive real axis – an upper edge ($\arg(z) = 0$) and a lower edge ($\arg(z) = 2\pi$). Our log function is the ordinary real-valued natural log function along the upper edge of the cut, and is this plus $2\pi i$ along the lower edge of the cut. We then define

$$z^{t-1} = e^{(t-1)\log z} \quad \text{for} \quad z \neq 0.$$

In the next theorem, we compute the integral (5.4.3) by using path integration on the function $f(z)z^{t-1}$.

**Theorem 5.4.2.** *Let $f$ be a function analytic on $\mathbb{C}$, except at finitely many singularities, none of which are on the positive real axis, and let $z_1, z_2, \cdots, z_n$ be the non-zero singularities of $f$. Suppose $t$ is not an integer and the function $|f(z)||z|^t$ has limit 0 as $z \to 0$ and as $z \to \infty$. Then the Mellin transform of $f$ at $t$ exists and is given by*

$$\int_0^\infty f(x)x^{t-1}\,dx = -\pi\frac{e^{-\pi t i}}{\sin(\pi t)}\sum_{j=1}^{n}\operatorname{Res}(g, z_j), \tag{5.4.4}$$

*where $g(z) = f(z)z^{t-1}$ on $\mathbb{C}\setminus[0,+\infty)$ and $z^{t-1}$ is defined as above.*

**Proof.** The path $\gamma$ for our path integral is the path which goes along the upper edge of the cut from $\epsilon > 0$ to $R$, then along the circle $\gamma_R$ of radius $R$ centered at the origin, then along the lower edge of the cut from $R$ to $\epsilon$ and finally around the origin along the circle $\gamma_\epsilon$ of radius $\epsilon$ (see Figure 5.4.1).

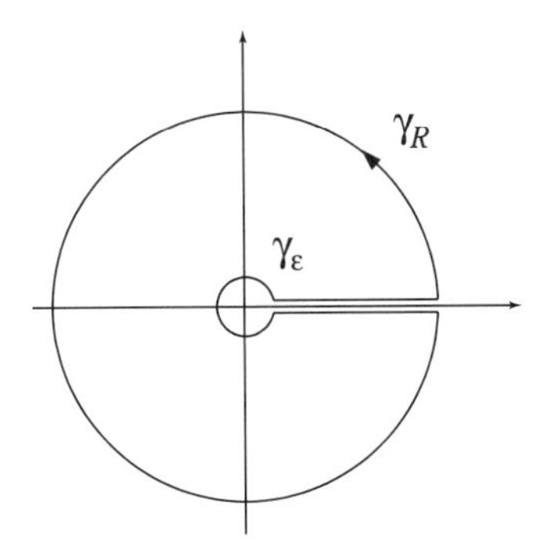


**Figure 5.4.1.** Path for Computing the Mellin Transform.

Let $z_1, z_2, \cdots, z_n$ be the non-zero singularities of $f$ and assume that $R$ is large enough and $\epsilon$ small enough that all of these singularities lie inside the circle of radius $R$ and outside the circle of radius $\epsilon$. Then $\gamma$ has index 1 about each $z_j$. If we set

$$g(z) = f(z)z^{t-1} \quad \text{for} \quad z \notin [0, +\infty),$$

then, by the Residue Theorem, we have

$$\begin{aligned} 2\pi i \sum_{j=1}^{n} \operatorname{Res}(g, z_j) &= \int_\gamma f(z)z^{t-1}\,dz \\ &= \int_\epsilon^R f(x)x^{t-1}\,dx + \int_{\gamma_R} f(z)z^{1-t}\,dz \\ &\quad - e^{2\pi i(t-1)} \int_\epsilon^R f(x)x^{t-1}\,dx - \int_{\gamma_\epsilon} f(z)z^{t-1}\,dz. \end{aligned} \tag{5.4.5}$$

If, as $R \to \infty$ and $\epsilon \to 0$, the integrals around $\gamma_R$ and $\gamma_\epsilon$ have vanishing limits, then

$$2\pi i \sum_{j=1}^{n} \operatorname{Res}(g, z_j) = (1 - e^{2\pi i t}) \int_0^\infty f(x)x^{t-1}\,dx.$$

If $t$ is not an integer, we may solve for the integral. This yields

$$\int_0^\infty f(x)x^{t-1}\,dx = -\pi \frac{e^{-\pi i t}}{\sin(\pi t)} \sum_{j=1}^{n} \operatorname{Res}(g, z_j). \tag{5.4.6}$$

It remains to show that the conditions on $f(z)z^{t-1}$ ensure the integrals around the two circles in (5.4.5) have limits 0 as $R \to \infty$ and $\epsilon \to 0$. If $M(R)$ is the maximum value of $|f(z)|$ on the circle of radius $R$, then

$$\left| \int_{\gamma_R} f(z)z^{t-1}\,dz \right| \le 2\pi M(R)R^t.$$

Hence, this integral will have limit 0 as $R \to \infty$ if

$$\lim_{z \to \infty} |f(z)||z|^t = 0.$$

Similarly, if $M(\epsilon)$ is the maximum value of $f$ on the circle of radius $\epsilon$, then

$$\left|\int_{\gamma_\epsilon} f(z)z^{t-1}\,dz\right| \le 2\pi M(\epsilon)\epsilon^t,$$

and so this integral will have limit 0 as $\epsilon \to 0$ if

$$\lim_{z\to 0} |f(z)||z|^t = 0.$$

This completes the proof. □

The calculation of the Mellin transform for the function $\dfrac{1}{1+x}$ is a key step in the developement of the properties of the gamma function in Chapter 9.

**Example 5.4.3.** Show that the Mellin transform of $f(x) = \dfrac{1}{1+x}$ is $\dfrac{\pi}{\sin(\pi t)}$.

**Solution:** According to the preceding theorem, the Mellin transform of this function $f$ will exist if $0 < t < 1$, since it is for these values of $t$ that

$$\lim_{z\to 0} \frac{|z|^t}{|1+z|} = 0 = \lim_{z\to\infty} \frac{|z|^t}{|1+z|}.$$

The only singularity of $f(z)$ occurs when $z = -1$, and the corresponding residue, $\mathrm{Res}(g, -1)$ for $g(z) = f(z)z^{t-1}$, is

$$(-1)^{t-1} = e^{\pi(t-1)i} = -e^{\pi t i}.$$

Thus, by Theorem 5.4.2, if $0 < t < 1$, then

$$\int_0^\infty \frac{x^{t-1}}{1+x}\,dx = \frac{\pi}{\sin(\pi t)}.$$

## Exercise Set 5.4

1. Find the Laplace transforms of the functions $f(x) = 1$ and $g(x) = x$. On what subset of $\mathbb{C}$ are these defined?
2. Find the Laplace transform of the function $f$ of Example 5.4.1. Does it coincide with the function $F$ of this example?
3. Find the inverse Laplace transform of $F(x) = \dfrac{\lambda}{(\lambda^2+4)}$.
4. Find the inverse Laplace transform of $F(x) = \dfrac{1}{1+\lambda^2}$.
5. If $f$ is continuous on $[0,\infty)$, differentiable on $(0,\infty)$, $f(0) = a$, and the Laplace transform $F$ of $f$ exists for $\mathrm{Re}(\lambda) > u$, what is the Laplace transform of $f'$ in terms of $F$?
6. If $f$ satisfies the differential equation $f'' + af' + bf = g$ on $(0,\infty)$ and $f$ and $f'$ are continuous on $[0,\infty)$, describe the Laplace transform $F$ of $f$ in terms of the Laplace transform $G$ of $g$ and the initial values ($f(0)$ and $f'(0)$) of $f$ and $f'$. Hint: Use the result of the previous exercise.
7. Can a non-zero constant function be the Laplace transform of a function $f$ on $[0,\infty)$?

8. Find the Mellin transform of $f(x) = \dfrac{1}{1+x^2}$.
9. Find the Mellin transform of $f(x) = \dfrac{1}{1+x^3}$.
10. Find $\displaystyle\int_0^\infty \frac{x^{2/3}}{1+x}\,dx$.
11. Show that if $h(y) = f(e^y)$, then the Mellin transform of $f$ can be expressed as $\displaystyle\int_{-\infty}^{\infty} h(y)\,e^{ty}\,dy$. Note the similarity to the inverse Fourier transform.
12. Verify the second equality in (5.4.5).

## 5.5. Summing Infinite Series

In this section we demonstrate a method for using the Residue Theorem to calculate the sum of an infinite series of the form

$$\sum_{n=-\infty}^{\infty} f(n), \tag{5.5.1}$$

where $f$ is an analytic function with isolated singularities in the complex plane.

The idea is this: If we can find an analytic function $g$ such that $g$ has a simple pole with residue 1 at each integer $n$, then $fg$ will have a simple pole with residue $f(n)$ at each integer $n$ (see Exercise 5.1.13). If we can also choose an expanding sequence of simple closed paths $\{\gamma_N\}$ such that each singularity of $fg$ is contained in $\gamma_N$ for sufficiently large $N$, and such that the integral of $fg$ around $\gamma_N$ has limit 0 as $N \to \infty$, then the Residue Theorem will imply that the sum of all residues of $fg$ constitutes a convergent series with sum 0. If $f$ has only finitely many singularities and none of them are integers, then the sum (5.5.1) will necessarily converge to the negative of the sum of the residues of $fg$ at the singularities of $f$. If some of the singularities of $f$ occur at integers, then the result will need to be adjusted accordingly.

This program actually works under reasonable hypotheses on $f$, if $g$ is chosen well. A particularly good choice is $g(z) = \pi\cot(\pi z)$.

**Some Properties of the Cotangent.** Since $\sin(\pi z)$ has a zero of order one at every integer and no other zeroes on $\mathbb{C}$, the meromorphic function

$$\cot(\pi z) = \frac{\cos(\pi z)}{\sin(\pi z)}$$

has a simple pole at each integer and no other poles. The residue of this function is $1/\pi$ at each $n$ (Exercise 5.1.5). Thus,

**Lemma 5.5.1.** *The function $\pi\cot(\pi z)$ has a simple pole with residue 1 at each integer, and no other poles.*

The function $\pi\cot(\pi z)$ is obviously not bounded near $\infty$, since it has a pole at each integer. However, it turns out that there is a sequence of closed paths, converging to $\infty$, on which this function is bounded. The paths in question are the

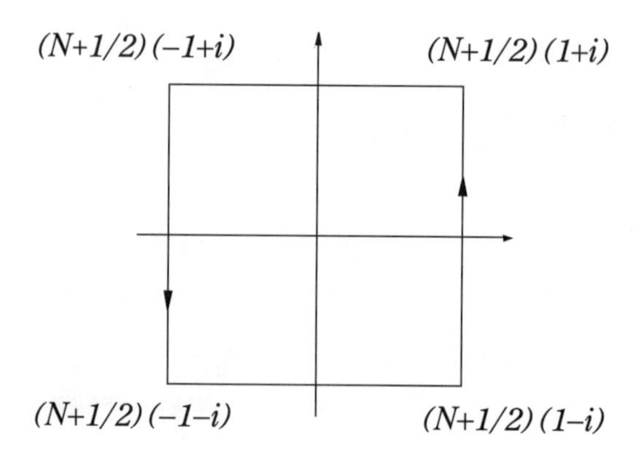


**Figure 5.5.1.** Path $\gamma_N$.

paths $\gamma_N$, where $N$ is a positive integer and $\gamma_N$ traces once in the positive direction around the square with vertices at

$$(N+1/2)(1+i),\ (N+1/2)(-1+i),\ (N+1/2)(-1-i),\ \text{and } (N+1/2)(1-i)$$

(see Figure 5.5.1).

**Lemma 5.5.2.** *There is a positive number $R$ such that $|\cot(\pi z)| \leq 2$ on $\gamma_N(I)$ for each $N \geq R$.*

**Proof.** By Definition 1.3.8, we have

$$\begin{aligned}\cot(\pi z) &= \frac{\cos(\pi z)}{\sin(\pi z)} = i\frac{e^{\pi iz} + e^{-\pi iz}}{e^{\pi iz} - e^{-\pi iz}} \\ &= i\frac{e^{2\pi iz} + 1}{e^{2\pi iz} - 1} = i\frac{e^{2\pi ix}\, e^{-2\pi y} + 1}{e^{2\pi ix}\, e^{-2\pi y} - 1},\end{aligned}$$

where $z = x + iy$. Thus,

$$|\cot(\pi z)| = \left|\frac{e^{2\pi ix}\, e^{-2\pi y} + 1}{e^{2\pi ix}\, e^{-2\pi y} - 1}\right|.$$

If $x = N + 1/2$ for an integer $N$, then $e^{2\pi ix} = -1$ and $|\cot(\pi z)| \leq 1$. Thus, $|\cot(\pi z)| \leq 1$ on the vertical sides of each $\gamma_N$. On the horizontal sides, where $y$ is fixed at $\pm(N+1/2)$, the maximum of $|\cot(\pi z)|$ occurs when $e^{2\pi ix} = 1$, that is, at integer values of $x$. Thus, the maximum is

$$\left|\frac{e^{\pm(2N+1)\pi} + 1}{e^{\pm(2N+1)\pi} - 1}\right|.$$

Since this expression has limit 1 as $N \to \infty$, there is an $M$ such that it is less than 2 when $N > M$. □

**Lemma 5.5.3.** *With $\gamma_N$ as above,*

$$\int_{\gamma_N} \frac{\pi \cot(\pi z)}{z} = 0 \tag{5.5.2}$$

*for each positive integer $N$.*

**Proof.** The function $h(z) = \dfrac{\pi \cot(\pi z)}{z}$ is analytic inside $\gamma_N$ except at integer points, where it has poles. By the Residue Theorem

$$\int_{\gamma_N} \frac{\pi \cot(\pi z)}{z} = \sum_{n=-N}^{N} \operatorname{Res}(h, n). \tag{5.5.3}$$

Since $h$ is an even function, its residue at 0 is 0 (Exercise 5.1.15 ). The poles of $h$ at non-zero integers are simple poles and may be calculated using Corollary 5.1.6. The result is that the residue of $h$ at $n$ is $1/n$. Since this is an odd function of $n$ and the range of summation in (5.5.3) is symmetric about the origin, the integral in (5.5.2) is 0. □

**A Summation Theorem.** We now have enough information about the function $g(z) = \pi \cot(\pi z)$ to prove the following summation theorem.

**Theorem 5.5.4.** *Suppose $f$ is analytic on $\mathbb{C}$ except at a finite set*

$$E = \{z_1, z_2, \cdots, z_n\}$$

*of isolated singularities. Also suppose that there exist positive constants $R$ and $M$ such that*

$$|f(z)| \leq \frac{M}{|z|} \quad \text{for} \quad |z| \geq R. \tag{5.5.4}$$

*Then*

$$\lim_{N \to \infty} \sum_{n \in \mathbb{Z}_0^N} f(n) = -\sum_{j=1}^{m} \operatorname{Res}(\pi f(z) \cot(\pi z), z_j),$$

*where $Z_0^N$ is the set of integers in $[-N, N]$ which do not belong to $E$.*

**Proof.** The singularities of $\pi f(z) \cot(\pi z)$ occur at the integers and at the points $z_1, \cdots, z_n$. At an integer $n$ not in $E$, this function has residue $f(n)$, since $\cot(\pi z)$ has a simple pole with residue 1 at $n$ and $f$ has no singularity at $n$. If $N$ is large enough so that the $z_j$ are all inside $\gamma_N$, then the Residue Theorem implies

$$\frac{1}{2\pi i} \int_{\gamma_N} \pi f(z) \cot(\pi z)\, dz = \sum_{n \in \mathbb{Z}_0^N} f(n) + \sum_{j=1}^{m} \operatorname{Res}(\pi f(z) \cot(\pi z), z_j). \tag{5.5.5}$$

Thus, to prove the theorem, we simply need to show that the integral on the left converges to 0 as $N \to \infty$.

Choose $R$ and $M$ large enough that (5.5.4) holds, the inequality in Lemma 5.5.2 holds, and $|z_j| \leq R$ for each $j$. Then $f$ is analytic in the annulus

$$A = \{z : R < |z| < \infty\}$$

and vanishes at infinity. It therefore has a Laurent expansion, in this annulus, involving only negative powers of $z$. Say,

$$f(z) = \frac{c_{-1}}{z} + \frac{c_{-2}}{z^2} + \cdots + \frac{c_{-n}}{z^n} + \cdots.$$

By Lemma 5.5.3,

$$\int_{\gamma_N} \frac{\pi \cot(\pi z)}{z}\, dz = 0.$$

and so

$$\frac{1}{2\pi i}\int_{\gamma_N} \pi f(z)\cot(\pi z)\,dz = \frac{1}{2\pi i}\int_{\gamma_N} \pi\left(f(z) - \frac{c_{-1}}{z}\right)\cot(\pi z)\,dz. \tag{5.5.6}$$

However, the function

$$f(z) - \frac{c_{-1}}{z} = \frac{c_{-2}}{z^2} + \cdots + \frac{c_{-n}}{z^n} + \cdots$$

has the form $\frac{q(z)}{z^2}$ in $A$, where $q$ is analytic in $A$ and has limit $c_{-2}$ at infinity. This implies that, for some $R_1 > R$, the function $|q|$ is bounded by some positive constant $M_1$ on $\{z : |z| \geq R_1\}$. Then, by Lemma 5.5.2, the integrand of the integral on the right in (5.5.6) is bounded above by $2MM_1/|z|^2$. Since $|z| > N$ on $\gamma_N$ and the length of the path $\gamma_N$ is $8N + 4$, we have that

$$\left|\frac{1}{2\pi i}\int_{\gamma_N} \pi f(z)\cot(\pi z)\,dz\right| \leq \frac{2MM_1(8N+4)}{N^2},$$

which converges to 0 as $N \to \infty$. This completes the proof. □

**Example 5.5.5.** Prove that $\displaystyle\sum_{n=1}^{\infty}\frac{1}{n^2} = \frac{\pi^2}{6}$.

**Solution:** We apply the preceding theorem with $f(z) = 1/z^2$. It clearly satisfies the hypothesis of the theorem. The only singularity of this function is the one at 0, and so the theorem tells us that

$$\sum_{n\in\mathbb{Z}_0}\frac{1}{n^2} = -\operatorname{Res}\left(\frac{\pi\cot(\pi z)}{z^2}, 0\right),$$

where $\mathbb{Z}_0$ is the set of non-zero integers. The residue of $\frac{\cot z}{z^2}$ at 0 is $-1/3$ by Example 5.1.9. Since the coefficient of $z$ in the Laurent expansion of $\pi\cot(\pi z)$ is $\pi^2$ times the coefficient of $z$ in the Laurent expansion of $\cot z$, we conclude that $\frac{\pi\cot(\pi z)}{z^2}$ has residue $\pi^2/3$ at 0. Then,

$$\sum_{n=1}^{\infty}\frac{1}{n^2} = \frac{1}{2}\sum_{n\in\mathbb{Z}_0}\frac{1}{n^2} = \frac{\pi^2}{6}.$$

## Exercise Set 5.5

1. Prove that, like $\pi\cot(\pi z)$, the function $\dfrac{2\pi i}{e^{2\pi i z} - 1}$ has a simple pole with residue 1 at each integer and it has no other poles.
2. Find $\displaystyle\sum_{n=1}^{\infty}\frac{1}{1+n^2}$.
3. Find $\displaystyle\sum_{n=1}^{\infty}\frac{1}{(n-i)^2}$.

4. Find $\sum_{n=1}^{\infty} \frac{1}{n^4}$.

5. Prove that if $w \in \mathbb{C} \setminus \mathbb{Z}$, then $\pi \cot \pi w = \lim_{N \to \infty} \sum_{n=-N}^{N} \frac{1}{n+w}$.

6. Assuming the result of the previous exercise, prove that if $w \in \mathbb{C} \setminus \mathbb{Z}$, then

$$\pi \cot \pi w = \frac{1}{w} + \sum_{n=1}^{\infty} \frac{1}{w^2 - n^2}.$$

7. Prove that the function $\frac{\pi}{\sin(\pi z)}$ has a pole at each integer $n$ with residue $(-1)^n$ and no other poles.

8. Derive a method for summing a series of the form $\sum_{-\infty}^{\infty} (-1)^n f(n)$, where $f$ is a meromorphic function with a finite number of poles. Hint: Use the preceding exercise and the method of Theorem 5.5.4.

Chapter 6

# Conformal Mappings

If $U$ and $V$ are open subsets of the complex plane, then just how different are $U$ and $V$ from the standpoint of the study of analytic functions? If there is a analytic map $h : U \to V$ with an analytic inverse $h^{-1} : V \to U$, then the answer is, "they are not different at all." In this case, we shall say that $h$ is a *conformal equivalence* between $U$ and $V$, and that $U$ and $V$ are *conformally equivalent.*

If $h : U \to V$ is a conformal equivalence, then each analytic function $g$ on $V$ gives rise to an analytic function $g \circ h$ on $U$, and each analytic function $f$ on $U$ gives rise to an analytic function $f \circ h^{-1}$ on $V$. This establishes a one-to-one correspondence between the analytic functions on $U$ and those on $V$. This correspondence preserves sums and products and exponentials and logarithms and roots. The same formulas define one-to-one correspondences between meromorphic functions on $U$ and on $V$, and between functions with isolated singularities on $U$ and on $V$. Similarly, paths, closed paths, and homotopies in $U$ all have corresponding objects in $V$. Indices of paths are preserved by this correspondence.

Clearly then, it is important to know when two open sets $U$ and $V$ are conformally equivalent. That is the subject of this chapter. The main theorem of the chapter is the *Riemann Mapping Theorem*, which says that every connected, simply connected, proper open subset of $\mathbb{C}$ is conformally equivalent to the open unit disc.

## 6.1. Definition and Examples

The word *conformal* refers to maps between regions of the plane that are angle preserving, in a sense to be defined below. It turns out these are exactly the maps which are analytic with a non-vanishing derivative.

**Angle Preserving Maps.** Let $U$ and $V$ be open sets in $\mathbb{C}$. A mapping $h : U \to V$ is said to be *conformal* if it preserves angles between smooth curves at each point of $U$. We will make this statement precise below.

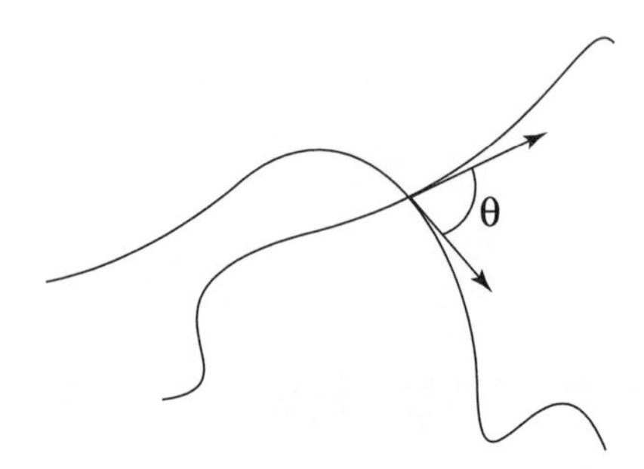


**Figure 6.1.1.** The Angle $\theta$ Between Two Paths.

If $h(x+iy) = u(x,y) + iv(x,y)$, for $z = x + iy \in U$, then we may consider $h$ to be an $\mathbb{R}^2$-valued function of two real variables $x$ and $y$ with coordinate functions $u(x,y)$ and $v(x,y)$. Let $z_0 = (x_0, y_0)$ be a point of $U$. If $h$ is differentiable at $z_0$, as a function in this sense, then its differential $dh(z_0)$ is the linear transformation from $\mathbb{R}^2$ to $\mathbb{R}^2$ whose matrix is

$$dh = \begin{pmatrix} u_x & u_y \\ v_x & v_y \end{pmatrix} \tag{6.1.1}$$

evaluated at $z_0$. If $\gamma(t) = (x(t), y(t))$ is a differentiable curve in $U$ with $\gamma(t_0) = z_0$, then $\gamma'(t_0) = (x'(t_0), y'(t_0))$ is the *tangent vector* to the curve at the point $z_0$. The chain rule implies that

$$(h \circ \gamma)'(t_0) = dh(z_0) \cdot \gamma'(t_0),$$

where $(h \circ \gamma)'(t_0)$ is the tangent vector to the curve $h \circ \gamma$ at $h(z_0)$, and the product on the right represents the linear transformation $dh(z_0)$ applied to the vector $\gamma'(t_0)$. Using the matrix for $dh$, this is given in terms of vector-matrix multiplication as

$$(h \circ \gamma)' = \begin{pmatrix} u_x & u_y \\ v_x & v_y \end{pmatrix} \begin{pmatrix} x' \\ y' \end{pmatrix},$$

where the partial derivatives are evaluated at $z_0$ while $(h \circ \gamma)'$, $x'$ and $y'$ are evaluated at $t_0$. Thus, the matrix $dh(z_0)$ transforms the tangent vector to $\gamma$ at $z_0$ into the tangent vector to the image curve $h \circ \gamma$ at $h(z_0)$.

**Definition 6.1.1.** We say a mapping $h$, as above, is conformal at $z_0$ if its differential $dh$ at $z_0$ is non-singular and angle preserving, in the sense that if $a$ and $b$ are any two non-zero vectors in $\mathbb{R}^2$, then the angle between $dh(z_0)a$ and $dh(z_0)b$ is the same as the angle between $a$ and $b$. If $h : U \to V$ is conformal at each point of $U$ and maps onto $V$, then we say it is a conformal map of $U$ onto $V$.

**Theorem 6.1.2.** *A complex-valued function $h$, defined in a neighborhood of $z_0 \in \mathbb{C}$, is conformal at $z_0$ if and only if it has a complex derivative at $z_0$ and this derivative is non-zero.*

**Proof.** If $h$ has a non-zero complex derivative at $z_0$, then its partial derivatives satisfy the Cauchy-Riemann equations at $z_0$ and are not all 0. Hence, the differential

(6.1.1) has the form

$$dh = \begin{pmatrix} u_x & u_y \\ -u_y & u_x \end{pmatrix},$$

with the partial derivatives evaluated at $z_0$ and the vector $(u_x, u_y)$ non-zero. If we put this vector in polar form $(r\cos\theta, r\sin\theta)$, with $r \neq 0$, then this matrix becomes

$$\begin{pmatrix} r\cos\theta & r\sin\theta \\ -r\sin\theta & r\cos\theta \end{pmatrix} = \begin{pmatrix} r & 0 \\ 0 & r \end{pmatrix} \begin{pmatrix} \cos\theta & \sin\theta \\ -\sin\theta & \cos\theta \end{pmatrix}. \tag{6.1.2}$$

That is, if $h$ has a complex derivative at $z_0$, then the differential of $h$ at $z_0$ is rotation through an angle $\theta$ followed by expansion or contraction by a positive factor $r$. A linear transformation of this form is clearly non-singular and preserves the angle between any two vectors, since both vectors get rotated by the same angle $\theta$. Hence, $h$ is conformal at $z_0$.

Conversely, suppose $h$ is conformal at $z_0$. Let $\theta$ be the angle from the vector $(1, 0)$ to its image under $dh(z_0)$. Let

$$A = \begin{pmatrix} \cos\theta & \sin\theta \\ -\sin\theta & \cos\theta \end{pmatrix}$$

be the matrix which rotates every vector through an angle $\theta$. Then $A^{-1} \cdot dh(z_0)$ is an angle preserving matrix which sends $(1, 0)$ to a positive multiple $(r, 0)$ of itself. Since it preserves angles between vectors, it must also send every other vector to a multiple of itself. We claim that, in fact, it multiplies each vector by $r$. Suppose $(0, 1)$ is sent to $(0, t)$. Then the vector $(1, 1)$ is sent to $(r, t)$ and, if this is to be a multiple of $(1, 1)$, we must have $r = t$. This implies that $A^{-1} \cdot dh(z_0)$ is $r$ times the identity matrix. Hence, $dh(z_0)$ has the form (6.1.2). This implies that $h$ satisfies the Cauchy-Riemann equations at $z_0$ and, hence, has a complex derivative at $z_0$. This derivative is non-zero because $dh(z_0)$ is non-singular. □

Note that a conformal map $h$ from $U$ onto $V$ need not be one-to-one. In fact, the exponential function $e^z$ is a conformal map of $\mathbb{C}$ onto $\mathbb{C}$ which is not one-to-one. However, the following is true, based on the Inverse Mapping Theorem (Theorem 4.5.5).

**Theorem 6.1.3.** *If $h$ is a one-to-one conformal map of $U$ onto $V$, then $h$ has an inverse function $h^{-1} : V \to U$ which is also a conformal map.*

**Proof.** The function $h : U \to V$ has a well-defined inverse function $h^{-1} : V \to U$ because $h$ is one-to-one and onto. By Theorem 4.5.5, since $h$ is analytic and has non-vanishing derivative at each point of $U$, its inverse function is analytic with non-vanishing derivative on all of $V$. □

As mentioned in the introduction, an analytic map from $U$ to $V$ with an analytic inverse is called a conformal equivalence between $U$ and $V$. By the preceding theorem, a map of $U$ onto $V$ is a conformal equivalence if and only if it is a one-to-one conformal map.

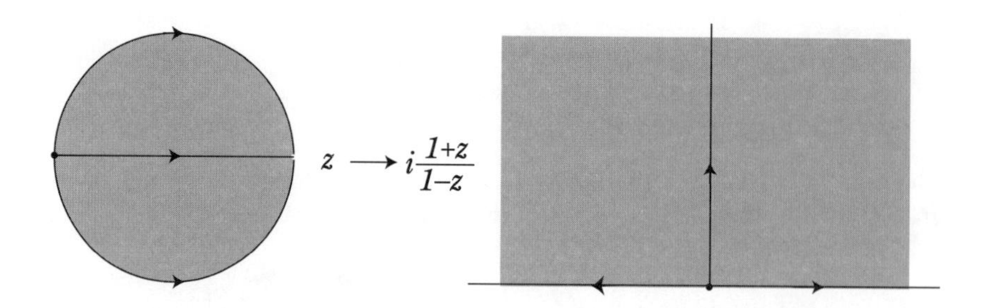


**Figure 6.1.2.** Conformal Map from Disc to Upper Half-Plane.

**Examples of Conformal Maps.** It can be useful to have a substantial inventory of conformal maps between various regions in the plane. Here we describe a short list of such maps. Many others can be constructed using similar ideas or by combining some of these.

**Example 6.1.4.** Find a conformal equivalence between the unit disc and the right half-plane.

**Solution:** Let

$$g(z) = \frac{1+z}{1-z}.$$

Then $g$ is defined and analytic on $\mathbb{C} \setminus \{1\}$. If $w \neq -1$, the equation

$$w = \frac{1+z}{1-z}$$

can be uniquely solved for $z$. The solution is

$$z = \frac{w-1}{w+1}.$$

It follows that $g$ is a one-to-one analytic map of the set $\mathbb{C}\setminus\{1\}$ onto the set $\mathbb{C}\setminus\{-1\}$ with an analytic inverse function

$$g^{-1}(w) = \frac{w-1}{w+1}.$$

If we multiply both numerator and denominator of the fraction defining $g$ by $1-\overline{z}$, we see that the real part of $g(z)$ is $\dfrac{1-|z|^2}{|1-z|^2}$. Hence, $g$ has positive real part if and only if $|z| < 1$. Thus, $g$ maps the unit disc onto the right half-plane. Since it has an inverse function, it is a conformal equivalence from the unit disc onto the right half-plane.

**Example 6.1.5.** Find a conformal equivalence between the unit disc and the upper half-plane.

**Solution:** Since multiplication by $i$ has the effect of rotating points of the plane by $\pi/2$ in the counterclockwise direction, we obtain a conformal equivalence $h$ of the unit disc to the upper half-plane by multiplying the function $g$ of the previous example by $i$. Thus,

$$h(z) = i\frac{1+z}{1-z}$$

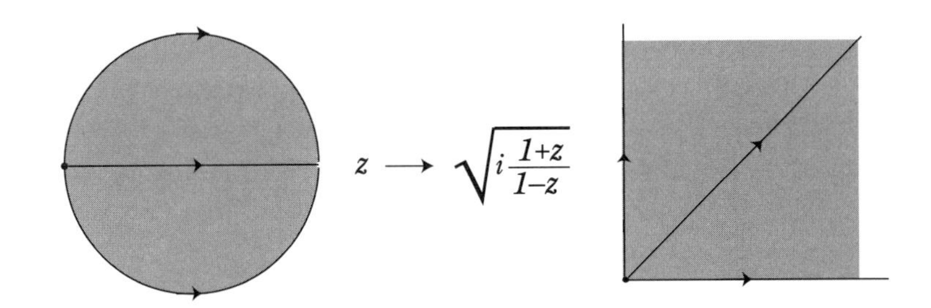


**Figure 6.1.3.** Conformal Map from Unit Disc to 1st Quadrant.

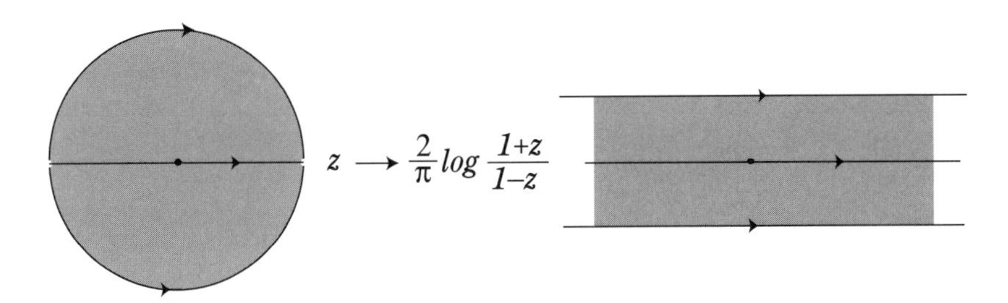


**Figure 6.1.4.** Conformal Map from Disc to $-1 < \text{Im}(z) < 1$.

is the equivalence we seek (see Figure 6.1.2).

**Example 6.1.6.** Find a conformal equivalence from the unit disc to the first quadrant of the plane.

**Solution:** The upper half-plane consists of all points which, in polar coordinates, have the form $z = r\,e^{i\theta}$ with $r > 0$ and $0 < \theta < \pi$, while the open first quadrant consists of points of this form with $0 < \theta < \pi/2$. It follows that the principal branch of the square root function (which sends $r\,e^{i\theta}$ to $\sqrt{r}\,e^{i\theta/2}$ for $-\pi < \theta < \pi$) is a conformal equivalence from the upper half-plane to the first quadrant with inverse function $w \to w^2$. It follows from this and the previous example that

$$h(z) = \sqrt{i\frac{1+z}{1-z}}$$

is a conformal equivalence from the unit disc to the first quadrant, provided we use the principal branch of the square root function.

**Example 6.1.7.** Find a conformal equivalence from the unit disc onto the strip $S = \{z : -1 < \text{Im}(z) < 1\}$.

**Solution:** The principal branch of the log function is a conformal equivalence of the open right half-plane onto the strip $\{z : -\pi/2 < \text{Im}(z) < \pi/2\}$. Thus, we

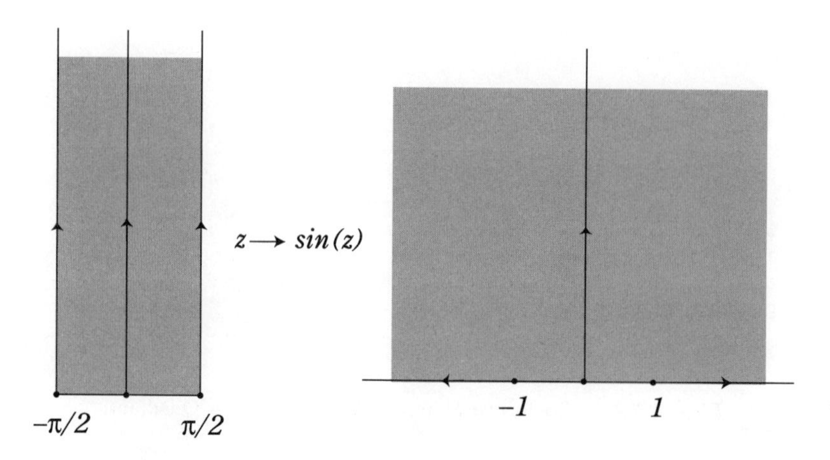


**Figure 6.1.5.** Conformal Map from Half-Strip to Half-Plane.

can construct a conformal equivalence from the unit disc to $S$ by composing the following series of conformal maps: First, we map the unit disc to the right half-plane using $z \to \dfrac{1+z}{1-z}$, as in Example 6.1.4; we follow this by the map log from the right half-plane to the strip $\{z : -\pi/2 < \mathrm{Im}(z) < \pi/2\}$; and finally, we divide by $\pi/2$. The composition of these maps is

$$h(z) = \frac{2}{\pi} \log \frac{1+z}{1-z},$$

which is the required conformal equivalence.

**Example 6.1.8.** Find a conformal equivalence from the set

$$A = \{z : -\pi/2 < \mathrm{Re}(z) < \pi/2,\ \mathrm{Im}(z) > 0\}$$

to the upper half-plane.

**Solution:** Recall from Exercise 1.3.14 that the function $\sin z$ can be written as

$$\sin(x + iy) = \sin x \cosh y + i \cos x \sinh y, \tag{6.1.3}$$

while its derivative $\cos z$ can be written as

$$\cos(x + iy) = \cos x \cosh y - i \sin x \sinh y. \tag{6.1.4}$$

The first of these formulas can be used to prove that sin is one-to-one on the set $A$ (Exercise 6.1.9), while the second can be used to prove that its derivative, $\cos z$, is non-vanishing on $A$ (Exercise 6.1.10). Therefore, sin is a one-to-one conformal map of $A$ onto some open set $B$.

To identify $B$ we note that sin takes the interval $[-\pi/2, \pi/2]$ on the real line to the interval $[-1, 1]$. It takes the half-lines $\{z : \mathrm{Re}(z) = \pm\pi/2, \mathrm{Im}(z) > 0\}$ to the half-lines $[1, \infty)$ and $[-\infty, -1]$. Hence, it must take $A$ to either the upper or lower half-plane. Since it takes $i$ to $i \sinh 1$ and $\sinh 1 > 0$, it must take $A$ to the upper half-plane. Hence, sin is the required conformal equivalence.

## Exercise Set 6.1

1. If $r\,e^{i\theta} \in \mathbb{C}$ and the map $z \to r\,e^{i\theta}\,z$ is considered as a linear transformation from $\mathbb{R}^2$ to $\mathbb{R}^2$, then what is its corresponding matrix?
2. Find a conformal equivalence from the half-plane $\{z : \mathrm{Re}(z) > 1\}$ to the unit disc $D$.
3. If $D$ is the unit disc, find a conformal equivalence from $D \setminus \{0\}$ to $\mathbb{C} \setminus \overline{D}$.
4. Find a conformal equivalence from the open unit disc to the set
$$W = \{z : 0 < \arg(z) < \pi/4\}.$$
5. Show that the image of the upper half-disc $\{z : |z| < 1,\ \mathrm{Im}(z) > 0\}$, under the mapping
$$z \to \frac{1+z}{1-z}$$
of Example 6.1.4, is the open first quadrant.
6. Find a conformal equivalence from the upper half-disc of the previous exercise to the unit disc.
7. For each $a$ with $0 < a < 2\pi$, find a conformal equivalence from the set $\{z : 0 < \arg z < a\}$ to the upper half-plane.
8. Find a conformal equivalence from the upper half-plane to $\mathbb{C} \setminus (-\infty, 1]$.
9. Prove that $\sin z$ is one-to-one on the strip $A$ of Example 6.1.8. Hint: Show that sinh and cosh are increasing functions on $\{y \in \mathbb{R} : y > 0\}$ and then use (6.1.3).
10. Prove $\cos z$ is never 0 on the strip $A$ of Example 6.1.8. Hint: Show that sinh and cosh are non-vanishing functions on $\{y \in \mathbb{R} : y > 0\}$ and then use (6.1.4).
11. Prove that the composition $h \circ g$ of a conformal map $g : U \to V$ with a conformal map $h : V \to W$ is conformal.
12. Prove that an analytic function $f$ on an open set $U$ in the plane is a conformal map from $U$ to $f(U)$ if and only if it has a local analytic inverse (in the sense of Definition 4.5.4) at each point of $U$.

## 6.2. The Riemann Sphere

In Lemma 4.3.1 we saw how the change of variables $z = 1/w + z_0$ transforms a function which is analytic in the exterior of a disc centered at 0, and has a limit at $\infty$, into a function which is analytic in a disc centered at $z_0$. This suggests that, in some sense, $\infty$ can be made to play the same role that ordinary points of $\mathbb{C}$ play in analytic function theory. In fact, it turns out that the set $\mathbb{C} \cup \{\infty\}$ can be treated as a space with both a topological and an analytic structure in which $\infty$ is in no way different from other points. This space is called the *Riemann Sphere*. It is the proper setting for the theory of conformal mappings and so it is time for us to introduce it and discuss its properties.

**Construction of the Riemann Sphere.** The Riemann Sphere $S^2$ has several descriptions. We begin with the simplest of these.

We denote by $S^2$ the set $\mathbb{C} \cup \{\infty\}$, where $\infty$ denotes an abstract point that is adjoined to $\mathbb{C}$. We want to emphasize that $\infty$ is to be thought of as a point of this space no different than any other. In fact, we will work with two coordinate systems on $S^2$. In one of these, the coordinate function is $z$ – a function analytic at every point except $z = \infty$. In the other, the coordinate function is $w = 1/z$ – a function analytic at every point except $z = 0$, including the point at $\infty$ (this will follow once we define what it means for a function to be analytic at $\infty$).

An open disc in $S^2$, centered at $\infty$, is defined to be the set of all points with $w$ coordinate satisfying $|w| < r$ for a fixed $r > 0$. Equivalently, it is the set of all points with $z$ coordinate satisfying $|z| > 1/r$. An open set in $S^2$ is a set $U$ such that every point of $U$ is the center of some open disc contained in $U$. This is the familiar definition of open set in case $U \subset \mathbb{C}$. It now also makes sense if $U$ contains $\infty$, since we have defined the notion of an open disc centered at $\infty$. With this definition, a *neighborhood of* $\infty$ is the union of $\{\infty\}$ with an open subset of $\mathbb{C}$ which contains the complement of a closed bounded disc centered at 0.

With the topology (the collection of open sets) defined, we can talk about limits of functions from a subset of $S^2$ to a subset of $S^2$, and about the continuity of such functions. These notions are defined using open sets (i.e., neighborhoods of points) rather than epsilons and deltas, as suggested by Theorem 2.1.10 and Definition 2.1.11. Thus, if $f : U \to S^2$ is a function defined on an open set $U \subset S^2$, then

$$\lim_{z \to z_0} f(z) = L$$

if and only if for each neighborhood $W$ of $L$, there is a deleted neighborhood $V$ of $z_0$ such that $V \subset U$ and $f(V) \subset W$. This leads to definitions for the meaning of the statements $\lim_{z \to z_0} f(z) = \infty$ and $\lim_{z \to \infty} f(z) = L$, that are consistent with the meanings attached to them in preceding chapters.

A function $f$, defined on an open set $U \subset S^2$, with values in $S^2$, is continuous at $z_0 \in U$ if and only if $\lim_{z \to z_0} f(z) = f(z_0)$. As in Theorem 2.1.13, $f$ is continuous at every point of its domain $U$ if and only if the inverse image, under $f$, of every open set is open.

**Analytic Functions on the Riemann Sphere.** In what follows, the phrase *analytic at* $z_0$, applied to a function on a subset of $S^2$, will mean *defined and analytic in a neighborhood of* $z_0$.

We know what it means for a function to be analytic at a finite point of $S^2$. What does it mean for a function $f(z)$ to be analytic at $\infty$? We make sense of this by, again, using the coordinate function $w = 1/z$, which is 0 at $z = \infty$.

**Definition 6.2.1.** We will say that a complex-valued function $f(z)$ is *analytic at* $\infty$ if $f(1/w)$ is analytic at $w = 0$ (meaning defined and analytic in a deleted neighborhood of 0 with a removable singularity at 0).

Thus, a function $f(z)$ is analytic at $\infty$ if $f(1/w)$, when given the appropriate value at 0, is complex differentiable with respect to $w$ for $w$ in a neighborhood of 0. Note that, at points where $z$ and $w = 1/z$ are both finite, differentiability with respect to $z$ and differentiability with respect to $w$ are equivalent, although the two derivatives are not the same (Exercise 6.2.1).

**Example 6.2.2.** Show that $f(z) = \dfrac{1-z}{1+z}$ is analytic at $\infty$ if we set $f(\infty) = -1$.

**Solution:** According to the above definition, we must show that $f(1/w)$ is analytic at $w = 0$. However,

$$f(1/w) = \frac{1 - 1/w}{1 + 1/w} = \frac{w-1}{w+1}$$

and this is, indeed, analytic at $w = 0$.

A way to think of $S^2$ that is suggested by the use of the $z$ and $w$ coordinate functions is to regard it as two copies of $\mathbb{C}$ glued together along $\mathbb{C} \setminus \{0\}$ by identifying the point $z$ in one copy of $\mathbb{C} \setminus \{0\}$ with the point $1/z$ in the other copy.

**Analytic Functions from $S^2$ to $S^2$.** We will now discuss analytic functions defined on an open subset of $S^2$ with values in $S^2$ rather than in $\mathbb{C}$. There is some question as to what this means. That is, we know what an analytic function from an open subset of $S^2$ to $\mathbb{C}$ is, but what does it mean for a function to be analytic at a point where it has the value $\infty$? This is only a problem if we insist on using the $z$ coordinate function on the image space of the function $f$. If we use the $w$ coordinate function, instead, then it is clear that $f$ should be considered analytic at a point $z_0$ where $f(z_0) = \infty$ if and only if $1/f$ is analytic at $z_0$.

**Example 6.2.3.** Show that the function $\dfrac{1-z}{1+z}$ is analytic at $z = -1$ as a function from $S^2$ to $S^2$.

**Solution:** Since $f(-1) = \infty$, to show that $f$ is analytic at $-1$ as an $S^2$-valued function, we must show that $1/f$ is analytic at $-1$. In fact,

$$\frac{1}{f(z)} = \frac{1+z}{1-z}$$

and this is analytic at $z = -1$.

The function $1/z$ that has played a major role in the discussion to this point, can not only be thought of as a new coordinate function on $S^2$, it can also be thought of as a mapping of $S^2$ to itself which interchanges 0 and $\infty$. We will call this mapping $s$. Thus,

$$s(z) = 1/z,$$

for all $z \in S^2$ (including 0 and $\infty$, where, of course, $1/0 = \infty$ and $1/\infty = 0$).

According to our definitions, a function $f$ is analytic at infinity if and only if $f \circ s$ is analytic at 0, while $f$ is analytic at a point $z_0$ where it has the value $\infty$ if and only if $s \circ f$ is analytic at $z_0$.

**Theorem 6.2.4.** *The mapping $s$ is a conformal equivalence of $S^2$ onto itself, that is, it is a one-to-one analytic map of $S^2$ onto itself with an analytic inverse map.*

**Proof.** Clearly $s$ is one-to-one and onto and is its own inverse function, since $s \circ s(z) = z$. It is also analytic at every point, with the possible exceptions of 0 and $\infty$. However, $s(1/w) = w$ is analytic at $w = 0$ and so $s$ is analytic at $z = \infty$. Similarly, $1/s(z) = z$ is analytic at $z = 0$ and so $s$ is analytic at 0. This completes the proof. $\square$

A meromorphic function $f(z)$ on $\mathbb{C}$ is said to have a pole of order $n$ at $\infty$ if $f(1/w)$ has a pole of order $n$ at 0 (we include the case where the pole is of order 0 – i.e. the case of a removable singularity). The next theorem shows that the meromorphic functions on $\mathbb{C}$, with poles at $\infty$, are the analytic functions from $S^2$ to $S^2$.

**Theorem 6.2.5.** *Each meromorphic function on $\mathbb{C}$ with a pole at $\infty$ defines an analytic function $\tilde{f}$ from $S^2$ to $S^2$, and each analytic function from $S^2$ to $S^2$ arises in this way.*

**Proof.** Certainly a meromorphic function with a pole at $\infty$ can be considered as a function $\tilde{f}$ from $S^2$ to $S^2$ – we let $\tilde{f}(z) = f(z)$ at points of $\mathbb{C}$ which are not poles of positive order and assign the value $\infty$ to $\tilde{f}$ at each point (including $\infty$) at which $f$ has a pole of positive order.

If $f(z)$ has a pole of order 0 at $\infty$, then $f(1/w)$ has a removable singularity at 0 and we assign $\tilde{f}(\infty)$ the value which makes $f(1/w)$ analytic at $w = 0$. The resulting function $\tilde{f} : S^2$ to $S^2$ is analytic at each point of $\mathbb{C}$ where $f$ is analytic. At a pole $z_0 \in \mathbb{C}$ of $f(z)$, $1/f(z)$ is analytic, and this implies that $\tilde{f}(z)$ is analytic.

At the point $\infty$, $f(z)$ has a pole as a meromorphic function on $\mathbb{C}$ if and only if $f(1/w)$ has a pole at 0. This happens if and only if $\tilde{f}$ is analytic at $\infty$ as a function from $S^2$ to $S^2$.

Thus, each meromorphic function $f$ on $\mathbb{C}$ with a pole at $\infty$ extends to be an analytic function $\tilde{f}$ from $S^2$ to $S^2$. It remains to show that every analytic function from $S^2$ to $S^2$ arises in this way.

If $g$ is an analytic function from $S^2$ to $S^2$, then at points $z_0$ of $\mathbb{C}$ where $g(z_0) \neq \infty$, $g$ is an analytic complex-valued function. At a point $z_0$ of $\mathbb{C}$ where $g(z_0) = \infty$, $1/g(z)$ is analytic and, hence, $g(z)$ has a pole. Therefore, the restriction $f$ of $g$ to $\mathbb{C}$ is a meromorphic function. At $\infty$, $g(1/w)$ is either analytic or has a pole and this implies that $f$ has a pole at $\infty$. This completes the proof. □

**Complex Projective Space.** The Riemann Sphere can also be described in a way that treats all points alike from the beginning, so that there is no special point $\infty$. This is its description as *Complex Projective Space* of dimension 1 as described below.

Let $\mathbb{C}^2$ denote the complex two-dimensional vector space consisting of all ordered pairs $(z, w)$ of complex numbers. We define a space $P^1(\mathbb{C})$ by starting with the space of non-zero vectors in $\mathbb{C}^2$ and then identifying any two of these that differ only by a scalar factor. That is, we identify $(z_1, w_1)$ and $(z_2, w_2)$ if there is a complex scalar $\lambda \neq 0$ such that $(z_2, w_2) = (\lambda z_1, \lambda w_1)$.

The set $\{(z, 0) : 0 \neq z \in \mathbb{C}\}$ is a single point $p$ of $P^1(\mathbb{C})$, since any two vectors in this set are scalar multiples of one another. Similarly, the set $\{(0, w) : 0 \neq w \in \mathbb{C}\}$ is a single point $q$ of $P^1(\mathbb{C})$. If we remove $p$ from $P^1(\mathbb{C})$, then there is a well-defined map

$$\phi : P^1(\mathbb{C}) \setminus \{p\} \to \mathbb{C}, \quad \text{where} \quad \phi(z, w) = \frac{z}{w}.$$

Note that $\phi$ defines a well-defined map on $P^1(\mathbb{C}) \setminus \{p\}$ because it sends equivalent points $(z, w)$ and $(\lambda z, \lambda w)$ to the same point $z/w$. Similarly, there is a well-defined map

$$\psi : P^1(\mathbb{C}) \setminus \{q\} \to \mathbb{C}, \quad \text{where} \quad \psi(z,w) = \frac{w}{z}.$$

Both $\phi$ and $\psi$ are one-to-one and onto. The inverse map for $\phi$ sends the complex number $z$ to the equivalence class containing $(z,1)$, while the inverse map for $\psi$ sends $w$ to the equivalence class containing $(1,w)$. Then, on $\mathbb{C}\setminus\{0\}$, the composition $\psi \circ \phi^{-1}$ is defined and it is, in fact,

$$\psi \circ \phi^{-1}(z) = \frac{1}{z}.$$

This means that $P^1(\mathbb{C})$ may also be thought of as two copies of $\mathbb{C}$ glued together along $\mathbb{C} \setminus \{0\}$ by identifying the point $z$ in one copy of $\mathbb{C} \setminus \{0\}$ with the point $1/z$ in the other copy. Definitions of continuity and complex differentiability for functions on $P^1(\mathbb{C})$ may then be defined at points on each copy of $\mathbb{C}$. These definitions will agree on the overlap because $1/z$ is an analytic function of $z$ on $\mathbb{C} \setminus \{0\}$. The space $P^1(\mathbb{C})$, with topology and notion of analytic function as defined above, is one-dimensional complex projective space.

It is apparent that our description of $P^1(\mathbb{C})$ is just another description of the Riemann Sphere $S^2$.

**Stereographic Projection.** Why do we call $S^2$ the Riemann *Sphere*? It does not appear to look like a sphere at all. However, there is a standard construction that shows how a plane with the point at infinity adjoined can be made to look like a sphere.

Consider a (hollow) sphere $S$ in $\mathbb{R}^3$. If one point is removed from $S$, the result is a space which can be mapped in a one-to-one fashion onto the plane. To see this, imagine $S$ sits in $\mathbb{R}^3 = \{(x,y,t) : x,y,t \in \mathbb{R}\}$, above the $xy$-plane and is tangent to it at the origin. Thus, the $t$-axis meets the sphere at the origin and one other point $p$. Then every line through $p$, that is not tangent to the sphere, meets the sphere $S$ at exactly one point $u$, other than $p$, and also meets the $xy$-plane at exactly one point $v$. The function $\rho$ which sends the point $u$ to the point $v$ is a one-to-one map of the sphere with $p$ removed onto the $xy$-plane. This function is called the *stereographic projection* of the sphere to the plane. The standard topology on the sphere is such that $\rho$ and its inverse function are both continuous. This means that a set is open in $S \setminus \{p\}$ if and only if its image in the plane is open. A one-to-one onto map with this property is called a *homeomorphism* between the two spaces involved, and its existence means that the two spaces are equivalent as topological spaces.

Note that, if $S$ is thought of as a globe representing the earth and $p$ is thought of as the north pole, then the spherical open discs on the sphere, centered at $p$, are just the sets consisting of all points above a certain north latitude. Each such disc is taken by the stereographic projection $\rho$ onto the exterior of a closed disc, centered at 0, in the $xy$-plane.

This discussion suggests that the sphere $S$, as a topological space, is homeomorphic to a space constructed from the plane by adding a single point and then defining the topology in such a way that exteriors of closed discs centered at 0 are deleted neighborhoods of $\infty$. This is how we constructed the Riemann Sphere.

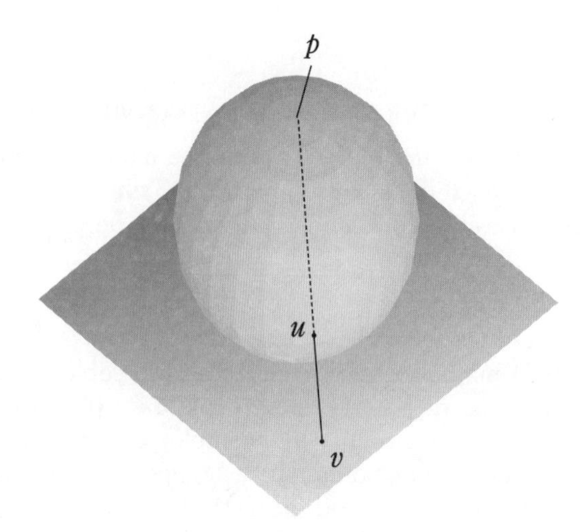


**Figure 6.2.1.** Stereographic Projection.

## Exercise Set 6.2

1. Prove that if $f$ is defined in a neighborhood of $z_0 \in \mathbb{C} \setminus \{0\}$ , then $f(z)$ is differentiable with respect to $z$ at $z_0$ if and only if $f(1/w)$ is differentiable with respect to $w$ at $w_0 = 1/z_0$. What is the relationship between the two derivatives?
2. Show that the function $f(z) = \dfrac{1}{z^2+5}$ is analytic at $\infty$ if it is given the value 0 at $\infty$.
3. Show that the function $f(z) = 2z - 1$ is not analytic at $\infty$ as a complex-valued function, but it is analytic at $\infty$ as an $S^2$-valued function.
4. Show that $e^z$ is not analytic at $\infty$ – either as a complex-valued function or as an $S^2$-valued function.
5. If $f(z) = \dfrac{z^4}{z^2-1}$ is considered as an analytic function from $S^2$ to $S^2$, where are the poles of $f$ and what are their orders?
6. Prove that a polynomial of degree $n$ determines an analytic function from $S^2$ to $S^2$ which has a pole of degee $n$ at $\infty$ and no other poles.
7. Show that if $f$ is an analytic function from $S^2$ to $S^2$ which has no poles on $\mathbb{C}$, then $f$ is a polynomial.
8. Find a conformal equivalence from $S^2$ to $S^2$ which sends 1 to $\infty$, $-2$ to 0, and 0 to 6.
9. Show that every open cover of $S^2$ has a finite subcover – that is, show that $S^2$ is compact as a topological space (see Section 2.5).
10. Show that if $f$ is a non-constant analytic function from $S^2$ to $S^2$ and $w \in S^2$, then $f$ takes on the value $w$ at most finitely many times. If $w = \infty$, this means that $f$ has only finitely many poles. If $w \neq \infty$, this means that $f(z) - w$ has only finitely many zeroes.

11. Can we consider $\sin z$ as an analytic function from $S^2$ to $S^2$? Justify your answer.
12. Show that if $f$ is a non-constant analytic function from $S^2$ to $S^2$, then $f$ is onto. That is, $f$ takes on each value $p \in S^2$ at least once.
13. Show that $S^2$ is connected. That is, show that $S^2$ cannot be expressed as the disjoint union of two non-empy open subsets.
14. If $f : S^2 \to S^2$ is an analytic function which has no pole or zero at $\infty$, then show that, for sufficiently large $R$, $\int_{|z|=R} f(z)\, dz = 0$.
15. If $f : S^2 \to S^2$ is a non-constant analytic function, $w \in S^2$, and $n(f, w)$ is the number of zeroes of $f(z) - w$, counting multiplicity, if $w \neq \infty$, and is the number of poles of $f$, counting multiplicity, if $w = \infty$, then prove that $n(f, w)$ is constant as a function of $w$. Hint: Use the result of previous exercise and Rouché's Theorem.
16. Prove that every analytic function from $S^2$ to $S^2$ is a rational function – that is, a function of the form $\dfrac{p(z)}{q(z)}$, where $p$ and $q$ are polynomials in $z$.

## 6.3. Linear Fractional Transformations

A *linear fractional transformation* or *Möbius transformation* is a mapping of the form

$$h(z) = \frac{az+b}{cz+d}, \tag{6.3.1}$$

where $ad - bc \neq 0$. Note that, if the condition $ad - bc \neq 0$ is not satisfied, then $h$ is the constant function $b/d$.

The right way to think about a linear fractional transformation is to regard it as a map from $S^2$ to $S^2$, where $\infty$ is taken to $\lim_{z\to\infty} h(z) = a/c$ and the point $-d/c$ is taken to $\infty$. In fact, as we shall see below, the linear fractional transformations are exactly the conformal equivalences from $S^2$ to itself.

**Conformal Automorphisms of $S^2$.** By a *conformal automorphism* of an open subset $U$ of $S^2$ we mean a one to one conformal map of $U$ onto itself. The composition of two conformal automorphisms of $U$ is clearly also a conformal automorphism of $U$, as is the inverse function of a conformal automorphism of $U$. Thus, the conformal automorphisms of an open set $U$ form a group under the operation of composition. We are particularly interested in the group of conformal automorphisms of $S^2$ itself.

We will show that the conformal automorphisms of $S^2$ are exactly the linear fractional transformations – that is, the transformations of the form (6.3.1). There are two types of linear fractional transformations that will be particularly useful in the argument. These are the *affine transformations*, that is, transformations of the form

$$L(z) = az + b,$$

and the inversion transformation $s$, given by

$$s(z) = \frac{1}{z}.$$

**Theorem 6.3.1.** *Each linear fractional transformation is a conformal automorphism of $S^2$. Conversely, each conformal automorphism of $S^2$ is either an affine transformation or a composition $L_1 \circ s \circ L_2$, where $L_1$ and $L_2$ are affine transformations. Hence, each conformal automorphism of $S^2$ is a linear fractional transformation.*

**Proof.** We show that each linear fractional transformation $h$ has an inverse function which is also a linear fractional transformation. In fact, if

$$h(z) = \frac{az+b}{cz+d},$$

and we solve the equation $w = h(z)$ for $z$, the result is

$$z = \frac{-dw+b}{cw-a} = g(w).$$

This suggests that $g$ is an inverse function for $h$. In fact, if we calculate $h \circ g$, then for all $w \in S^2$ except $w = \infty$ and $w = -a/c$ we get

$$h \circ g(w) = \frac{(ab-cd)w}{ab-cd}.$$

Since $ab - cd \neq 0$, we conclude that $h \circ g(w) = w$. The cases $w = \infty$ and $w = -a/c$ can be checked separately (Exercise 6.3.1), and so $h \circ g(w) = w$ for all $w \in S^2$. A similar calculation shows that $g \circ h(z) = z$ for all $z \in S^2$. Hence, $h$ has an inverse function $h^{-1} = g : S^2 \to S^2$, and it is also a linear fractional transformation. Since $h$ has an analytic inverse function, defined on all of $S^2$, it is a conformal automorphism of $S^2$. Thus, each linear fractional transformation is a conformal automorphism of $S^2$.

To prove the converse, we let $f$ be a conformal automorphism of $S^2$ and proceed to show that it has the form claimed in the theorem. We first consider the case where $f(\infty) = \infty$. By Theorem 6.2.4, $s(z) = 1/z$ is a conformal automorphism of $S^2$. Thus, the composition

$$q(z) = s \circ f \circ s(z) = \frac{1}{f(1/z)}$$

is also a conformal automorphism of $S^2$, and it takes 0 to 0. Thus, $q$ is a one-to-one analytic function with a zero at 0. This zero must be of order one since a conformal map has non-vanishing derivative at each point. This implies that $f(1/z)$ has a pole of order 1 at $z = 0$ and is analytic and finite at every other point of $S^2$. Then $f(z)$ has a pole of order 1 at $\infty$ and is analytic and finite everywhere on $\mathbb{C}$. It follows that $z^{-1}(f(z) - f(0))$ is analytic and finite everywhere on $S^2$. Hence, it is constant by Liouville's Theorem. If this constant is $a$ and $b = f(0)$, then $f$ is the affine transformation $f(z) = az + b$. Such a transformation is of the form (6.3.1), with $c = 1$ and $d = 0$, and, hence, is a linear fractional transformation.

We next consider the case where $f(\infty) = k \neq \infty$. If we let

$$p(z) = \frac{1}{z-k},$$

then $p$ is a linear fractional transformation and, hence, a conformal automorphism of $S^2$. The composition

$$p \circ f(z) = \frac{1}{f(z) - k}$$

is then a confromal automorphism of $S^2$ which takes $\infty$ to $\infty$. By the previous paragraph, it has the form $az + b$. This implies that

$$f(z) = \frac{1}{az + b} + k = \frac{akz + bk + 1}{az + b}.$$

This is the composition of the affine transformation $z \to az + b$, followed by $s$, followed by the affine transformation $z \to z + k$. Since this composition is also a linear fractional transformation, the proof is complete. $\square$

**Lines and Circles.** By a circle in $S^2$ we mean either a circle in $\mathbb{C}$ or a line in $\mathbb{C}$ together with the point $\{\infty\}$. That is, lines are to be thought of as circles in $S^2$ which pass through $\infty$.

**Theorem 6.3.2.** *Each linear fractional transformation transforms a circle in $S^2$ to a circle in $S^2$.*

**Proof.** By the previous theorem, each linear fractional transformation is either affine or the composition of affine transformations and $s$. Thus, it suffices to show that $s$ and each affine transformation map circles in $S^2$ to circles in $S^2$.

Each affine transformation $L(z) = az + b$ obviously takes lines to lines and circles to circles, since, if $a = r\, e^{i\theta}$, then $L$ is just rotation through an angle $\theta$, followed by an expansion or contraction by a factor $r$, followed by a translation by $b$. To finish the proof, we must show that $s$ also takes circles in $S^2$ to circles in $S^2$.

Every line or circle in the plane is the solution set to an equation of the form

$$a|z|^2 + \overline{w}z + w\overline{z} + b = 0, \tag{6.3.2}$$

where $a$ and $b$ are real numbers and $w$ is a complex number. Conversely, every equation of this form has a line, circle, point, or the empty set as its set of solutions (Exercise 1.1.18). Given a line or circle with equation (6.3.2), the transformation $s$ takes a point on this line or circle to a point $z$ which satisfies the equation

$$a|z|^{-2} + \overline{w}z^{-1} + w\overline{z}^{-1} + d.$$

On multiplying through by $|z|^2$, this becomes

$$a + \overline{wz} + wz + d|z|^2,$$

which is another equation of the form (6.3.2). Hence, its solution set is a line, circle, point, or the empty set. Clearly, only the first two are possible images of a line or circle under $s$. This completes the proof. $\square$

The preceding result can be very useful in making a quick determination of the image of a set under a linear fractional transformation.

**Example 6.3.3.** Find the image of the unit disc under the transformation

$$h(z) = \frac{2z}{z - i}.$$

**Solution:** We have $h(1) = 1 + i$, $h(-1) = 1 - i$, and $h(i) = \infty$. Since $h$ must take the unit circle to a line or a circle, it evidently takes the unit circle to the vertical line $\text{Re}(z) = 1$. Since the unit circle is the boundary of the open unit disc, $h$ must take this disc to a set having the line $\text{Re}(z) = 1$ and the point $\infty$ as its boundary. Thus, the image must be one of the half-planes $\{z : \text{Re}(z) < 1\}$, $\{z : \text{Re}(z) > 1\}$. Since $h(0) = 0$, the image must be $\{z : \text{Re}(z) < 1\}$.

A linear fractional transformation $h$ is determined by the images of any three distinct points in $S^2$ under $h$.

**Theorem 6.3.4.** *Given two ordered triples $\{w_1, w_2, w_3\}$ and $\{z_1, z_2, z_3\}$ of distinct points of $S^2$, there is exactly one linear fractional transformation $h$ such that $h(w_j) = z_j$, $j = 1, 2, 3$.*

**Proof.** We first show that there is a linear fractional tranformation that takes a given ordered triple of complex numbers $\{w_1, w_2, w_3\}$ to $\{0, 1, \infty\}$. In fact,

$$h(z) = \frac{(w_2 - w_3)(z - w_1)}{(w_2 - w_1)(z - w_3)}$$

does the job.

Now suppose we have a triple $\{w_1, w_2, w_3\}$ of distinct points of $S^2$, where one of the points is $\infty$. Then we can still find an $h$ taking this triple to $\{0, 1, \infty\}$. For example, if $w_3 = \infty$, then

$$h(z) = \frac{z - w_1}{w_2 - w_1}$$

will suffice.

Now to get a linear fractional transformation $h$ which takes $\{w_1, w_2, w_3\}$ to $\{z_1, z_2, z_3\}$, we choose $h_1$ taking $\{w_1, w_2, w_3\}$ to $\{0, 1, \infty\}$ and $h_2$ taking $\{z_1, z_2, z_3\}$ to $\{0, 1, \infty\}$ and then set $h = h_2^{-1} \circ h_1$.

It remains to prove that $h$ is unique. Suppose $g$ is another linear fractional transformation that takes $\{w_1, w_2, w_3\}$ to $\{z_1, z_2, z_3\}$. With $h = h_2^{-1} \circ h_1$, as in the previous paragraph, we have that $f = h_2 \circ g \circ h_1^{-1}$ is a linear fractional transformation that takes $\{0, 1, \infty\}$ to $\{0, 1, \infty\}$. If

$$f(z) = \frac{az + b}{cz + d},$$

then $f(0) = 0$ implies that $b = 0$, $f(\infty) = \infty$ implies that $c = 0$, and $f(1) = 1$ implies that $a/d = 1$. We conclude that $f$ is the identity transformation – that is, $f(z) = z$. This, in turn, implies that

$$g = h_2^{-1} \circ f \circ h_1 = h_2^{-1} \circ h_1 = h. \qquad \square$$

**Automorphisms of the Unit Disc.** We next determine the conformal automorphisms of the open unit disc and study their properties. These maps will play a key role in our proof of the Riemann Mapping Theorem. We begin by identifying which linear fractional transformations map the unit disc to itself.

Let $D = D_1(0)$ be the open unit disc and let $w$ be a point of $D$. Then the linear fractional transformation

$$h_w(z) = \frac{z - w}{1 - \overline{w}z} \tag{6.3.3}$$

maps $w$ to 0, and 0 to $-w$. It also maps the unit circle to itself since, if $|u| = 1$, then $u^{-1} = \overline{u}$, and so

$$\left|\frac{u-w}{1-\overline{w}u}\right| = \left|\frac{u-w}{u^{-1}-\overline{w}}\right| = \left|\frac{u-w}{\overline{u}-\overline{w}}\right| = 1.$$

Since $h_w$ maps the unit circle to itself and a point, $w$, inside the unit circle to another point, 0, inside the unit circle, it must map the unit disc $D$ onto itself (it has to map it to some open set in $S^2$ which has the unit circle as boundary and there are only two choices: $D$ and the complement of its closure). It follows that $h_w$ is a conformal automorphism of $D$.

Suppose that $h$ is some other conformal automorphism of $D$ that takes $w$ to 0. Then $h \circ h_w^{-1}$ is a conformal automorphism of $D$ that takes 0 to 0. By Theorem 3.5.6 there is a complex number $u$ of modulus one such that $h \circ h_w^{-1}(z) = uz$. On replacing $z$ by $h_w(z)$, this implies that $h(z) = uh_w(z)$. Thus, we have proved the following theorem.

**Theorem 6.3.5.** *The conformal automorphisms of the unit disc are the linear fractional transformation of the form*

$$h(z) = u\frac{z-w}{1-\overline{w}z} = uh_w(z),$$

*where* $|u| = 1$ *and* $|w| < 1$.

The following is an elementary calculation and is left as an exercise (Exercise 6.3.8). It will be used in the proof of the Riemann Mapping Theorem.

**Theorem 6.3.6.** *With* $w \in D$ *and* $h_w$ *as above,*

$$h_w'(0) = 1 - |w|^2 \quad \text{and} \quad h_w'(w) = \frac{1}{1-|w|^2}.$$

## Exercise Set 6.3

1. Prove that $h \circ g(w) = w$ for the transformations $h$ and $g$ of Theorem 6.3.1 in the exceptional cases $w = \infty$ and $w = a/c$.
2. Find a linear fractional transformation that takes the line $\mathrm{Re}(z) = 1$ to the circle of radius one centered at $-1$.
3. What is the image of the unit disc under the linear fractional transformation
$$h(z) = \frac{2iz}{z-1}?$$
4. What is the image of the disc of radius 1, centered at $1+i$, under the linear fractional transformation
$$h(z) = \frac{1+z}{1-z}?$$
5. What is the image of the disc of radius 1/2, centered at 1/2, under the linear fractional transformation of the previous exercise?

6. The linear fractional transformation $h(z) = \dfrac{1+z}{1-z}$ takes the open unit disc $D$ to the right half-plane. Show that $h$ sends the set consisting of those points in $D$ that lie outside another circle which intersects the unit circle at right angles and passes through the point $z = 1$ to a set of the form

$$\{z : \text{Re}(z) > 0,\ \text{Im}(z) < a\} \quad \text{or} \quad \{z : \text{Re}(z) > 0,\ \text{Im}(z) > -a\}$$

for some positive $a$. What is $a$?

7. A certain linear fractional transformation $h$ takes the ordered triple of points $\{a_1, b_1, c_1\}$ on circle number one to the ordered triple $\{a_2, b_2, c_2\}$ on circle number two. Just by looking at these points, how can you tell whether $h$ maps the inside of circle one to the inside or to the outside of circle two?
8. Prove Theorem 6.3.6.
9. Prove that, for each $w \in D$, the mapping $h_w$ of (6.3.3) has $h_{-w}$ as its inverse transformation.
10. Find a conformal automorphism of $D$ which takes $1/2$ to 0 and has derivative $3i/4$ at $z = 0$.
11. Find a conformal automorphism of $D$ which takes $1/2$ to $i/3$ and has positive derivative at $1/2$.
12. If $w \in D$, prove that among all analytic functions that map $D$ into $D$, map $w$ to 0, and have a positive derivative at $w$, the one with the largest derivative is the function $h_w$ of (6.3.3). Hint: Use Schwarz's Lemma (Lemma 3.5.5).
13. Each non-singular $2 \times 2$ complex matrix

$$A = \begin{pmatrix} a & b \\ c & d \end{pmatrix}$$

determines a linear fractional transformation

$$\phi_A(z) = \frac{az+b}{cz+d}.$$

Show that this correspondence $A \to \phi_A$ is a group homomorphism – that is, show that it takes the product $AB$ of two matrices $A$ and $B$ to the composition $\phi_A \circ \phi_B$ of the corresponding linear fractional transformations.

14. What is the *kernel* of the map $A \to \phi_A$ of the preceding exercise – that is, for which matrices $A$ is $\phi_A$ the identity transformation $z \to z$?

## 6.4. The Riemann Mapping Theorem

In this section we prove the Riemann Mapping Theorem, which says that every proper, simply connected, open subset of $\mathbb{C}$ is conformally equivalent to the unit disc.

The proof proceeds as follows: Given a proper, simply connected, open subset $U \subset \mathbb{C}$, we fix a point $z_0 \in U$ and consider the family $\mathcal{F}$ of all one-to-one conformal maps of $U$ into the unit disc $D$ which map $z_0$ to 0. We first show that this set is non-empty. We then show that if a given $f \in \mathcal{F}$ does not map $U$ onto $D$, then there is a $g \in \mathcal{F}$ with $|g'(z_0| > |f'(z_0)|$. Finally, we show that there is an $h \in \mathcal{F}$ whose derivative at $z_0$ has maximum absolute value. Such an $h$ must map $U$ onto $D$.

The last step in this program – showing that there is an $h \in \mathcal{F}$ with derivative of maximal modulus at $z_0$ – involves showing that any sequence in $\mathcal{F}$ has a subsequence which converges to a function which is also in $\mathcal{F}$. This means we must prove that $\mathcal{F}$ is a *normal family*. We take up this task first.

**Normal Families.**

**Definition 6.4.1.** Let $U$ be an open subset of $\mathbb{C}$. A collection $\mathcal{F}$ of analytic functions defined on $U$ is called a *normal family* if each sequence in $\mathcal{F}$ either converges uniformly to infinity on each compact subset of $U$ or has a subsequence which converges uniformly on each compact subset of $U$ to an analytic function.

Normal families containing sequences which converge to infinity will not occur in this chapter, but they will play a role in the next chapter in the proof of the Big Picard Theorem.

Here we encounter another one of the amazing properties of analytic functions: It takes very little for a family of analytic functions to be a normal family. In fact, it is enough that the family be uniformly bounded, where $\mathcal{F}$ is said to be *uniformly bounded* on $U$ if there is an $M$ such that $|f(z)| \le M$ for every $z \in U$ and every $f \in \mathcal{F}$.

**Theorem 6.4.2** (Montel's Theorem). *Given an open set $U \subset \mathbb{C}$, each uniformly bounded sequence of analytic functions on $U$ has subsequence which converges uniformly on each compact subset of $U$. Hence, each uniformly bounded family of functions on $U$ is a normal family.*

**Proof.** Let $\{f_n\}$ be a uniformly bounded sequence of functions on $U$. We must show that this sequence has a subsequence that converges uniformly on each compact subset of $U$.

We begin by choosing an enumeration $z_1, z_2, \cdots, z_n, \cdots$ of the points of $U$ with rational coordinates. Since $\mathcal{F}$ is uniformly bounded, there is an $M > 0$ such that $|f(z)| \le M$ for all $z \in U$ and all $f \in \mathcal{F}$. Then $\{f_n(z_1)\}$ is a sequence of complex numbers, bounded in modulus by $M$. It therefore has a convergent subsequence. We denote by $\{f_{1n}\}$ the corresponding subsequence of $\{f_n\}$. Then $\{f_{1n}(z_2)\}$ is a bounded sequence, and so it too has a convergent subsequence. We denote the corresponding subsequence of $\{f_{1n}\}$ by $\{f_{2n}\}$. Continuing in this way we inductively construct a sequence of subsequences of $\{f_n\}$:

$$\begin{array}{ccccc} f_{11}, & f_{12}, & \cdots & f_{1n}, & \cdots \\ f_{21}, & f_{22}, & \cdots & f_{2n}, & \cdots \\ . & . & \cdots & . & \cdots \\ . & . & \cdots & . & \cdots \\ f_{k1}, & f_{k2}, & \cdots & f_{kn}, & \cdots \\ . & . & \cdots & . & \cdots \\ . & . & \cdots & . & \cdots \end{array} \tag{6.4.1}$$

with each sequence a subsequence of the preceding sequence, and $\{f_{kn}(z_j)\}_n$ a convergent sequence of numbers for each $k$ and each $j \le k$. Then the diagonal sequence $\{f_{nn}\}$ is a subsequence of $\{f_n\}$ which converges at every $z_j$.

Let $g_n = f_{nn}$. We will show that $\{g_n\}$ converges uniformly on every compact subset of $U$ – not just pointwise at each rational point $z_j$.

If $w \in U$, we choose $r > 0$ such that the closed disc $\overline{D}_{2r}(w)$ is contained in $U$. Then for each $z \in \overline{D}_r(w)$ we have

$$\overline{D}_r(z) \subset \overline{D}_{2r}(w) \subset U.$$

By Cauchy's estimates, each $f \in \mathcal{F}$ satisfies

$$|f'(z)| \leq \frac{M}{r}.$$

Then, for any two points $z, z' \in \overline{D}_r(w)$ we have

$$|f(z) - f(z')| = \left| \int_{z'}^{z} f'(\lambda)\, d\lambda \right| \leq \frac{M}{r}|z - z'|. \tag{6.4.2}$$

Note that the bound on the right in this estimate is independent of the function $f \in \mathcal{F}$. In particular, it holds for each function in the sequence $\{g_n\}$.

Given $\epsilon > 0$, we choose $\delta = \dfrac{r\epsilon}{3M}$. If $z \in D_r(w)$, there are points with rational coefficients arbitrarily close to $z$. Let $z_j$ be one such point with $|z - z_j| < \delta$. By (6.4.2),

$$|g_n(z) - g_n(z_j)| < \frac{M}{r}\frac{r\epsilon}{3M} = \frac{\epsilon}{3}.$$

We next choose $N$ such that

$$|g_n(z_j) - g_m(z_j)| < \frac{\epsilon}{3} \quad \text{whenever} \quad n, m \geq N.$$

We can do this because $\{g_n(z_j)\}$ is a convergent, hence Cauchy, sequence for each $j$. Then

$$|g_n(z) - g_m(z)| \leq |g_n(z) - g_n(z_j)| + |g_n(z_j) - g_m(z_j)| + |g_m(z_j) - g_m(z)| < \epsilon,$$

if $n, m \geq N$. This proves that $g_n$ is uniformly Cauchy on the disc $D_r(w)$ and, hence, converges uniformly on this disc.

If $K$ is a compact subset of $U$, then the discs of the form $D_r(w)$, where $w \in K$ and $r$ is chosen so that $D_{2r}(w) \subset U$, form an open cover of $K$. Since $K$ is compact, it is covered by finitely many of these discs. Since $\{g_n\}$ converges uniformly on each of these discs, it also converges uniformly on $K$. □

**The Riemann Mapping Theorem.** Let $U$ be a non-empty, connected, proper subset of $\mathbb{C}$ with the property that every non-vanishing analytic function on $U$ has an analytic square root. This property is satisfied, for example, if $U$ is simply connected (Theorem 4.6.16). We fix a point $z_0 \in U$ and let $\mathcal{F}$ denote the set of one-to-one conformal maps $f$ of $U$ into the open unit disc $D$, such that $f(z_0) = 0$. The proof of the Riemann Mapping Theorem consists of several steps leading to the conclusion that there is an element of $\mathcal{F}$ which maps onto $D$.

**Lemma 6.4.3.** *The set $\mathcal{F}$ is non-empty.*

**Proof.** Since $U$ is a proper subset of $\mathbb{C}$, there is a point $\lambda \in \mathbb{C}$ which is not in $U$. Then the function $f(z) = z - \lambda$ is non-vanishing on $U$ and, hence, has an analytic square root $g$. Since $f$ is one-to-one and non-vanishing, so is $g$ and, hence, $g$ is a one-to-one conformal map whose image does not contain 0.

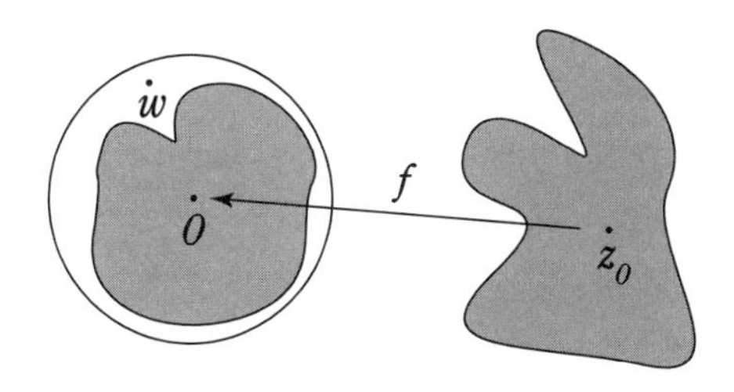


**Figure 6.4.1.** The Mapping $f$ of Lemma 6.4.4.

Since $g$ is analytic, its image $g(U)$ is open by the Open Mapping Theorem (Theorem 4.5.8). It follows that we can find a closed disc $\overline{D}_r(w_0) \subset g(U)$ with $0 < r < \infty$. Since $g^2 = f$ and $f$ is a one-to-one function, the image of $g$ cannot contain a non-zero point and its negative, since then these two points would be taken to the same point by $g^2$. Thus no point of the reflection through the origin, $\overline{D}_r(-w_0)$, of $\overline{D}_r(w_0)$ is contained in $g(U)$. It follows that

$$|g(z) + w_0| > r$$

for all $z \in U$, and this means that the function

$$p(z) = \frac{r}{g(z) + w_0}$$

is a one-to-one conformal map of $U$ into the unit disc. If $p(z_0) = w$, then we can compose $p$ with the conformal automorphism $h_w$ of the unit disc (6.3.3) to obtain a one-to-one conformal map $h_w \circ p$ of $U$ into $D$ which takes $z_0$ to 0. Thus, $\mathfrak{F}$ is not empty. □

**Lemma 6.4.4.** *Let $U$, $z_0$, and $\mathfrak{F}$ be as above. If $f \in \mathfrak{F}$ and $f$ does not map $U$ onto $D$, then there is a $g \in \mathfrak{F}$ with*

$$|g'(z_0)| > |f'(z_0)|.$$

**Proof.** Let $w \in D$ be a point which is not in the image of $f$ (see Figure 6.4.1). Recall from the previous section that the map $h_w$ of (6.3.3) is a conformal automorphism of the unit disc which sends $w$ to 0. Thus, $h_w \circ f(z) \neq 0$ for all $z \in U$. As in the proof of the previous lemma, this means that $h_w \circ f$ has an analytic square root $q$. If $q(z_0) = \lambda$, then $\lambda^2 = w$, $q^2 = h_w \circ f$, and

$$q'(z_0) = \frac{h_w'(0)}{2q(z_0)} f'(z_0) = \frac{(1 - |w|^2)}{2\lambda} f'(z_0) = \frac{(1 - |\lambda|^4)}{2\lambda} f'(z_0).$$

The function $q$ is a one-to-one conformal map of $U$ into the unit disc, but $q(z_0)$ is not 0 – it is $\lambda$. To get an element of $\mathfrak{F}$ we compose $q$ with $h_\lambda$. The resulting map $g = h_\lambda \circ q$ belongs to $\mathfrak{F}$ and satisfies

$$g'(z_0) = h_\lambda'(\lambda) q'(z_0) = \frac{(1 - |\lambda|^4)}{2\lambda(1 - |\lambda|^2)} f'(z_0) = \frac{(1 + |\lambda|^2)}{2\lambda} f'(z_0). \tag{6.4.3}$$

Since $0 < (1 - |\lambda|)^2 = 1 - 2|\lambda| + |\lambda|^2$, we have $2|\lambda| < 1 + |\lambda|^2$. Since a conformal map has non-vanishing derivative, $|f'(z_0)| > 0$. It then follows from (6.4.3) that $|g'(z_0)| > |f'(z_0)|$. □

**Theorem 6.4.5** (Riemann Mapping Theorem). *Let $U$ be a proper, connected open subset of $\mathbb{C}$ with the property that every non-vanishing analytic function on $U$ has an analytic square root. Then there is a conformal equivalence of $U$ onto the unit disc $D$.*

**Proof.** Let $z_0 \in U$ and $\mathcal{F}$ be as above. We know that $\mathcal{F}$ is non-empty by Lemma 6.4.3. We set

$$m = \sup\{|f'(z_0)| : f \in \mathcal{F}\}.$$

Then $m$ is either $+\infty$ or a positive number, since every conformal map has non-vanishing derivative. We will show that there is an element $h$ of $\mathcal{F}$ such that $|h'(z_0)| = m$. Clearly, in view of the previous lemma, such an $h$ maps $U$ onto $D$.

We choose a sequence $\{f_n\}$ of elements of $\mathcal{F}$ such that

$$\lim_{n\to\infty} |f_n'(z_0)| = m.$$

Since $\mathcal{F}$ is a uniformly bounded family of analytic functions, it is a normal family by Theorem 6.4.2. Hence, there is a subsequence of $\{f_n\}$ which converges uniformly on compact subsets of $U$ to a function $h$. It follows from Cauchy's estimates that the derivatives of the functions in this subsequence converge to $h'$ (Exercise 3.2.15), and so $|h'(z_0)| = m$. Since $m \neq 0$, $h$ is not a constant. Since each $f_n$ is one-to-one, it follows that $h$ is also one to one (Exercise 4.5.5). Since $f_n(z_0) = 0$ for every $n$, we also have $h(z_0) = 0$. Hence, $h$ is an element of $\mathcal{F}$ whose derivative at $z_0$ has modulus $m$. By the previous lemma, $h$ must map $U$ onto $D$. Hence, $h$ is a conformal equivalence of $U$ to $D$. □

A conformal equivalence $h$ from an open set $U$ to an open set $V$ is, in particular, a homeomorphism – that is, a one-to-one continuous function from $U$ onto $V$ with a continuous inverse function. Such a function $h$ takes a curve $\gamma$ in $U$ to a curve $h \circ \gamma$ in $V$. Its inverse function $h^{-1}$ takes curves in $V$ to curves in $U$. Clearly homotopies are taken to homotopies by these maps. It follows that if there is a homeomorphism between two open sets, then one is simply connected if and only if the other is also. This implies the following corollary of the Riemann Mapping Theorem.

**Corollary 6.4.6.** *If $U$ is a connected open set in the plane with the property that every non-vanishing analytic function on $U$ has an analytic square root, then $U$ is simply connected.*

This corollary completes the proof of Theorem 4.6.16.

## Exercise Set 6.4

1. Is the set of all conformal automorphisms of the unit disc a normal family? If so, then every sequence of such maps has a convergent subsequence. Is the limit necessarily also a conformal automorphism of the unit disc?

2. If $U$ is a connected open subset of $\mathbb{C}$ and $z_0$ is a point of $U$, prove that the set of all analytic functions $f$ on $U$ with positive real part and with $f(z_0) = 1$ is a normal family.
3. Suppose $U$ is a connected open set and $V$ is a simply connected open set which is not the whole plane. If $z_0 \in U$ and $w_0 \in V$ are fixed, let $\mathcal{F}$ be the family of all analytic functions from $U$ to $V$ which send $z_0$ to $w_0$. Prove that $\mathcal{F}$ is a normal family.
4. Show by example that it is not true that every uniformly bounded family $\mathcal{F}$ of continuous functions on an interval on the line has the property that every sequence in $\mathcal{F}$ has a convergent subsequence.
5. Prove that the complex plane $\mathbb{C}$ is not conformally equivalent to the unit disc even though it is simply connected. Why doesn't this contradict the Riemann Mapping Theorem?
6. Is an open rectangle in the plane conformally equivalent to the open unit disc $D$?
7. Is the open right half-plane, with the interval $[1, \infty)$ removed, conformally equivalent to the unit disc $D$?
8. Is the open right half-plane, with the interval $[1, 2]$ removed, conformally equivalent to the unit disc $D$?
9. Is the punctured disc $D_1(0) \setminus \{0\}$ conformally equivalent to the right half-plane $\{z : \mathrm{Re}(z) > 0\}$?
10. Given one conformal equivalence $h$ from an open set $U$ to the unit disc $D$, can you describe all the others in terms of $h$?
11. With $\mathcal{F}$ as in the proof of Theorem 6.4.5, is it necessarily true that a conformal equivalence $h$ from $U$ to $D$ that takes $z_0 \in U$ to 0 has the property that $|h'(z_0)|$ is maximal among the numbers $|f'(z_0)|$ for $f \in \mathcal{F}$?
12. Prove that if $f$ is analytic and satisfies $|f(z)| < 1$ on the right half-plane, and if $f(1) = 0$, then
$$|f(z)| \leq \left|\frac{z-1}{z+1}\right|$$
for all $z$ in the right half-plane.

## 6.5. The Poisson Integral

Let $U$ be a bounded open set in the plane and let $g$ be a continuous function on $\partial U$. The *Dirichlet problem* for $U$ and $g$ is the problem of finding a function $u$ which is continuous on $\overline{U}$, harmonic on $U$, and equal to $g$ on $\partial U$. The function $g$ is called the *boundary function* for the problem. A uniqueness theorem for solutions to the Dirichlet problem follows easily from the results of Section 3.5 (see Exercise 6.5.2):

**Theorem 6.5.1.** *If $U$ is a bounded open subset of $\mathbb{C}$ and $g$ is a continuous function on $\partial U$, then there is at most one solution to the Dirichlet problem for $U$ and $g$.*

If $U$ is simply connected and has a simple closed curve as its boundary, then the Dirichlet problem always has a solution. One proof that this is so, uses the Riemann Mapping Theorem. This proof proceeds as follows: We exhibit an explicit solution to the Dirichlet problem for the unit disc $D$ and any continuous function on its boundary. The Riemann Mapping Theorem tells us there is a conformal equivalence $h : U \to D$. We use this map to show that the Dirichlet problem on $U$ is equivalent to the Dirichlet problem on $D$.

The difficulty with the above approach is that it requires that the conformal equivalence $h : U \to D$ extends to a one-to-one continuous map $\overline{U} \to \overline{D}$ with a continuous inverse. In fact, this is true and we will state it as a theorem, but we do not include a proof. A proof can be found in Walter Rudin's text [**8**].

**Theorem 6.5.2.** *If $U$ is a simply connected open set with a simple closed curve as boundary, and if $h : U \to D$ is a conformal equivalence, then $h$ extends to a continuous function from $\overline{U}$ to $\overline{D}$ with a continuous inverse.*

Assuming this theorem, we will proceed with the proof that the Dirichlet problem always has a solution for a set $U$ as above. In this section we deal with the first step – the case where $U$ is the unit disc $D$. The next section is devoted to showing how to use the Riemann Mapping Theorem to attack the problem on more complicated sets.

**The Poisson Kernel.** Recall the linear fractional transformation from Example 6.1.4:

$$\frac{1+z}{1-z}. \tag{6.5.1}$$

This is a one-to-one conformal map of the unit disc onto the right half-plane. The *Poisson kernel* is constructed from the real part of this function. This was calculated in Example 6.1.4 by multiplying numerator and denominator in (6.5.1) by the conjugate of the denominator. The result is

$$\operatorname{Re}\left(\frac{1+z}{1-z}\right) = \frac{1-|z|^2}{|1-z|^2}. \tag{6.5.2}$$

If $z = r\,e^{i\theta}$ is the polar form of $z$, then

$$|1-z|^2 = |1-r\,e^{i\theta}|^2 = 1 - 2r\cos\theta + r^2$$

and so the expression on the right in (6.5.2) becomes

$$P_r(\theta) = \frac{1-r^2}{1-2r\cos\theta + r^2}. \tag{6.5.3}$$

This may also be written as

$$P_r(\theta) = \sum_{-\infty}^{\infty} r^{|n|}\,e^{in\theta}$$

by Exercise 6.5.1.

The Poisson kernel is the function

$$\begin{aligned} P_r(\theta - t) &= \frac{1-r^2}{1-2r\cos(\theta-t)+r^2} \\ &= \operatorname{Re}\left(\frac{1+e^{-it}z}{1-e^{-it}z}\right) = \operatorname{Re}\left(\frac{e^{it}+z}{e^{it}-z}\right). \end{aligned} \tag{6.5.4}$$

It produces a harmonic function on $D$ when it is integrated with respect to $t$ against a continuous function on the boundary of $D$.

**Theorem 6.5.3.** *If $D$ is the open unit disc, $g$ is a continuous real-valued function on $\partial D$, and $u$ is defined by*

$$u(r\,e^{i\theta}) = \frac{1}{2\pi}\int_0^{2\pi} g(e^{it})P_r(\theta - t)\,dt,$$

*then $u$ is a harmonic function of $z = r\,e^{i\theta}$ on $D$.*

**Proof.** Since $g$ is real-valued, the function $u$ is the real part of

$$f(z) = \frac{1}{2\pi}\int_0^{2\pi} g(e^{it})\frac{e^{it}+z}{e^{it}-z}\,dt. \tag{6.5.5}$$

This function is analytic on the complement of the unit circle (Exercise 3.2.16) and, in particular, on $D$. Hence, its real part $u$ is harmonic on $D$. □

The integral defining the harmonic function $u$ in the above theorem is called the *Poisson integral.*

**Boundary Values.** We will show that the function $u$, defined in the above theorem, has the property that

$$\lim_{z\to w} u(z) = g(w)$$

for each $w$ on the unit circle. This implies that we get a continuous extension of $u$ to the closed unit disc $\overline{D}$ if we define $u(w)$ to be $g(w)$ at each $w \in \partial D$. Before proving this, we need to establish some properties of the function $P_r(\theta)$.

**Lemma 6.5.4.** *The Poisson kernel $P_r(\theta - t)$ of (6.5.4) satisfies*

$$\frac{1}{2\pi}\int_{-\pi}^{\pi} P_r(\theta - t)\,dt = 1$$

*for every $r < 1$ and every $\theta$.*

**Proof.** We have

$$\frac{1}{2\pi}\int_{-\pi}^{\pi} P_r(\theta - t)\,dt = \operatorname{Re}\left(\frac{1}{2\pi}\int_{-\pi}^{\pi}\frac{e^{it}+z}{e^{it}-z}\,dt\right). \tag{6.5.6}$$

We can easily evaluate this using residue theory. In fact,

$$\begin{aligned}\frac{1}{2\pi}\int_{-\pi}^{\pi}\frac{e^{it}+z}{e^{it}-z}\,dt &= \frac{1}{2\pi i}\int_{-\pi}^{\pi}\frac{1+e^{-it}z}{e^{it}-z}\,i\,e^{it}\,dt\\ &= \frac{1}{2\pi i}\int_{|w|=1}\frac{1+z/w}{w-z}\,dw.\end{aligned}$$

The last expression can be evaluated by applying the Residue Theorem to the function

$$\frac{1+z/w}{w-z} = \frac{w+z}{w(w-z)}.$$

This has residue $-1$ at $w = 0$ and residue 2 at $w = z$. Hence,

$$\frac{1}{2\pi}\int_{-\pi}^{\pi}\frac{e^{it}+z}{e^{it}-z}\,d\theta = 2 - 1 = 1.$$

Thus, its real part is also 1. In view of (6.5.6), the proof is complete. □

**Lemma 6.5.5.** *If $g$ is a continuous function on the unit circle, then*

$$\lim_{r\to 1-}\frac{1}{2\pi}\int_{-\pi}^{\pi} g(e^{it})P_r(\theta-t)\,dt = g(e^{i\theta}). \tag{6.5.7}$$

*Furthermore, the convergence is uniform in $\theta$.*

**Proof.** The idea of the proof is this: When $r$ is near 1, $P_r(\theta - t)$ is nearly 0 for all $t \in [-\pi, \pi]$ except near $\theta$, where the function has a large spike. This means that only values of $g(e^{it})$ for $t$ near $\theta$ contribute significantly to the integral (6.5.7). The proof below makes this claim precise.

The function $P_r(\theta)$ is a positive, even function of $\theta$. Since its denominator is increasing on the interval $[0, \pi]$, this function is decreasing as $\theta$ moves away from 0 in either direction. It follows that, if $\eta$ is a fixed number between 0 and $\pi$, then

$$P_r(\theta-t) \le P_r(\eta) \quad \text{when} \quad \eta \le |\theta - t| \le \pi. \tag{6.5.8}$$

Since $g$ is continuous on the unit circle, which is compact, $g$ is uniformly continuous. Thus, given $\epsilon > 0$, we may choose $\eta > 0$, which does not depend on $\theta$, such that

$$|g(e^{it}) - g(e^{i\theta})| < \epsilon/2 \quad \text{whenever} \quad |\theta - t| < \eta. \tag{6.5.9}$$

It follows from Lemma 6.5.4 that

$$\frac{1}{2\pi}\int_{-\pi}^{\pi} g(e^{it})P_r(\theta-t)\,dt - g(e^{i\theta}) = \frac{1}{2\pi}\int_{-\pi}^{\pi}(g(e^{it}) - g(e^{i\theta}))P_r(\theta-t)\,dt.$$

The absolute value of this expression is less than or equal to

$$\frac{1}{2\pi}\int_{-\pi}^{\pi}|g(e^{it}) - g(e^{i\theta})|P_r(\theta-t)\,dt.$$

We split this integral into the sum of an integral over $[\theta-\eta, \theta+\eta]$ and integrals over $[-\pi, \theta-\eta]$ and $[\theta+\eta, \pi]$. The first summand may be estimated using (6.5.9) and Lemma 6.5.4:

$$\frac{1}{2\pi}\int_{\theta-\eta}^{\theta+\eta}|g(e^{it}) - g(e^{i\theta})|P_r(\theta-t)\,dt < \frac{\epsilon}{2}\frac{1}{2\pi}\int_{\theta-\eta}^{\theta+\eta}P_r(\theta-t)\,dt \le \frac{\epsilon}{2}. \tag{6.5.10}$$

The integrals over $[-\pi, -\theta_0]$ and $[\theta_0, \pi]$ are estimated using (6.5.8). If $M$ is an upper bound for $g$ on the circle, then

$$\begin{aligned} &\frac{1}{2\pi}\int_{\theta+\eta}^{\pi}|g(e^{it}) - g(e^{i\theta})|P_r(\theta-t)\,dt \\ &\qquad < P_r(\eta)\frac{1}{2\pi}\int_{\theta+\eta}^{\pi}|g(e^{it}) - g(e^{i\theta})|\,dt \le MP_r(\eta), \end{aligned} \tag{6.5.11}$$

and the integral over $[-\pi, \theta-\eta]$ satisfies the same estimate.

Since $0 < \eta < \pi$,

$$\lim_{r\to 1-} P_r(\eta) = \lim_{r\to 1-}\frac{1-r^2}{1-2r\cos\eta + r^2} = 0,$$

because the numerator has limit 0 and the denominator has limit $2 - 2\cos\eta \ne 0$.

We choose $\delta > 0$ such that

$$|P_r(\eta)| < \frac{\epsilon}{4M} \quad \text{if} \quad 1-\delta < r < 1.$$

Then (6.5.11) implies that

$$\frac{1}{2\pi}\int_{\theta+\eta}^{\pi} |g(e^{it}) - g(e^{i\theta})| P_r(\theta - t)\, dt \le M\frac{\epsilon}{4M} = \frac{\epsilon}{4}. \tag{6.5.12}$$

Putting (6.5.10) together with (6.5.12) and the analogous estimate for the integral on $[-\pi, \theta - \eta]$, we conclude that

$$\left|\frac{1}{2\pi}\int_{-\pi}^{\pi} g(e^{i\theta})P_r(\theta)\, d\theta - g(e^{i\theta})\right| < \epsilon$$

for $1-\delta < r < 1$ and for every $\theta \in [-\pi, \pi]$. This proves the Theorem. □

The next theorem tells us that the Poisson integral yields a solution to the Dirichlet problem for the unit disc and an arbitrary continuous boundary function $g$.

**Theorem 6.5.6.** *If $g$ is a real-valued continuous function on the unit circle and $u$ is defined on the closed unit disc $\overline{D}$ by*

$$u(r\,e^{i\theta}) = \frac{1}{2\pi}\int_0^{2\pi} g(e^{it})P_r(\theta - t)\, dt \tag{6.5.13}$$

*if $r < 1$ and by $u(e^{i\theta}) = g(e^{i\theta})$ on the unit circle, then $u$ is continuous on $\overline{D}$ and harmonic on $D$.*

**Proof.** In view of Theorem 6.5.3, the only thing left to prove is that

$$\lim_{z \to e^{i\theta_0}} u(z) = g(e^{i\theta_0}) \quad \text{for all} \quad \theta_0. \tag{6.5.14}$$

The previous lemma almost says this, but not quite. The limit there is taken only along radial lines, and this is not exactly what is meant by (6.5.14). We need to do a bit more work.

Because the convergence in Lemma 6.5.5 is uniform in $\theta$, given $\epsilon > 0$ we may choose a $\delta_1 > 0$ such that

$$|u(r\,e^{i\theta}) - g(e^{i\theta})| \le \frac{\epsilon}{2}$$

for all $\theta$ and all $r$ with $1 - \delta_1 < r < 1$. We may also choose $\delta_2$ such that

$$|g(e^{i\theta}) - g(e^{i\theta_0})| < \frac{\epsilon}{2}$$

whenever $|e^{i\theta} - e^{i\theta_0}| < \delta_2$. If $\delta = \min\{\delta_1, \delta_2\}$, then $|r\,e^{i\theta} - e^{i\theta_0}| < \delta$ implies both $|r\,e^{i\theta} - e^{i\theta}| < \delta_1$ and $|e^{i\theta} - e^{i\theta_0}| < \delta_2$, from which it follows that

$$|u(r\,e^{i\theta}) - g(e^{i\theta_0})| \le |u(r\,e^{i\theta}) - g(e^{i\theta})| + |g(e^{i\theta}) - g(e^{i\theta_0})| < \epsilon.$$

This proves (6.5.14). □

**Poisson's Formula.** Suppose $g$ is a continuous function defined on the circle of radius $R$ centered at $z_0$ and we wish to solve the Dirichlet problem for $D_R(z_0)$ and $g$. By using the conformal map $z \to z_0 + Rz$, which takes $D$ to $D_R(z_0)$, we can transform this problem into a Dirichlet problem on the unit disc. The solution given by Theorem 6.5.3 may then be pulled back to $D_R(z_0)$ using the inverse map $z \to (z - z_0)/R$. The result is a solution to the Dirichlet problem for $D_R(z_0)$ and $g$ given by

$$u(r\,\mathrm{e}^{i\theta}) = \frac{1}{2\pi}\int_0^{2\pi} g(z_0 + R\,\mathrm{e}^{it})\frac{(R^2 - r^2)}{R^2 - 2rR\cos(\theta - t) + r^2}\,dt. \tag{6.5.15}$$

We leave the details to Exercise 6.5.8.

This result may be used to prove Poisson's Integral Formula for a harmonic function on an open set $U$:

**Theorem 6.5.7.** *Let $u$ be a real-valued harmonic function defined on an open set $U$ in the plane. If $\overline{D}_R(z_0) \subset U$, then*

$$\begin{aligned} u(z) &= \frac{1}{2\pi}\int_0^{2\pi} u(z_0 + R\,\mathrm{e}^{it})\frac{(R^2 - r^2)}{R^2 - 2Rr\cos(\theta - t) + r^2}\,dt \\ &= \mathrm{Re}\left(\frac{1}{2\pi}\int_0^{2\pi} u(z_0 + R\,\mathrm{e}^{it})\frac{R\,\mathrm{e}^{it} + z}{R\,\mathrm{e}^{it} - z}\,dt\right) \end{aligned}$$

*for all $z = z_0 + r\,\mathrm{e}^{i\theta} \in D_R(z_0)$.*

The proof is left as an exercise (Exercise 6.5.9)

Note that the second equality in the above theorem expresses $u$ as the real part of an analytic function in the disc $D_R(z_0)$.

## Exercise Set 6.5

1. Prove that the Poisson kernel may also be written as

$$P_r(\theta) = \sum_{-\infty}^{\infty} r^{|n|}\,\mathrm{e}^{in\theta}.$$

2. Prove the uniqueness theorem (Theorem 6.5.1) for solutions of the Dirichlet problem.
3. By inspection, find a solution to the Dirichlet problem on $D$ if the boundary function $g$ is $g(\mathrm{e}^{it}) = \cos t$. Use your answer to evaluate

$$\frac{1}{2\pi}\int_0^{2\pi} \frac{(1 - r^2)\cos t}{1 - 2r\cos(\theta - t) + r^2}\,dt.$$

4. Show that the Poisson kernel $P_r(\theta - t)$ is a harmonic function of $z = r\,\mathrm{e}^{i\theta}$ on the open unit disc.
5. Equation (6.5.5) expresses the harmonic function $u$, given by the Poisson integral, as the real part of an analytic function on $D$. Use this equation to find an integral formula for a harmonic conjugate to $u$.

6. Show that if $g$ is a positive, continuous function on the unit circle, then the analytic function $f$ defined by equation (6.5.5) has positive real part and its imaginary part vanishes at $z = 0$.
7. Let $f$ be analytic on the open unit disc $D$. If the real part of $f$ is positive and extends to be continuous on $\overline{D}$, then prove that $f$ is a function of the form given in the previous exercise plus an imaginary constant.
8. Supply the details of the proof that (6.5.15) gives a solution to the Dirichlet problem for $D_R(z_0)$ with boundary function $g$.
9. Prove Theorem 6.5.7.
10. Show that if $u$ is a real-valued, non-negative harmonic function on an open set containing $\overline{D}_R(z_0)$, then

$$u(z_0)\frac{R-r}{R+r} \le u(z) \le u(z_0)\frac{R+r}{R-r}$$

for $0 \le r = |z - z_0| < R$. This is Harnack's Inequality. Hint: Use Theorem 6.5.7.
11. If $u$ is harmonic and non-negative on $D$ and $u(0) = 1$, give upper and lower bounds for $u(1/2)$. Hint: Use the result of the previous exercise.
12. Prove that if $u_n$ is a sequence of positive harmonic functions on a connected open set $U$, and if $u_n(z_0) \to 0$ at one point $z_0 \in u$, then $u_n(z) \to 0$ at every point $z$ of $U$. Hint: Use the result of Exercise 10.

## 6.6. The Dirichlet Problem

The Dirichlet problem is of great importance in physics and engineering. Here we briefly discuss some of the applications that make this so. We then move on to showing how to use conformal equivalence to reduce the Dirichlet problem for a set $U$ and boundary function $g$ to a simpler problem which, hopefully, we know how to solve.

**Heat Flow.** The termperature $T$ at points of a flat plate represented by a region $U$ in the plane satisfies the *heat equation*

$$\frac{\partial T}{\partial t} = k\Delta T = k\left(\frac{\partial^2 T}{\partial^2 x} + \frac{\partial^2 T}{\partial^2 y}\right). \tag{6.6.1}$$

If the set $U$ is bounded and if each point $w$ of its boundary is maintained at a constant temperature $T_0(w)$, then, over time, the temperature distribution on the entire plate will approach an equilibrium state $\tilde{T}$ – one that does not change with time. This means $\partial\tilde{T}/\partial t = 0$. The heat equation (6.6.1) then implies that $\tilde{T}$, as a function of $x$ and $y$ satisfies Laplace's equation. Hence, $\tilde{T}$ is a harmonic function on $U$ which agrees with the boundary function $T_0$ on $\partial U$. Thus, to find the equilibrium temperature on $U$ that will result from maintaining the temperature on the boundary at $T_0$ means solving the Dirichlet problem for $U$ and $T_0$.

**Electrostatics.** An electric field on a plane region $U$ is a vector function

$$E(z) = (E_1(z), E_2(z))$$

which, for each $z \in U$, describes the force that would be exerted on a charged particle located at $z$. This vector function has the form

$$E = -\nabla\phi = -\left(\frac{\partial\phi}{\partial x}, \frac{\partial\phi}{\partial y}\right)$$

for a real-valued function $\phi$ on $U$ called the *potential function* for the electric field $E$. A static electric field (one that is not changing in time) satisfies

$$\text{div } E = \frac{\partial E_1}{\partial x} + \frac{\partial E_2}{\partial x} = 0.$$

This means its potential function $\phi$ satisfies Laplace's equation and is, therefore, harmonic.

If the boundary of $U$ is maintained at a potential given by the boundary function $\psi$ on $\partial U$, then the problem of finding the resulting potential at points of $U$ is the Dirichlet problem for $U$ with boundary function $\psi$.

**Hydrodynamics.** In the study of two-dimensional fluid flow for non-viscous fluids, the velocity of the fluid at a given point $z$ is a vector $q(z) = (q_1(z), q_2(z))$. If the fluid is incompressible and the flow is irrotational, so that

$$\text{div } q = 0 \quad \text{and} \quad \text{curl } q = 0,$$

then the vector $q$ is the gradient $\nabla\phi$ of a potential function $\phi$ which satisfies Laplace's equation and, hence, is a harmonic function. A typical fluid flow problem is not, however, a Dirichlet type problem. The typical boundary condition for fluid flow is that the flow along the walls of the container should be parallel to the walls; that is, the component $\partial\phi/\partial n$ of $\nabla\phi$ normal to the wall should be 0. The problem of finding a harmonic function which satisfies this condition on a given region is called the *Neumann problem*.

**Existence of Solutions.** Given a bounded simply connected open set $U$ with a simple closed curve for its boundary, the Riemann Mapping Theorem and Theorem 6.5.2 tell us that there is a homeomorphism $h : \overline{U} \to \overline{D}$ which, on $U$, is a conformal map of $U$ onto $D$. This allows us to prove the existence of a solution to the Dirichlet problem for $U$ in the following way.

If $g$ is a continuous real-valued function on $\partial U$, then $g \circ h^{-1}$ is a continuous real valued function on $\partial D$. By Theorem 6.5.6, $g \circ h^{-1}$ is the restriction to $\partial D$ of a function $u$ which is continous on $\overline{D}$ and harmonic on $D$. Then $u \circ h$ is a continuous function on $\overline{U}$ which equals $g$ on $\partial U$. It is also harmonic on $U$, since the composition of an analytic function followed by a harmonic function is harmonic (Exercise 6.6.11). This proves that the Dirichlet problem for $U$ and $g$ has a solution for every continuous function $g$ on $\partial U$.

The Dirichlet problem also makes sense for unbounded open sets. Suppose $U$ is an unbounded, simply connected, open subset of $\mathbb{C}$ with closure $\overline{U}$ a proper subset of $C$. If we choose a point $w \in \mathbb{C} \setminus \overline{U}$, then the linear fractional transformation

$$h(z) = \frac{1}{z - w}$$

is a conformal automorphism of $S^2$ which takes $U$ to a bounded simply connected open subset $V$ of $\mathbb{C}$. If $V$ has a simple closed curve as boundary, then we will say that $U$ has a simple closed curve as boundary in $S^2$. If this is the case, then the Dirichlet problem can be solved for $V$ and any continuous function on $\partial V$. The map $h$ is a homeomorphism from the closure of $U$ in $S^2$ to the closure of $V$ and so it can be used to transform a continuous function $g$ on the boundary of $U$ in $S^2$ to a continuous function $g \circ h^{-1}$ on the boundary of $V$. A solution to the Dirichlet problem for $V$ and $g \circ h$ may then be transformed back to a solution to the Dirichlet problem for $U$ and $g$ by composing with $h$.

The following theorem summarizes the preceding discussion.

**Theorem 6.6.1.** *Let $U$ be a proper simply connected open subset of $\mathbb{C}$ with a simple closed curve as its boundary in $S^2$. Then each continous function on the boundary of $U$ in $S^2$ is the restriction of a continuous function on the closure of $U$ in $S^2$ which is harmonic on $U$.*

**Remark 6.6.2.** The hypothesis that $g$ is continuous in the above theorem is not really necessary if one is willing to accept somewhat less in the conclusion. For example, if $g$ is bounded and continuous except at a finite set of discontinuities, then Lemma 6.5.3 continues to hold, while in Lemma 6.5.5 and Theorem 6.5.6, the conclusions about limits and continuity of $u$ at boundary points continue to hold at the points where $g$ is continuous. This implies that the previous theorem can be modified to cover the case where $g$ is bounded and continuous except at a finite set of discontinuities. The conclusion is then that $g$ is the restriction to $\partial U$ of a function which is harmonic on $U$ and continuous on $\overline{U}$ except at those points of $\partial U$ where $g$ is not continuous.

**Finding Explicit Solutions.** The previous theorem tells us that solutions to the Dirichlet problem exist if $U$ is simply connected with a simple closed curve as boundary, but it does not give us an explicit expression for the solution in terms of elementary functions. It is too much to hope for this in general, but there are situations where a conformal equivalence between $U$ and some other set $V$ reduces the problem to one where we can find an explicit solution, often by inspection. This is another way in which conformal maps aid in the study of the Dirichlet problem. We will demonstrate this with a series of examples.

**Example 6.6.3.** Let $U$ be the strip $\{z : -1 < \mathrm{Im}(z) < 1\}$. Find a solution to the Dirichlet problem for $U$ and the boundary function which is 0 on the line $\{z : \mathrm{Im}(z) = -1\}$ and 2 on the line $\{z : \mathrm{Im}(z) = 1\}$.

**Solution:** This one is easy. The function $1 - iz$ is analytic on $\mathbb{C}$ and so its real part

$$1 + \mathrm{Im}(z) = 1 + y$$

is harmonic on $\mathbb{C}$. The restriction of this function to $\{z : -1 \le \mathrm{Im}(z) \le 1\}$ is continuous, is harmonic on $\{z : -1 < \mathrm{Im}(z) < 1\}$, and has the required values on the boundary of this set.

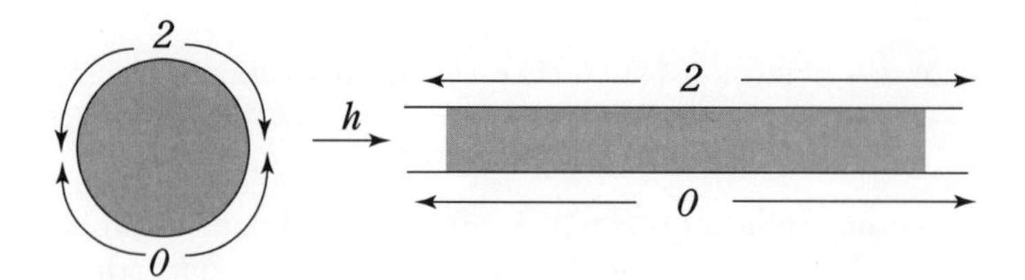


**Figure 6.6.1.** The Picture for Example 6.6.4.

**Example 6.6.4.** Find a function on the closed unit disc which is continuous except at $\pm 1$, is harmonic on the open unit disc, and is 0 on the lower half of the unit circle and 2 on the upper half of the unit circle.

**Solution:** In Example 6.1.7 we showed that the function

$$h(z) = \frac{2}{\pi} \log\left(\frac{1+z}{1-z}\right)$$

is a conformal equivalence from the unit disc to the strip $\{z : -1 < \operatorname{Im}(z) < 1\}$. This map takes the upper semicircle to the line $\operatorname{Im}(z) = 1$ and the lower semicircle to the line $\operatorname{Im}(z) = -1$. It follows that a solution to the problem may be obtained by composing the solution to the previous problem with $h$ (see Figure 6.6.1). This yields

$$u(z) = 1 + \operatorname{Im}(h(z)) = 1 + \frac{2}{\pi} \arg\left(\frac{1+z}{1-z}\right),$$

so that

$$u(x,y) = 1 + \frac{2}{\pi} \tan^{-1}\left(\frac{2y}{1-x^2-y^2}\right)$$

is our solution. Here, for points on the unit circle $x^2 + y^2 = 1$, we interpret this expression to mean its limit as the point $(x,y)$ is approached from within the unit disc. Note that the resulting function $u$ is, indeed, continuous on the closed disc, except at the points $\pm 1$ where the boundary function fails to be continuous.

**Example 6.6.5.** Let $g$ be the function on $\mathbb{R}$ which is $\sqrt{1-x^2}$ for $-1 < x < 1$ and is 0 for all other values of $x$. Find a harmonic function on the upper half-plane which has $g$ as its boundary function and which has limit 0 as $z \to \infty$ in the upper half-plane.

**Solution:** We use the results of Example 6.1.8. There we showed that $\sin z$ is a conformal equivalence of the strip $A = \{z : -\pi/2 < \operatorname{Re}(z) < \pi/2,\ \operatorname{Im}(z) > 0\}$ onto the upper half-plane. If we compose $g$ with $\sin z$, we get the function $q$ on $\partial A$ which is $\cos x$ on the real line between $-\pi/2$ and $\pi/2$ and is 0 on the vertical lines $\operatorname{Re}(z) = \pm\pi/2$. The real part $\mathrm{e}^{-y}\cos x$ of $\mathrm{e}^{iz}$ is a harmonic function on $A$ which has $q$ as boundary function and has limit 0 as $z \to \infty$ in $A$. Hence, we obtain a solution $u$ to our problem by setting

$$f(z) = \mathrm{e}^{i \sin^{-1}(z)} \quad \text{and} \quad u(z) = \operatorname{Re}(f(z))$$

for $z$ in the upper half-plane. We can simplify this.

If $w = \sin^{-1}(z)$, then $\sin w = z$ and $\cos^2 w = 1 - z^2$. Since, for every $z$ in the open upper half-plane, $1 - z^2$ is in the domain of the principal branch of the square

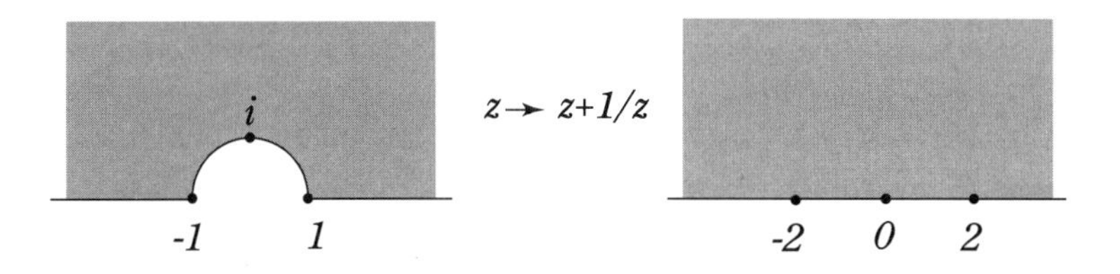


**Figure 6.6.2.** The Picture for Example 6.6.6.

root function (denoted $\sqrt{\cdot}$ ), we have $\cos w = \pm\sqrt{1-z^2}$. Since $\cos w$ has positive real part if $w \in A$ (see (6.1.4)), and the principal branch of the square root function takes values in the right half-plane, we conclude that $\cos w = \sqrt{1-z^2}$. Then

$$e^{iw} = \cos w + i\sin w = \sqrt{1-z^2} + iz,$$

and so

$$f(z) = \sqrt{1-z^2} + iz.$$

Then our solution $u$ is

$$\begin{aligned} u(z) &= \operatorname{Re}\left(\sqrt{1-z^2}+iz\right) \\ &= -y + \sqrt{\frac{1}{2}\left(1-x^2+y^2+\sqrt{(1-x^2+y^2)^2+4x^2y^2}\right)}. \end{aligned} \tag{6.6.2}$$

This last equality is left as an exercise (Exercise 6.6.13).

**Example 6.6.6.** Let $B$ be the open upper half-plane with the upper half of the closed unit disc removed. That is,

$$B = \{z : |z| > 1,\ \operatorname{Im}(z) > 0\}.$$

Solve the Neumann problem for $B$: Find a function $u$ on $\overline{B}$ which is harmonic on $B$ and has normal derivative 0 at every boundary point of $B$.

**Solution:** The same problem on the upper half-plane has an easy solution: the function $\operatorname{Re}(z) = x$ is harmonic on the upper half-plane and its gradient $(1, 0)$ is parallel to the real axis at each point. Thus, its normal derivative on the real axis is 0. To get a solution to our original problem, we need only compose this solution with a conformal equivalence of $B$ to the upper half-plane. In fact

$$z \to z + \frac{1}{z}$$

is such a map (Exercise 6.6.9). Thus, our solution is

$$\begin{aligned} u(z) &= \operatorname{Re}\left(z+\frac{1}{z}\right) \\ &= x + \frac{x}{x^2+y^2}. \end{aligned}$$

Any constant multiple of this solution is also a solution. If this solution represents the potential function for a fluid flow problem, then the corresponding velocity vector field for the flow would be its gradient

$$\nabla u(x,y) = \left(1 + \frac{y^2 - x^2}{(x^2+y^2)^2}, -\frac{2xy}{(x^2+y^2)^2}\right).$$

## Exercise Set 6.6

1. Solve the Dirichlet problem for the upper half-plane and the boundary function which is 1 on the positive real axis and 0 on the negative real axis. Hint: Think about the function $\arg(z)$ on the upper half-plane.
2. Use the result of the preceding exercise to find a different way of solving the problem in Example 6.6.4 .
3. Solve the Dirichlet problem on the first quadrant with boundary function $g$ which is 1 on the real axis and 0 on the imaginary axis.
4. Solve the Dirichlet problem on the half-disc $\{z : |z| < 1, \ \mathrm{Im}(z) > 0\}$ with boundary function which is 0 on the line $[-1, 1]$ and 1 on the semicircle where $|z| = 1$. Hint: Use the result of Exercise 6.1.5.
5. Solve the Dirichlet problem for the open set consisting of the points inside the unit circle and outside a circle $C$ which intersects the unit circle at right angles at the points 1 and $i$. The boundary function is the function which is 1 on the part of the boundary which lies on the unit circle and is 0 on the part which lies on $C$. Hint: see Exercise 6.3.6.
6. Show that $\dfrac{1}{1+x^2}$, for $x \in \mathbb{R}$, is the real part of $\dfrac{1}{1-ix}$. Use this to solve the Dirichlet problem on the upper half-plane with boundary function $\dfrac{1}{1+x^2}$ on the real line and limit 0 as $z \to \infty$.
7. Modify the approach used in the preceding problem to find a solution to the Dirichlet problem on the lower half-plane for the boundary function $\dfrac{1}{1+x^2}$.
8. For each $w \in \mathbb{C}$ show that the equation $w = z + 1/z$ has two solutions for $z$, one is the inverse of the other, and they are distinct if and only if $w \neq \pm 2$. Show that the solution with $|z| \geq 1$ has imaginary part with the same sign as $\mathrm{Im}(w)$, while if $|z| < 1$, then $z$ and $w$ have imaginary parts with opposite sign.
9. Show that the transformation $z \to z + 1/z$ is a conformal equivalence of the exterior of the unit circle onto the complex plane with the interval $[-2, 2]$ removed. Also show that the restriction of this transformation to

$$\{z : |z| > 1, \ \mathrm{Im}(z) > 0\}$$

is a conformal equivalence from this set to the upper half-plane. Hint: Use the result of the previous exercise.
10. Find a solution to the Neumann problem for the set consisting of the exterior of the unit disc. Hint: Use the result of Exercise 9.
11. Let $U$ and $V$ be open sets in the plane and let $h : U \to V$ be analytic. Prove that $u \circ h$ is harmonic on $U$ for every harmonic function $u$ on $V$.
12. Suppose $U$ is a bounded open set and $g$ is a continous function on $\partial U$. Prove that if $u$ is a solution to the Dirichlet problem on $U$ with boundary function

$g$, then, for any real constants $a$ and $b$, $au + b$ is a solution to the Dirichlet problem on $U$ with boundary function $ag + b$.

13. Prove the second equality in (6.6.2). Hint: Express $1 - z^2$ in polar form:

$$1 - z^2 = r\,e^{i\theta} = r(\cos\theta + i\sin\theta),$$

with $r = |1 - z^2|$ and $\theta = \pm\cos^{-1}(\mathrm{Re}(1 - z^2)/r)$.

Chapter 7

# Analytic Continuation and the Picard Theorems

It often happens in solving complex variables problems, particularly where the solution is obtained as the sum of a series, that we are led to an analytic function $f$ defined, as far as we know, just on a certain open set $U$. Yet it may turn out that $f$ can be extended to be an analytic function on a much larger open set. For example, the power series

$$\sum_{n=0}^{\infty} z^n$$

converges on the open unit disc $D$ to a function analytic on $D$. In fact, this function is

$$f(z) = \frac{1}{1-z},$$

on $D$, and it can obviously be extended to a function, given by the same algebraic expression, which is analytic on the much larger set $\mathbb{C} \setminus \{1\}$.

According to the Inverse Function Theorem, a function $f$ which is analytic on an open set $U$ and which has a non-zero derivative at a point $z_0 \in U$ has an analytic inverse function defined in some neighborhood $W$ of $f(z_0)$. Although the theorem assures us of an inverse function just in some possibly very small neighborhood of $f(z_0)$, we would naturally like to know on just how large a set can the inverse function be defined.

There are many problems of this nature, problems where a solution in the form of an analytic function is known or is known to exist locally, in a neighborhood of a point, and the question arises as to just how much we can enlarge the domain on which this function is defined, is analytic, and is still a solution to the original problem. Questions of this type are questions of *analytic continuation*.

One method of analytic continuation, using the Schwarz Reflection Principle, will at the end of the chapter lead to the existence of a special analytic function (an

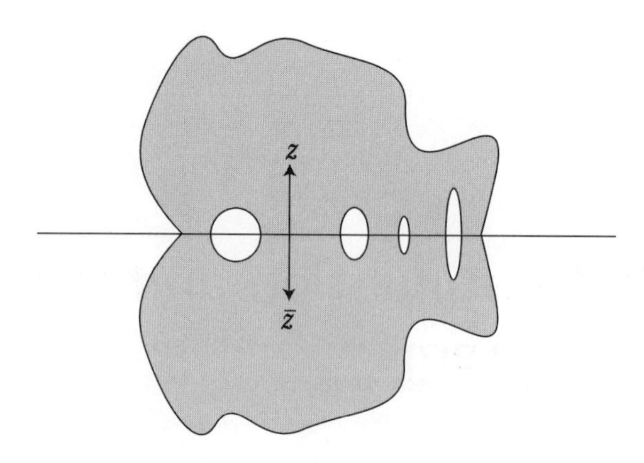


**Figure 7.1.1.** Continuation by Reflection.

analytic covering map) from the unit disc onto $\mathbb{C} \setminus \{0, 1\}$. Using this function, one can turn an entire function with two distinct points of the plane not in its range into an entire function with range contained in the unit disc. Such a function must be constant. This leads directly to the proofs of the Big and Little Picard theorems.

## 7.1. The Schwarz Reflection Principle

One way to extend an analytic function to a larger domain, is to take advantage of some kind of symmetry operation. The Schwarz Reflection Principle does exactly this by making use of the symmetry operation $z \to \overline{z}$, which sends a point to its reflection through the $x$ axis.

**Schwarz Reflection.** Figure 7.1.1 illustrates the situation described in the next theorem. In what follows, if $V$ is a subset of $\mathbb{C}$, then $V^-$ will denote $\{\overline{z} : z \in V\}$.

**Theorem 7.1.1** (Schwarz Reflection Principle). *Let $U$ be an open set which is symmetric about the $x$-axis $\mathbb{R}$. Let $A = U \cap \mathbb{R}$ and let $V$ be the part of $U$ in the open upper half-plane. If $f$ is a function which is continuous on $V \cup A$, analytic on $V$, and real-valued on $A$, then $f$ can be continued to a function analytic on all of $U$ by setting its continuation equal to $\overline{f(\overline{z})}$ on $V^-$.*

**Proof.** The function $g$, defined by

$$g(z) = \overline{f(\overline{z})},$$

is analytic on the set $V^-$ (Exercise 2.2.15).

Note that $V^-$ is the part of $U$ which lies in the open lower half-plane, since $U$ is symmetric about the $x$-axis. Note also that the function $g$ is continuous on $V^- \cup A$, since $f$ is continuous on $V \cup A$.

The fact that $f$ is real on $A$ implies that $f$ and $g$ agree on $A$ and, hence, that they define a single continuous function on $U = V \cup A \cup V^-$. This function is analytic on $U$, except possibly at points of $A$.

Now $A$ is an open subset of the real line and, hence, is a union of open line segments. Let $L$ be one of these line segments. Then the set $W = V \cup L \cup V^-$ is an open subset of $U$ and $f$ is continuous on $W$ and analytic on $W$ except possibly at points of $L$. However, it is a simple consequence of Morera's Theorem that a function which is continuous on an open set $W$ and analytic on $W$ except possibly at the points of a line segment in $W$ is actually analytic on all of $W$ (Exercise 3.2.14). Since $f$ is analytic on each of the open sets $W = V \cup L \cup V^-$ for $L$ one of the line segments making up $A$, it is anaytic on the union of these sets, and $U$ is this union. This completes the proof. □

**Continuation Across a Simple Analytic Curve.** Recall from our discussion of the Riemann Sphere $S^2$ in Section 6.3 that lines in $\mathbb{C}$ can be thought of as circles in $S^2$ that pass through the point at infinity. In fact, the set of all lines and circles in $\mathbb{C}$ is just the set of all circles in $S^2$.

**Definition 7.1.2.** An analytic curve is the image under a conformal equivalence $h : W \to V$, between domains in $S^2$, of a set $I \subset W$ which is a circle or an open arc on a circle in $S^2$.

Of course, by using a linear fractional transformation to move $W$, we can always assume that the circle in the above definition is $\mathbb{R} \cup \{\infty\}$ and $I$ is either an open interval on the line or is $\mathbb{R} \cup \{\infty\}$ itself. In the latter case, the curve will be a simple closed curve.

In the proof of the Schwarz Reflection Principle, we used the fact that a function which is continuous on an open set $U$ and analytic on $U$ except possibly at the points of a line segment in $U$ is actually analytic on all of $U$ (Exercise 3.2.14). We can strengthen this as follows:

**Theorem 7.1.3.** *If $f$ is a function which is continuous on an open set $U$ and analytic on $U$ except possibly at the points of a simple analytic curve $C$ in $U$, then $f$ is actually analytic on all of $U$.*

**Proof.** By removing a point $p$ from $C$, if necessary, we obtain a simple analytic curve $C_1$ which is not a closed curve. Then there is a domain $V$ containing $C_1$, a domain $W$ which contains an open interval $L \subset \mathbb{R}$, and a conformal equivalence $h : W \to V$ which maps $L$ onto $C_1$. We may assume that $V \subset U$. Let $h^{-1} : V \to W$ denote the inverse function of $h$.

The composition $f \circ h$ is a continuous function on $W$ which is analytic on $W \setminus L$. It is, therefore, analytic on all of $W$. Since, on $V$,

$$f = f \circ h \circ h^{-1},$$

we conclude that $f$ is analytic on all of $V$. Since it is analytic on each of two open sets, $V$ and $U \setminus C$, whose union is $U \setminus \{p\}$, $f$ is actually analytic on all of $U \setminus \{p\}$. Since it is continuous at $p$, it is analytic on all of $U$, by Theorem 3.4.8. □

**Example 7.1.4.** Prove that a function $f$ which is continuous on an open set $U$ and analytic on $U \setminus C$, where $C$ is an open arc of a circle, is actually analytic on all of $U$.

**Solution:** An open arc of a circle is a simple analytic curve. Thus, by the above theorem, $f$ is analytic on all of $U$.

**Reflection Through an Analytic Curve.** The symmetry consisting of reflection through the $x$-axis played the key role in the Schwarz Reflection Principle. It turns out there is a similar reflection symmetry for each simple analytic curve and it leads to a similar reflection principle. To describe this requires the notion of a *conjugate analytic* function.

A function $f$ on an open set $U$ is said to be *conjugate analytic* if $\overline{f}$ is analytic on $U$, where

$$\overline{f}(z) = \overline{f(z)}.$$

**Definition 7.1.5.** Suppose $U$ is a domain in $\mathbb{C}$ and $C$ is a simple analytic curve in $U$ such that $U \setminus C$ has two connected components $V$ and $W$. If $\rho$ is a conjugate analytic function from $U$ to $U$ which fixes each point of $C$ and interchanges $V$ and $W$, and if

$$\rho \circ \rho(z) = z \quad \text{for every} \quad z \in U, \tag{7.1.1}$$

then $\rho$ is called a reflection through $C$ defined on $U$.

Note that (7.1.1) implies that $\rho$ is its own inverse function on $U$. If $\kappa(z) = \overline{z}$, then $\kappa \circ \rho$ and $\rho \circ \kappa$ are both analytic functions and are inverse functions of one another. This implies that $\kappa \circ \rho$ is a one-to-one conformal map. Thus, each reflection defined on $U$ is the conjugate of a conformal equivalence from $U$ to $U^-$.

**Theorem 7.1.6.** *If $C$ is any simple analytic curve in $\mathbb{C}$, then there is a reflection $\rho$ through $C$ defined on some domain $V$ containing $C$. This reflection is unique in the sense that another reflection through $C$, defined on a neighborhood $V_1$ of $C$, must be equal to $\rho$ on the connected component of $V \cap V_1$ which contains $C$.*

**Proof.** Since $C$ is a simple analytic curve, there is a conformal equivalence

$$h : W \to V,$$

where $W$ is a domain in $S^2$ which meets $\mathbb{R} \cup \{\infty\}$ in a set $L$ which is either $\mathbb{R} \cup \{\infty\}$ or a line segment in $\mathbb{R}$, $V$ contains $C$, and $C$ is the image of $L$ under $h$.

Note that $W \cap W^-$ also has $L$ as its intersection with $\mathbb{R} \cup \{\infty\}$. If $W \cap W^-$ is not connected, then exactly one of its connected components contains $L$. Thus, we may as well assume that $W^- = W$, since, otherwise, we can replace $W$ by the connected component of $W \cap W^-$ containing $L$ without affecting its intersection, $L$, with $\mathbb{R} \cup \{\infty\}$, or the fact that its image under $h$ is a domain containing $C$.

With the assumption that $W^- = W$, conjugation defines a reflection $\kappa$ through $L$ on $W$ ($\kappa(z) = \overline{z}$). We then get a reflection $\rho$ through $C$ on $V$ by setting

$$\rho = h \circ \kappa \circ h^{-1}$$

on $V$, where $h^{-1} : V \to W$ is the inverse function for $h$. The mapping $\rho$ is conjugate analytic because it is the composition of an analytic function $h$ with a conjugate analytic function $\kappa \circ h^{-1} = \overline{h^{-1}}$ (Exercise 7.1.7). It also satisfies

$$\begin{aligned} \rho \circ \rho(z) &= h \circ \kappa \circ h^{-1} \circ h \circ \kappa \circ h^{-1}(z) \\ &= h \circ \kappa \circ \kappa \circ h^{-1}(z) = z \end{aligned}$$

on $V$, and

$$\rho(z) = z$$

on $C$, because $\kappa(z) = z$ on $L$ and $h^{-1}$ maps $C$ to $L$. Thus, $\rho$ is a reflection through $C$ defined on $V$.

If $\sigma$ is any other reflection through $C$, defined on $V_1$, then $\sigma \circ \rho(z)$ is an analytic function of $z$ on $V \cap V_1$, since the composition of two conjugate analytic functions is analytic (Exercise 7.1.8). However, it is also equal to $z$ on $C$, because this is true of both $\sigma(z)$ and $\rho(z)$. But then it must be equal to $z$ on all of the connected component of $V \cap V_1$ containing $C$. This implies $\rho = \sigma^{-1} = \sigma$ on this component. Thus, $\rho$ is unique in the sense described in the theorem. □

It is now a simple matter to prove that a version of the Schwarz Reflection Principle holds for reflection through a simple analytic curve.

**Theorem 7.1.7.** *If $U$ is a domain, $C \subset U$ is a simple analytic curve with $U \setminus C$ having two components $V$ and $W$, and $\rho$ is a reflection through $C$ defined on $U$, then any function $f$ which is continuous on $V \cup C$, analytic on $V$, and real-valued on $C$ can be continued to an analytic function defined on all of $U$ by setting its continuation equal to $\overline{f(\rho(z))}$ on $W$.*

**Proof.** The proof is the same as the proof of the Schwarz Reflection Principle. We set

$$g(z) = \overline{f(\rho(z))}$$

on $W \cup C$ and note that $g$ is continuous on $W \cup C$ and analytic on $W$ because it is the composition of two conjugate analytic functions, $\overline{f}$ and $\rho$ (Exercise 7.1.8). The functions $g$ and $f$ agree on $C$ because $\rho$ fixes points of $C$ and $f$ and $g$ are real-valued on $C$. Hence, they define a single function on $U$ which is continuous on $U$ and analytic on $U \setminus C$. By Theorem 7.1.3, this function is analytic on all of $U$. Since it agrees with $f$ on $V \cup C$, it is an analytic continuation of $f$ to $U$. □

**Example 7.1.8.** What is the reflection through an arc on the unit circle and what does the previous theorem say about continuing an analytic function across such an arc?

**Solution:** The map

$$\rho(z) = \frac{1}{\overline{z}} \quad \text{or} \quad \rho(r\,e^{i\theta}) = \frac{1}{r}\,e^{i\theta}$$

is defined and conjugate analytic on $\mathbb{C} \setminus \{0\}$. It satisfies $\rho \circ \rho(z) = z$. It also fixes each point on the unit circle since, for $z$ on the unit circle, $|z|^2 = z\overline{z} = 1$, which implies $z = 1/\overline{z}$. It follows that, for any domain $U$ which meets the unit circle in an arc $C$ and is taken to itself by $\rho$, the unique reflection through $C$, defined on $U$, is the map $\rho$.

For such a $U$ and $C$, the previous theorem implies that any function analytic on the part $V$ of $U$ lying on one side of $C$, continuous on $V \cup C$, and real-valued on $C$ can be analytically continued to an analytic function on all of $U$.

## Exercise Set 7.1

1. The power series $\sum_{n=0}^{\infty}(n+1)z^n$ defines an analytic function $g$ on the open unit disc. What is the largest domain in the plane to which $g$ can be analytically continued?
2. Suppose that the hypotheses for the Schwarz Reflection Principle (Theorem 7.1.1) are satisfied for $f$, $U$, $V$ and $A$, except that the function $f$ is purely imaginary on $A$ rather than purely real. Is it still true that $f$ can be analytically continued to all of $U$? If so, prove it, and describe the formula for the continuation of $f$ to $V^-$.
3. If $L$ is the line with equation $ax + by = 0$, find a formula for the map which is reflection through $L$, and verify that this map is conjugate analytic.
4. The function $\sqrt{z^2 - 1}$ is defined and analytic on the set $V$ consisting of the open right half-plane with the interval $(0, 1]$ removed, provided we use the principal branch of the log function to define the square root (this is because $z^2 - 1$ lies in the complement of the non-negative real axis if $z \in V$). This function can be extended to be continuous on the closed right half-plane with the interval $[0, 1]$ removed. Show how to use the Schwarz Reflection Principle to extend this function to an analytic function on $\mathbb{C} \setminus [-1, 1]$.
5. Formulate and prove a version of the Schwarz Reflection Principle (Theorem 7.1.1) for meromorphic functions.
6. Formulate and prove a version of the Schwarz Reflection Principle for harmonic functions.
7. Prove that the composition $f \circ g$ of an analytic function $f$ with a conjugate analytic function $g$ is conjugate analytic.
8. Prove that the composition $f \circ g$ of two conjugate analytic functions is analytic.
9. Find a formula for the reflection through the circle of radius $r$ centered at $z_0$.
10. Prove that the curve in the plane with equation $y = x^2$ is a simple analytic curve.
11. Prove the following form of the Schwarz Reflection Principle: Suppose $C_1$ and $C_2$ are circles and the corresponding reflections through these circles are $\rho_1$ and $\rho_2$. If $U$ is a domain containing $C_1$ which is symmetric under $\rho_1$, and $V$ is the part of $U$ which lies on one side of $C_1$ (inside or outside), then any analytic function $f$ on $V$ which is continuous on $V \cup C_1$ and maps $C_1$ to $\mathbb{C}_2$ may be analytically continued to a function defined on all of $U$ by setting it equal to $\rho_2 \circ f \circ \rho_1$ on $\rho_1(V)$ (the part of $U$ which lies on the other side of $C_1$).
12. The power series $\sum_{n=0}^{\infty} z^{n!}$ defines an analytic function $f$ on the open unit disc. Prove that $f$ cannot be analytically continued to any larger open set. Hint: Consider the values of $f$ along rays of the form $z = r\,e^{2\pi i p/q}$, where $p$ and $q$ are integers.

## 7.2. Continuation Along a Curve

Suppose $f$ is a function which is analytic on a domain $U$. One way to try to extend $f$ to a larger domain is to consider its power series expansion about a point $z_0$ near

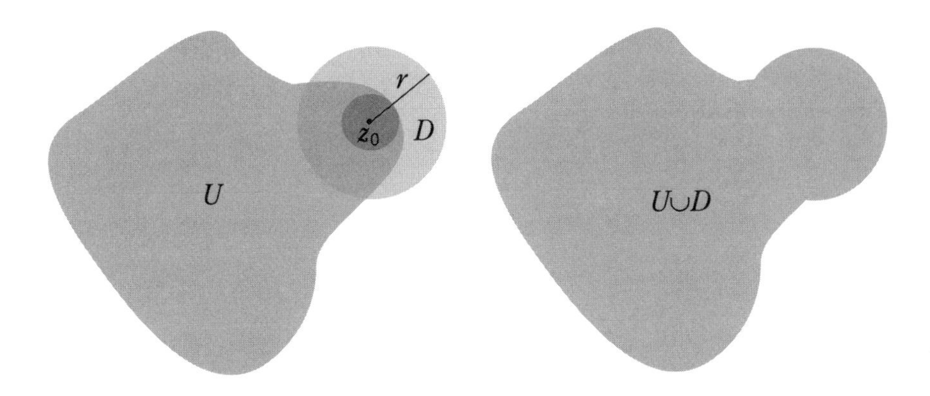


**Figure 7.2.1.** Continuation to a Disc Overlapping the Boundary.

the boundary of $U$. Suppose this power series converges on disc $D$, centered at $z_0$, of radius $r$. Then it defines an analytic function $g$ on $D$. Suppose that $r$ is greater than the distance from $z_0$ to the boundary of $U$, so that $D$ overlaps the boundary of $U$, and suppose that $U \cap D$ is connected. We know the power series expansion converges to $f$ on any disc centered at $z_0$ which is contained in $U$ and, hence, $g$ must agree with $f$ on such a disc. It follows from the Identity Theorem (Theorem 3.4.4) that $g$ agrees with $f$ on all of $U \cap D$. Thus, we succeed in extending $f$ to an analytic function on the larger domain $U \cup D$ by setting it equal to $g$ on $D$.

It may be possible to continue the above procedure, extending the domain on which $f$ is analytic in various directions to ever larger domains until something gets in the way. This could be a singularity where $f$ blows up or it could be a place where to extend the domain in a given direction would cause the function to be defined in a new way on a set where it was already defined. We can imagine extending along different "fingers" from $U$ and having these eventually overlap resulting in different definitions for $f$ on the overlap. It seems that this process has the potential to become quite complicated, maybe hopelessly so.

The concept of analytic continuation along a curve and the Monodromy Theorem serve to bring order to this process and to show that it is not so complicated as it seems. We will get to these ideas shortly, but first we need to introduce some useful terminology.

An *analytic function element*, $(f, D)$, is an analytic function $f$ defined on an open disc $D$. We will say that it is an analytic function element *at* $w$ if $w \in D$. Two analytic function elements $(f_1, D_1)$ and $(f_2, D_2)$ at $w$ are said to be *equivalent* at $w$ if $f = g$ on $D_1 \cap D_2$. In particular, an analytic function element $(f, D)$ at $w$ is equivalent to each function element $(f_1, D_1)$ where $w \in D_1 \subset D$ and $f_1$ is the restriction of $f$ to $D_1$. In other words, the disc in an analytic function element at $w$ is not important as long as it contains $w$ and the function $f$ is analytic on it.

**Definition 7.2.1.** Let $\gamma : [0, 1] \to \mathbb{C}$ be a curve and $(f_0, D_0)$ an analytic function element at $z_0 = \gamma(0)$. Suppose there exist

(a) a partition $0 = t_0 < t_1 < \cdots < t_{n+1} = 1$ of $[0, 1]$; and

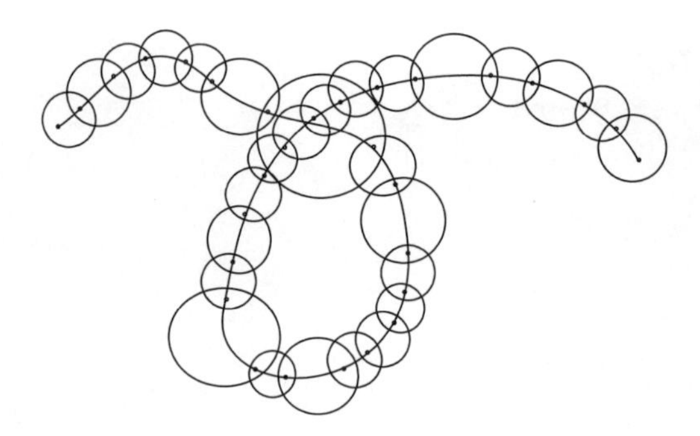

**Figure 7.2.2.** Continuation Along a Curve.

(b) a sequence of function elements

$$(f_1, D_1), (f_2, D_2), \cdots, (f_n, D_n),$$

with $\gamma([t_j, t_{j+1}]) \subset D_j$ for $j = 0, \cdots, n$ and $f_j = f_{j+1}$ on $D_j \cap D_{j+1}$ for $j = 0, \cdots, n-1$.

Then we will say that $(f_n, D_n)$ is an analytic continuation of $(f_0, D_0)$ along $\gamma$ (see Figure 7.2.2).

Although the function element $(f_n, D_n)$ of the above definition, seems to depend on the choice of partition $0 = t_0 < t_1 < \cdots < t_{n+1} = 1$ of $[0, 1]$ and sequence $(f_1, D_1), (f_2, D_2), \cdots, (f_n, D_n)$ of function elements, it turns out that, up to equivalence, it is actually independent of these choices.

**Theorem 7.2.2.** *Given a curve beginning at $z_0$ and ending at $w$ and an analytic function element $(f_0, D_0)$ at $z_0$, any two analytic continuations of $(f_0, D_0)$ along $\gamma$ are equivalent as analytic function elements at $w$.*

**Proof.** We first observe that, given an analytic continuation of $(f_0, D_0)$ along $\gamma$, involving a particular partition $0 = t_0 < t_1 < \cdots < t_{n+1} = 1$ of $[0, 1]$ and corresponding function elements $(f, D) = (f_1, D_1), (f_2, D_2), \cdots, (f_n D_n)$, we can replace the original partition with any given refinement $0 = s_0, s_1, \cdots, s_{m+1} = 1$ and modify the sequence of function elements so that this new set of data still determines the same analytic continuation of $(f_0, D_0)$. In fact, if $[t_j, t_{j+1}]$ is partitioned into $k$ subintervals in the new partition, we simply assign $(f_j, D_j)$ to each of these new subintervals (relabeling them in accordance with the labeling of the subintervals in the new partition). Obviously, this does not change the function element assigned to the last subinterval.

Given two analytic continuations of $(f_0, D_0)$ along $\gamma$, by passing to a common refinement of the corresponding partitions, we may assume that the two continuations are defined using the same partition

$$0 = s_0 < s_1 < \cdots < s_{m+1} = 1.$$

Suppose one of them is defined from the sequence of function elements

$$(f_1, D_1), \cdots, (f_m, D_m),$$

while the other is defined from

$$(\tilde{f}_1, \tilde{D}_1), \cdots, (\tilde{f}_m, \tilde{D}_m)$$

(we set $(\tilde{f}_0, \tilde{D}_0) = (f_0, D_0)$). If we set $V_j = D_j \cap \tilde{D}_j$ for $j = 1, \cdots, m$, then, for each $j$, $V_j$ is a connected (in fact, convex) open set containing $\gamma(t_j)$ and $\gamma(t_{j+1})$. To prove the theorem we need to show that $f_m = \tilde{f}_m$ on $V_m$. In fact, we will prove by induction on $j$ that $f_j = \tilde{f}_j$ on $V_j$ for each $j$.

We have $f_0 = \tilde{f}_0$ on all of $V_0$ by definition. Suppose $j < m$ and $f_j = \tilde{f}_j$ on $V_j$. We have $f_{j+1} = f_j$ on $D_j \cap D_{j+1}$ and $\tilde{f}_{j+1} = \tilde{f}_j$ on $\tilde{D}_j \cap \tilde{D}_{j+1}$. Since $\gamma(t_j)$ is in both of these open sets, it follows that $f_{j+1} = \tilde{f}_{j+1}$ in a neighborhood of $\gamma(t_j)$ and, hence, on all of $V_{j+1}$, since $V_{j+1}$ is connected. This completes the induction step and finishes the proof. □

The next theorem is almost obvious, but there is some work to be done to show the required chain of overlapping discs can be chosen as in Definition 7.2.1. We leave the details to the exercises (Exercise 7.2.1).

**Theorem 7.2.3.** *Suppose $(f_0, D_0)$ is a function element at $z_0$ and $\gamma$ is a curve joining $z_0$ to $w$. If there is an open set $U$, containing $D_0$ and $\gamma(I)$, and an analytic function $f$ on $U$ such that $f = f_0$ on $D_0$, then $(f_0, D_0)$ can be analytically continued along $\gamma$.*

The converse of the above theorem is not true. It may seem that it should be true, since the union of the overlapping discs in Definition 7.2.1 is an open set $U$ containing $\gamma(I)$, and it appears that the functions $f_j$ fit together to define a single function, on this union, that agrees with $f_0$ on $D_0$ . However, this is not the case, due to the fact that the curve $\gamma$ may cross itself, as in Figure 7.2.2, and the crossing point may be contained in two of the discs $D_j$ and $D_k$. The corresponding functions $f_j$ and $f_k$ may not be equivalent as function elements at this point. In fact, the curve might be a closed curve, so that $z_0 = \gamma(0)$ and $w = \gamma(1)$ are the same point. In this case the analytic continuation of $(f_0, D_0)$ along $\gamma$ leads to another function element $(f_n, D_n)$ at $z_0$. This function element need not be equivalent to $(f_0, D_0)$.

**Example 7.2.4.** Give an example of the phenomenon referred to in the previous paragraph.

**Solution:** Let log denote the principal branch of the log function and consider the analytic function element $(\log, D_1(1))$. If $\gamma$ is the unit circle traversed once in the counterclockwise direction beginning at $\gamma(0) = \gamma(1) = 1$, then $(\log, D_1(1))$ can be analytically continued along $\gamma$ (Exercise 7.2.2), but the resulting function element on the final disc will differ from that on the first disc by $2\pi i$.

There is no domain containing the unit circle to which $(\log, D_1(1))$ can be analytically continued. If there were such a domain $U$ and a continuation $f$ of $(\log, D_1(1))$ to $U$, then $f$ would have to agree with the principal branch of the log function on all of $U \setminus [-\infty, 0]$, but this function has a $2\pi i$ jump discontinuity across the half-line $[-\infty, 0]$ and certainly cannot be analytically continued across it.

The previous example is closely related to the following question: If $\gamma_1$ and $\gamma_2$ are two curves joining $z_0$ to $w$, $(f_0, D_0)$ is an analytic function element at $z_0$, and

$(f_0, D_0)$ can be analytically continued along $\gamma_1$ and along $\gamma_2$, then are the resulting function elements at $w$ equivalent? Not necessarily, as the following example shows:

**Example 7.2.5.** Can analytic continuations of a function element at $z_0$ along different paths from $z_0$ to $w$ lead to non-equivalent function elements at $w$?

**Solution:** Let the initial function element be $(\log, D_1(1))$, where log is the principal branch of the log function, $z_0 = 1$, and $w = -1$. Clearly this can be continued along the curve $\gamma_1$ consisting of the upper half of the unit circle traversed from 1 to $-1$ and along the curve $\gamma_2$ consisting of the lower half of the unit circle traversed from 1 to $-1$ . However, the resulting function elements at $-1 = \gamma_1(1) = \gamma_2(1)$ differ by $2\pi i$.

**The Monodromy Theorem.** The next theorem gives conditions under which we can be sure that the analytic continuations of a function along two curves from $z_0$ to $w$ are necessarily equivalent function elements, in contrast to the preceding example.

**Theorem 7.2.6** (Monodromy Theorem). *Let $U$ be a connected open set in $\mathbb{C}$, $z_0$ and $w$ points of $U$, and $(f_0, D_0)$ an analytic function element at $z_0$, with $D_0 \subset U$. Suppose*

(a) *$(f_0, D_0)$ can be analytically continued along every curve in $U$; and*
(b) *$\gamma_0$ and $\gamma_1$ are homotopic curves in $U$ joining $z_0$ to $w$.*

*Then the continuations of $f$ along $\gamma_0$ and $\gamma_1$ are equivalent function elements at $w$.*

**Proof.** We refer to terminology and notation developed in Section 4.6.

A homotopy from $\gamma_0$ to $\gamma_1$ in $U$ determines a continuous one-parameter family of curves $\{\gamma_s\}$ from $z_0$ to $w$ in $U$. The continuity of this family means that, given $\epsilon > 0$, there is a $\delta > 0$ such that

$$||\gamma_s - \gamma_r|| < \epsilon \quad \text{whenever} \quad |s - r| < \delta.$$

The analytic function element $(f_0, D_0)$ has an analytic continuation along each of the curves $\gamma_s$, by hypothesis. Denote the terminal function element for the continuation along $\gamma_s$ by $\phi_s$. We claim that, for each $r \in [0, 1]$, there is a $\delta > 0$ such that $\phi_s$ is equivalent to $\phi_r$ whenever $|s - r| < \delta$.

Let $0 = t_0 < t_1 < \cdots < t_{n+1} = 1$ be a partition and $(f_1, D_1), \cdots, (f_n, D_n)$ a sequence of function elements defining $\phi_r = (f_n, D_n)$ as an analytic continuation of $(f_0, D_0)$ along $\gamma_r$. Then

$$\gamma_r([t_j, t_{j+1}]) \subset D_j \quad \text{for} \quad j = 0, \cdots, n.$$

For each $j = 0, \cdots, n$, let $\epsilon_j$ be the distance from the compact set $\gamma_r([t_j, t_{j+1}])$ to the boundary of the disc $D_j$. If $||\gamma_s - \gamma_r|| < \epsilon_j$, then it will also be true that $\gamma_s([t_j, t_{j+1}]) \subset D_j$. Thus, if $\epsilon = \min\{\epsilon_0, \cdots, \epsilon_n\}$, and we choose $\delta > 0$ such that

$$||\gamma_s - \gamma_r|| < \epsilon \quad \text{whenever} \quad |s - r| < \delta,$$

then, for each $s$ with $|s - r| < \delta$, the partition $0 = t_0 < t_1 < \cdots < t_{n+1} = 1$ and the sequence of function elements $(f_1, D_1), \cdots, (f_n, D_n)$ also defines $(f_n, D_n)$ as an analytic continuation of $(f_0, D_0)$ along $\gamma_s$. Since, by the previous theorem, any other continuation of $(f_0, D_0)$ along $\gamma_s$ is equivalent to this one, we conclude that $\phi_r$ is equivalent to $\phi_s$. This proves that $\phi_s$ is equivalent to $\phi_r$ whenever $|r - s| < \delta$.

The remainder of the proof is a standard connectedness argument. We define a function $h$ on $[0,1]$ by setting $h(s) = 0$ if $\phi_s$ is equivalent to $\phi_0$, and $h(s) = 1$ otherwise. Then, by the result of the previous paragraph, $h$ is a continuous function on $[0,1]$. If $h(1) = 1$, then, since $h(0) = 0$, the Intermediate Value Theorem implies that $h$ must take on every value between 0 and 1. Since this is not the case, we conclude that $h(1) = 0$. This means that $\phi_1$ is equivalent to $\phi_0$. □

**Example 7.2.7.** Consider the function element $\phi$ consisting of the principal branch of the log function and the disc $D_1(1)$. Set

$$A = \{z \in \mathbb{C} : |z| > 0, \ -\pi/2 < \arg(z) < 3\pi/2\}.$$

Show that the analytic continuations along any two curves from 1 to $-1$ in $A$ have equivalent terminal function elements at $-1$.

**Solution:** The branch of the log function defined by restricting $\arg(z)$ to lie in the inteval $(-\pi/2, 3\pi/2)$ agrees with the principle branch of the log function on $D_1(1)$. By Theorem 7.2.3, $\phi$ can be analytically continued along any curve in $A$. The set $A$ is simply connected and so any two curves in $A$ from 1 to $-1$ are homotopic. In view of the Monodromy Theorem, continuations of $\phi$ along two such curves will have equivalent terminal function elements at $-1$.

The Monodromy Theorem allows us, in the next theorem, to conclude that if a function element can be analytically continued along every curve in an open set $U$ and if the open set is simply connected, then there is an analytic continuation of $f$ to all of $U$.

**Theorem 7.2.8.** *Suppose $U$ is a simply connected open set and $(f_0, D_0)$ is an analytic function element at $z_0 \in U$ (with $D_0 \subset U$). If $(f_0, D_0)$ can be analytically continued along every curve in $U$, then there is a function $f$ which is analytic on $U$ and equal to $f_0$ on $D_0$.*

**Proof.** If $z \in U$, then since any two curves from $z_0$ to $z$ in $U$ are homotopic in $U$, the Monodromy Theorem implies that any two terminal elements of analytic continuations of $(f_0, D_0)$ along curves from $z_0$ to $z$ in $U$ will be equivalent and, hence, will determine the same analytic function in some neighborhood of $z$. Hence all such analytic continuations determine the same function value at $z$. We define $f(z)$ to be this value.

Clearly $f_0(z) = f(z)$ on $D_0$. It remains to prove that $f$ is analytic on $U$. Let $w$ be a point of $U$, let $\gamma$ be a curve in $U$ joining $z_0$ to $w$, and let $(D_n, f_n)$ be the terminal function element of some continuation of $f$ along $\gamma$. If $z \in D_n$, then $(D_n, f_n)$ is also the terminal element of a continuation of $(f_0, D_0)$ along a curve $\gamma_1$ from $z_0$ to $z$ in $U$ – we simply extend $\gamma$ to a curve $\gamma_1$, ending at $z$, by joining $\gamma$ with the line segment from $w$ to $z$. It follows that $f(z) = f_n(z)$ on all of $D_n$, not just when $z = w$. Since $f_n$ is analytic on $D_n$, $f$ is also analytic on $D_n$. Since $w$ was an arbitrary point of $U$, we conclude that $f$ is analytic everywhere on $U$. □

## Exercise Set 7.2

1. Prove Theorem 7.2.3.
2. Prove that if log is the principal branch of the log function and $D_0$ is any disc centered at 1 and not containing 0, then $(\log, D_0)$ can be analytically continued along any curve $\gamma$ in $\mathbb{C} \setminus \{0\}$.
3. Prove that any analytic continuation of the element $(\log, D_0)$ of the previous exercise, along a curve $\gamma$ in $\mathbb{C} \setminus \{0\}$, yields a function element $(f, D)$ at $\gamma(1)$ which is some branch of the log function restricted to $D$. That is, prove that it satisfies $e^{f(z)} = z$ on $D$.

In the next four exercises $U$ is a domain and $(g_0, D_0)$ is a function element in $U$ which can be analytically continued along a curve $\gamma$ in $U$ to a function element $(g, D)$ in $U$.

4. If $f$ is an entire function, prove that $(f \circ g_0, D_0)$ can be analytically continued along $\gamma$ to $(f \circ g, D)$.
5. If $f$ is an entire function, $h$ is a function analytic on $U$ and $f \circ g_0 = h$ on $D_0$, then $f \circ g = h$ on $D$.
6. If a linear differential equation on $U$, with analytic coeficients, has $g_0$ for a solution on $D_0$, then it has $g$ for a solution on $D$.
7. If $u$ is a harmonic function on $U$ and $u = \mathrm{Re}(g_0)$ on $D_0$, then $u = \mathrm{Re}(g)$ on $D$.

In the next three exercises, use Theorem 7.2.8 to give a proof, different from the one in Theorem 4.6.16, that the indicated property holds for a simply connected domain $U$. In each case, show there is a solution in a disc $D_0$ centered at a point $z_0 \in U$ and then show that this solution can be analytically continued along any curve in $U$.

8. Suppose $g$ is a non-vanishing analytic function on $U$. Prove that $g$ has an analytic square root on $U$.
9. Prove that every analytic function on $U$ has an analytic antiderivative.
10. Prove that every real-valued harmonic function on $U$ is the real part of an analytic function on $U$.

## 7.3. Analytic Covering Maps

If $U$, $V$ and $W$ are open subsets of $\mathbb{C}$ and

$$h : V \to W \quad \text{and} \quad f : U \to W$$

are analytic maps, then we say that $f$ *lifts* through $h$ if there is an analytic map

$$g : U \to V \quad \text{with} \quad f = h \circ g$$

(see Figure 7.3.1). If $h$ is a conformal equivalence from $V$ to $W$, then $f$ trivially lifts through $h$. In this case, $h$ has an analytic inverse function $h^{-1} : W \to V$ and we can simply set

$$g = h^{-1} \circ f : U \to V.$$

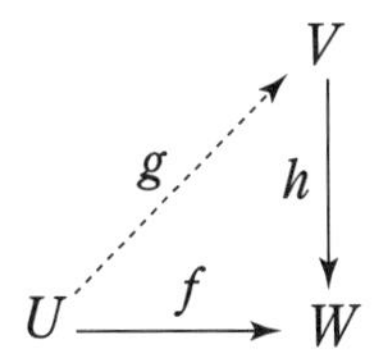


**Figure 7.3.1.** Lifting a Map $f$ Through Another Map $h$.

There is another class of maps $h$ for which a kind of lifting result holds. This is the class of *analytic covering maps.*

**Definition 7.3.1.** An analytic map $h : V \to W$ is called an *analytic covering map* if, for each $w_0 \in W$, there is a neighborhood $A$ of $w_0$, contained in $W$, such that $h^{-1}(A)$ is the disjoint union of a collection $\{B_j\}$ of open subsets of $V$ with the property that, for each $j$, $h$ is a conformal equivalence of $B_j$ onto $A$.

**Example 7.3.2.** Show that $z \to e^z$ is an analytic covering map from $\mathbb{C}$ to $\mathbb{C} \setminus \{0\}$.

**Solution:** The function exp, defined by $\exp(z) = e^z$, is certainly an analytic map of $\mathbb{C}$ onto $\mathbb{C} \setminus \{0\}$. We claim that each disc $D$ in $\mathbb{C} \setminus \{0\}$ has inverse image $\exp^{-1}(D)$ consisting of a disjoint union of open sets on each of which exp is a conformal equivalence onto $D$.

Given a disc $D \subset \mathbb{C} \setminus \{0\}$, there is a ray from 0 to $\infty$ disjoint from $D$ (e.g., take the ray from 0 to $\infty$ which passes through the center of $D$ and rotate it by the angle $\pi$). This means that there is an angle $\theta_0$ such that every $z$ for which $e^z \in D$ satisfies

$$\theta_0 + 2n\,\pi < \text{Im}(z) < \theta_0 + 2(n+1)\pi$$

for some integer $n$. Thus, $\exp^{-1}(D)$ is a disjoint union of open sets

$$B_n = \exp^{-1}(D) \cap \{z : \theta_0 + 2n\,\pi < \text{Im}(z) < \theta_0 + 2(n+1)\pi\}.$$

On each $B_n$, exp is a conformal equivalence from $B_n$ to $D$, with inverse function equal to the branch of the log function for which arg takes values between $\theta_0 + 2n\,\pi$ and $\theta_0 + 2(n+1)\pi$ (see Figure 7.3.2). Thus, by definition, exp is an analytic covering map.

An analytic covering map $h : V \to W$ is, in particular, a conformal map, since, for each point $z_0$ of $V$ it is a conformal equivalence of a neighborhood of $z_0$ onto a neighborhood of $h(z_0)$. However, not every conformal map is a covering map.

**Example 7.3.3.** Give an example of a conformal map $h$ of a region $V$ onto a region $W$ such that $h$ is not an analytic covering map.

**Solution:** We use the exponential map again. However, this time we restrict its domain to be the set

$$V = \{z : -\pi < \text{Im}(z) < 2\pi\}.$$

The image of this map is still $\mathbb{C} \setminus \{0\}$, and it is clearly a conformal map. However, any disc $D \subset \mathbb{C} \setminus \{0\}$, centered at 1, has an inverse image which is a disjoint union of two open sets – one which lies in the strip $-\pi < \text{Im}(z) < \pi$ and one which lies

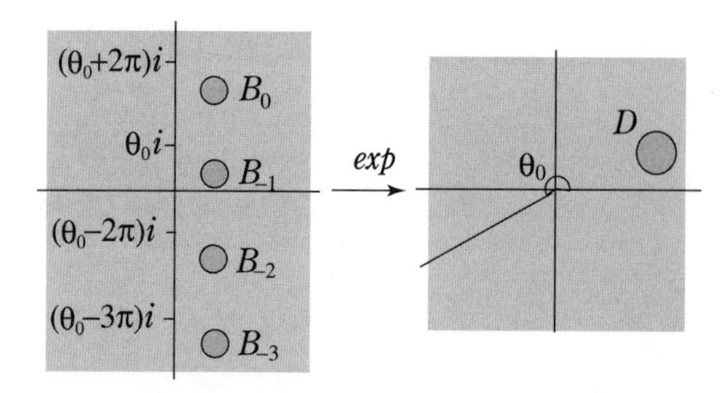


**Figure 7.3.2.** The Exponential as a Covering Map.

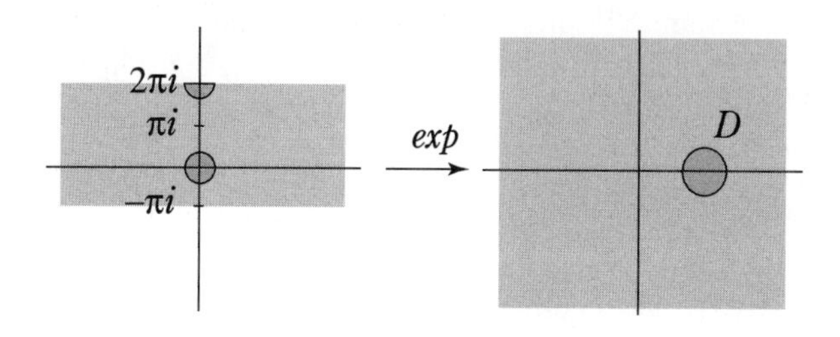


**Figure 7.3.3.** A Conformal Map Which is Not a Covering Map.

in the strip $\pi < \mathrm{Im}(z) < 2\pi$. The map exp is a conformal equivalence of the first of these onto $D$, but only maps the second one onto the lower half of $D$ (see Figure 7.3.3). Since this problem persists no matter how small a disc we choose centered at 1, the map exp is not an analytic covering map from $V$ to $\mathbb{C} \setminus \{0\}$.

### Lifting Through an Analytic Covering Map.

**Theorem 7.3.4.** *Let $h : V \to W$ be an analytic covering map and $U$ a simply connected domain in $\mathbb{C}$. Then each analytic function $f : U \to W$ can be lifted through $h$. That is, there is an analytic map $g : U \to V$ with $f = h \circ g$.*

**Proof.** We fix a $z_0 \in U$ and let $w_0 = f(z_0)$. We then choose a neighborhood $A_0$ of $w_0$ such that $h^{-1}(A_0)$ is a disjoint union of open sets on each of which $h$ is a conformal equivalence onto $A_0$. We choose one of these, call it $B_0$, and denote by $h_0^{-1}$ the inverse of $h : B_0 \to A_0$. We then choose an open disc $D_0$, centered at $z_0$ and contained in $f^{-1}(A_0)$. On this disc, we define a function $g_0$ by

$$g_0 = h_0^{-1} \circ f.$$

This function serves to lift $f$ through $h$, but only on the set $D_0$. The main work of the proof is to show that the analytic function element $(g_0, D_0)$ can be analytically continued along any curve in $U$ beginning at $z_0$. If we can do this, then the Monodromy Theorem implies that $(g_0, D_0)$ can be analytically continued to an analytic function defined on all of $U$ (since $U$ is simply connected). This function $g$ serves to lift $f$ through $h$ on all of $U$, since it satisfies the identity

$$f = h \circ g$$

on $D_0$ and, hence, on all of $U$, by Theorem 3.4.2.

So it remains to prove that $(g_0, D_0)$ can be analytically continued along any curve in $U$ beginning at $z_0$. Let $\gamma : [0,1] \to U$ be such a curve. Let $S$ be the subset of $I = [0,1]$ consisting of those points $s$ such that $(g_0, D_0)$ can be analytically continued along $\gamma$ as far as $s$. Clearly, if $s \in S$ and $0 \le s_1 \le s$, then $s_1 \in S$ as well. Also, if $s \in S$, then there is a chain of open discs along $\gamma$ restricted to $[0, s]$ that serves to analytically continue $(g_0, D_0)$ along this curve to $\gamma(s)$. If $s < 1$, then there is an $s_1 > s$ such that $s_1$ is also in the last disc in this chain of open discs. This implies that $(g_0, D_0)$ can also be analytically continued along $\gamma$ restricted to $[0, s_1]$ and, hence, that $s_1$ is also in $S$. We conclude that $S$ is either all of $[0,1]$, in which case the proof is complete, or it is a half-open interval of the form $[0, r)$ with $r = \sup S \le 1$.

If $S$ is a half-open interval $[0, r)$, we let $z_r = \gamma(r)$ and choose a neighborhood $A$ of $f(z_r)$ in $V$ such that $h^{-1}(A)$ is a disjoint union of open sets $B_j$, each of which is conformally equivalent to $A$ under $h$. We choose an open disc $D \subset U$ such that $f(D) \subset A$. This is possible because $f$ is continuous. We also choose an $s \in [0, r)$ such that $w = f(\gamma(s))$ belongs to $D$. This is possible because $D$ is open and $f \circ \gamma$ is continuous.

Because $s \in S$, the function element $(g_0, D_0)$ may be analytically continued along $\gamma$ to a function element $(g_n, D_n)$ with $\gamma(s) \in D_n$. Since $f = h \circ g_0$ on $D_0$, the same thing will be true of the function element $(g_n, D_n)$. That is,

$$f = h \circ g_n \quad \text{on} \quad D_n.$$

In particular, $w = f(\gamma(s)) = h(g_n(\gamma(s)))$. This means that $g_n(\gamma(s))$ belongs to $h^{-1}(A)$ and, hence, to exactly one of the sets $B_j$, call it $B_k$. We define a new function element $(g_{n+1}, D_{n+1})$ by choosing $D_{n+1} = D$ and setting $g_{n+1}$ equal to the composition of $f : D \to A$ with an inverse function for $h : B_k \to A$. This new function element certainly satisfies

$$f = h \circ g_{n+1} \quad \text{on} \quad D_{n+1},$$

but does it agree with $g_n$ on $D_n \cap D_{n+1}$? It does because, not only is the point $g_n(\gamma(s))$ in $B_k$, but all of $g_n(D_n \cap D_{n+1})$ is contained in $B_k$; otherwise, the connected open set $D_n \cap D_{n+1}$ would be separated by $g_n^{-1}(B_k)$ and the union of the sets $g_n^{-1}(B_j)$ for $j \neq k$. But this means that the two inverse functions for $h$ used in the definitions of $g_n$ and $g_{n+1}$ agree on $f(D_n \cap D_{n+1})$, which implies that $g_n$ and $g_{n+1}$ agree on $D_n \cap D_{n+1}$. But now, since $\gamma(r)$ is an interior point of $D_{n+1}$, this implies that $(g_0, D_0)$ can be analytically continued to $\gamma(r)$ and, hence, that $r \in S$. This contradicts the assumption that $S$ has the form $[0, r)$. The only other possiblility is that $S = [0,1]$, and this means that $(g_0, D_0)$ can be analytically continued along $\gamma$. □

The preceding theorem is, on the one hand, a powerful application of the Monodromy Theorem and, on the other hand, is the key ingredient in our proofs of the Picard theorems in the next section.

## Exercise Set 7.3

1. Is the function $h(z) = z^2$ an analytic covering map of $\mathbb{C}$ onto $\mathbb{C}$? Is it an analytic covering map of $\mathbb{C} \setminus \{0\}$ onto $\mathbb{C} \setminus \{0\}$? Justify your answers.
2. Give a proof, using Theorem 7.3.4, that if $U$ is a simply connected open set and $f$ is a non-vanishing analytic function on $U$, then there is an analytic logarithm of $f$ – that is, an analytic function $g$ on $U$ such that $f = e^g$.
3. Prove that if $h : V \to W$ is an analytic covering map from a connected open set $U$ to a simply connected open set $V$, then $h$ is a conformal equivalence.
4. Prove that if $h : V \to W$ is an analytic covering map and $U$ is any simply connected open subset of $W$, then $h^{-1}(U)$ is a disjoint union of open sets on each of which $h$ is a conformal equivalence onto $U$.
5. Is there an analytic covering map from the unit disc to $\mathbb{C} \setminus \{0\}$? Justify your answer.
6. Prove that if $h : V \to W$ is an analytic cover and $\gamma : [0, 1] \to W$ is a curve, then $\gamma$ can be lifted through $h$ in the sense that there is a curve $\lambda : [0, 1] \to V$ such that $\gamma = h \circ \lambda$.
7. Let $p(z)$ be a polynomial of degree $n$ in $z$ and let $w$ be a point of $\mathbb{C}$. Prove that the polynomial $p(z) - w$ fails to have $n$ distinct roots (that is, it has a repeated root) if and only if $w = p(z_0)$ for some point $z_0$ at which $p'(z_0) = 0$.
8. Let $p(z)$ be a polynomial of degree $n$ in $z$ and let $S$ be the set of all points $w \in C$ at which the polynomial $p(z) - w$ fails to have $n$ distinct roots. If $W = \mathbb{C} \setminus S$ and $V = p^{-1}(W)$, prove that $p : V \to W$ is an analytic covering map. Hint: Use the Inverse Mapping Theorem and the preceding exercise.
9. Prove that the polynomial $p(z) = z^3 - 3z$ is an analytic covering map from $\mathbb{C} \setminus \{-1, 1\}$ to $\mathbb{C} \setminus \{-2, 2\}$.
10. Use the previous exercise to prove that if $g$ is an analytic function on a simply connected domain $U$ and if $g$ does not take on the values $-2$ and 2, then there is an analytic function $f$ on $U$ which satisfies $f^3(z) - 3f(z) = g(z)$ on $U$.

## 7.4. The Picard Theorems

The Little Picard Theorem states that a non-constant entire function takes on every value in $\mathbb{C}$ except possibly one. This is a vast generalization of Liouville's Theorem and an impressive application of the analytic continuation techniques we have been developing in this chapter. The Big Picard Theorem states that an analytic function with an essential sigularity at $z_0$ takes on every complex value but one infinitely often in every neighborhood of $z_0$.

The strategy for proving the Picard theorems is to construct an analytic covering map $h$ from the unit disc to the plane with two points removed. The Little Picard Theorem follows easily from this since any analytic function from the plane to the plane with two points removed will lift through this covering map to an analytic function from the plane to the unit disc. By Liouville's Theorem the only such functions are constants. The Big Picard Theorem uses the existence of $h$ along with Montel's Theorem. The construction of $h$ relies heavily on analytic continuation

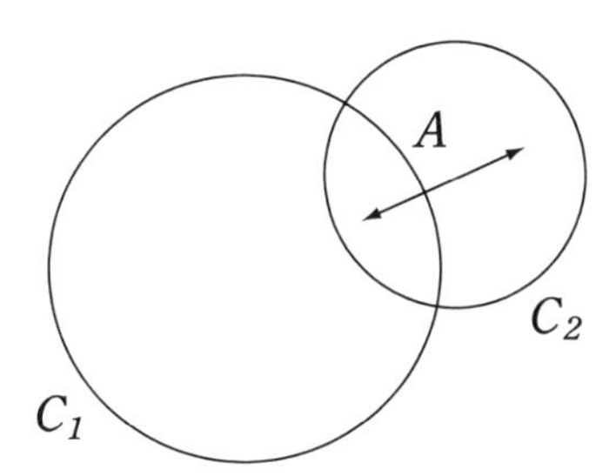


**Figure 7.4.1.** Reflection Through an Arc of a Perpendicular Circle.

by reflection through a circle (Example 7.1.8 – the Schwarz Reflection Principle for circles). We begin by proving a few simple facts about reflection through a circle.

**Reflection Through a Circle.** Suppose $C_1$ and $C_2$ are two circles in the plane that intersect one another in two points. If the tangents to the two circles are perpendicular at each of these two points, then we will say that the circles meet at right angles (Figure 7.4.1). Actually, it turns out that if the tangents are perpendicular at one of the points of intersection, then they are also perpendicular at the other point of intersection (Exercise 7.4.1).

**Theorem 7.4.1.** *If circles $C_1$ and $C_2$ meet at right angles, and $A$ is the arc of $C_1$ that lies inside $C_2$, then reflection through $A$ maps $C_2$ onto itself.*

**Proof.** There is a linear fractional transformation $h$ that maps $C_1$ onto $\mathbb{R} \cup \{\infty\}$ and maps the inside of $C_1$ onto the upper half-plane. We can choose which point on $C_1$ is sent to $\infty$ and we choose it so that it is not on $\overline{A}$; in particular, it is not one of the points where $C_1$ and $C_2$ intersect.

The fact that linear fractional transformations are conformal maps (angle preserving maps) and the fact that $C_1$ and $C_2$ meet at right angles mean that the image of $C_2$ under $h$ meets the real line at two points and it has vertical tangents at these points. The image of a circle under a linear fractional transformation is either a line or a circle, and so $h(C_2)$ must be a circle which meets the real axis at right angles. This implies that the real axis passes through a diameter of $h(C_2)$. It follows that reflection through the $x$-axis, maps the inside of the circle $h(C_2)$ onto itself.

Now reflection through the arc $A$ may be described as $h$ composed with reflection through the horizontal diameter of $h(C_2)$ followed by $h^{-1}$. Why? Well, this certainly describes a reflection through $A$ defined on the inside of $C_2$ and, by the uniqueness part of Theorem 7.1.6, this is the only reflection through $A$.

Since reflection through the $x$-axis takes the inside of $h(C_2)$ onto itself, reflection through $A$ takes the inside of $C_2$ onto itself. □

**Theorem 7.4.2.** *With $C_1$, $C_2$ as above, the reflection through $C_1$ takes any other circle $C$, which meets $C_2$ at right angles, to another circle which meets $C_2$ at right angles or to a line through the center of $C_2$.*

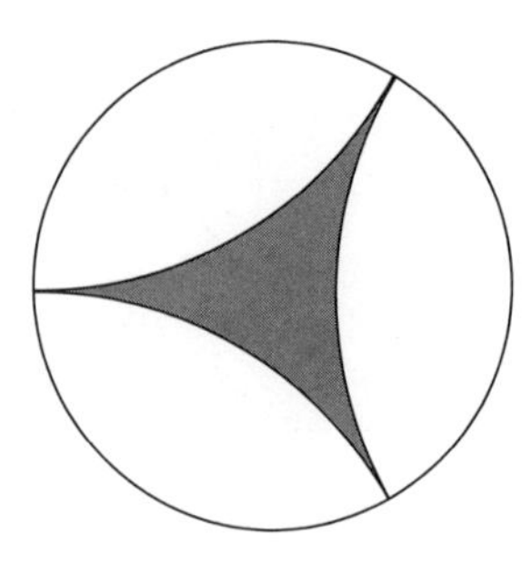

**Figure 7.4.2.** The Curvilinear Triangle $V_0$.

**Proof.** A reflection through a circle is a linear fractional transformation followed by conjugation (Exercise 7.1.9). It therefore takes a circle $C$ to either a circle or a line.

The reflection through $C_1$ takes the inside of $C_2$ onto itself. It follows that the image of $C$ under this reflection still meets $C_2$ in two points.

A conformal map preserves angles, while conjugation reverses the orientation of any angle – that is, it takes an angle to its negative. It follows that a reflection, which is a conformal map followed by conjugation, is also an angle reversing map. This implies that if two curves have perpendicular tangents at a point of intersection, then the same will be true of their images under a reflection. Therefore, reflection through $C_1$ takes $C$ to a circle or line which also meets $C_2$ at right angles. The only way the image of $C$ could be a line is if it passes through the center of $C_2$. Thus, the reflection of $C$ through $C_1$ is another circle which meets $C_2$ at right angles or is a line through the center of $C_2$. □

**An Analytic Covering Map.** The three points $-1$, $1/2 + i\sqrt{3}/2$, $1/2 - i\sqrt{3}/2$ are equidistant points on the unit circle $C$. According to Exercise 7.42, we can join each pair of these points with an arc of a circle that meets $C$ at right angles. These three arcs then bound an open curvilinear triangle $V_0$ (see Figure 7.4.2).

The open set $V_0$ is clearly simply connected. By the Riemann Mapping Theorem, there is a conformal equivalence $h : V_0 \to H$, where $H$ is the upper half-plane. By Theorem 6.5.2, the map $h$ extends to a continous map of the closure of $V_0$ to the closure of $H$ in $S^2$, which takes the boundary of $V_0$ to the boundary of $H$ in $S^2$. By composing with a linear fractional transformation, if necessary, we may choose this map in such a way that it takes the points $-1, 1/2 - i\sqrt{3}/2, 1/2 + i\sqrt{3}/2$ to the points $\infty, 0, 1$.

The function $h$ is real-valued on each of the three "edges" of $V_0$. By the circle version of the Schwarz Reflection Principle, $h$ can be continued by reflection across each of these edges. This results in $h$ being defined and analytic in the set $V_1$ described in Figure 7.4.3. Note that each of the three new cells on which $h$ is defined in this way is contained in the unit disc $D$ (Theorem 7.4.1) and is bounded by three circles that meet the unit circle $C$ in right angles (Theorem 7.4.2).

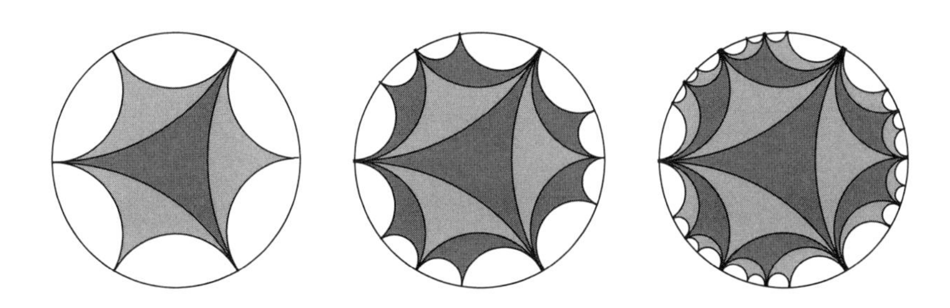

**Figure 7.4.3.** The Sets $V_1$, $V_2$ and $V_3$. of Theorem 7.4.3.

Note also that, while $h$ maps $V_0$ to the upper half-plane, it maps each of the new cells of $V_1$ to the lower half-plane, since each is a reflection of $V_0$ through one of its edges.

We can now analytically continue $h$ to a still larger domain $V_2$ by reflecting across each of the circular arcs that bound $V_1$. This results in $h$ being defined in the set $V_2$ represented in Figure 7.4.3. Note that $h$ now maps each of the new cells in $V_2$ into the upper half-plane. We can clearly continue this process by induction to create an increasing sequence $\{V_n\}$ of open sets to which $h$ may be analytically continued. The union of these open sets is $D$ (Exercise 7.4.5), and so the result is an analytic function $h$ which maps the open unit disc $D$ onto a subset of the plane which contains the upper and lower open half-planes and the intervals $(-\infty, 0)$, $(0, 1)$, and $(1, \infty)$ on the real line. The points $\infty$, 0, and 1 are not in the image because every triangular cell that occurs in the above construction has all of its vertices on the unit circle. Thus, $h$ is an analytic map of $D$ onto $\mathbb{C} \setminus \{0, 1\}$.

**Theorem 7.4.3.** *The analytic map $h : D \to \mathbb{C} \setminus \{0, 1\}$, described above, is an analytic covering map.*

**Proof.** Think of the disc $D$ as being partitioned into light cells and dark cells, separated by arcs of circles, as in Figure 7.4.3. The dark cells are mapped into the upper half-plane by $h$ and the light cells are mapped into the lower half-plane. Each arc separating two of these cells is mapped to one of the intervals $I_1 = (-\infty, 0)$, $I_2 = (0, 1)$, $I_3 = (1, \infty)$.

For $j = 1, 2, 3$, let $P_j$ be the union of the open upper and lower half-planes and the open interval $I_j$. Then each $P_j$ is an open set consisting of $S^2$ with a closed arc removed.

Let $\Delta$ be any open disc in $\mathbb{C} \setminus \{0, 1\}$. Then $\Delta$ lies entirely inside $P_j$ for at least one $j$. We fix a $j$ for which this is true. The inverse image of $P_j$ under $h$ consists of the union of all the light and dark open cells together with some of the arcs separating them – those arcs which $h$ maps to $I_j$. The collection of these arcs is pairwise disjoint, and each one of them separates exactly one light cell from one dark cell. The union of the arc and the two cells it separates is an open set on which $h$ is a conformal equivalence onto $P_j$. No two distinct open sets of this form can overlap since, if they did, the overlap would have to be a cell with boundary containing two arcs mapping to $I_j$ under $h$. Let $\{U_k\}_{k=1}^{\infty}$ be the collection of open sets of this form. This is a pairwise disjoint collection of open sets with union equal to $h^{-1}(P_j)$. Also,

$$h^{-1}(\Delta) \subset \bigcup_k U_k,$$

and so $h^{-1}(\Delta)$ is the disjoint union of the open sets $B_k = h^{-1}(\Delta) \cap U_k$. The map $h$ is a conformal equivalence of $B_k$ onto $\Delta$ for each $k$. It follows that $h : D \to \mathbb{C} \setminus \{0, 1\}$ is an analytic covering map. □

**The Little Picard Theorem.**

**Theorem 7.4.4** (Little Picard Theorem). *If $f$ is an entire function and there are two distinct points in the plane that are not in the image of $f$, then $f$ is constant.*

**Proof.** If $f$ does not take on the values $z_0$ and $z_1$, then the function

$$\frac{f(z) - z_0}{z_1 - z_0}$$

does not take on the values 0 and 1. Thus, we may as well assume that $f$ itself has this property. Then $f$ is an analytic function from the simply connected set $\mathbb{C}$ to $\mathbb{C} \setminus \{0, 1\}$. Since the map $h : D \to \mathbb{C} \setminus \{0, 1\}$ of the previous theorem is an analytic covering map, Theorem 7.3.4 implies that $f$ may be lifted through $h$ to an analytic function $g : \mathbb{C} \to D$ such that

$$f = h \circ g.$$

However, this means that $g$ is a bounded entire function and, hence, is a constant, by Liouville's Theorem. This, of course, forces $f$ to also be a constant. □

**The Big Picard Theorem.** The proof of the big Picard Theorem depends on the following theorem. Recall the definition of a normal family (Definition 6.4.1).

**Theorem 7.4.5.** *Let $U$ be a connected subset of the plane. Then the set of analytic functions on $U$ with values in the set $\mathbb{C} \setminus \{0, 1\}$ is a normal family.*

**Proof.** We will prove the theorem in the case where $U$ is an open disc $\Delta$. The proof that the theorem for general $U$ follows from this special case will be left to the exercises.

Let $\Delta$ be any open disc and let $z_0$ be its center. Let $\mathcal{F}$ denote the set of analytic functions on $U$ with values in $\mathbb{C} \setminus \{0, 1\}$. Let $\{f_n\}$ be a sequence in $\mathcal{F}$. We will show that this sequence converges uniformly to $\infty$ on compact subsets of $\Delta$ or it has a subsequence which converges uniformly on compact subsets of $\Delta$ to a function analytic on $\Delta$.

Suppose $\{f_n\}$ does not converge uniformly to $\infty$ on compact subsets of $\Delta$. Then there is a disc $\Delta_1$ with the same center and with $\overline{\Delta}_1 \subset \Delta$ and an $R > 0$ such that $f_n(\Delta_1)$ has elements with modulus less than $R$ for infinitely many $n$. This means that we can choose a subsequence of $\{f_n\}$ which itself has no subsequence converging uniformly to $\infty$ on $\Delta_1$. Without loss of generality, we may replace $\{f_n\}$ by this subsequence. That is, we may assume that $\{f_n\}$ has no subsequence converging uniformly to $\infty$ on $\Delta_1$.

Let $h : D \to \mathbb{C} \setminus \{0, 1\}$ be the analytic covering map of Theorem 7.4.3. Then, since $\Delta$ is simply connected, Theorem 7.3.4 implies each $f_n$ can be lifted through

$h$ to an analytic function $g_n : \Delta \to D$ with $f_n = h \circ g_n$. Furthermore, since the set $V_1$ in Figure 7.4.3 maps onto $\mathbb{C} \setminus \{0, 1\}$ under $h$, we can choose the $g_n$ in such a way that $g_n(z_0) \in V_1$.

The sequence $\{g_n\}$ is a uniformly bounded sequence of analytic functions on $\Delta$ and, hence, by Montel's Theorem (Theorem 6.4.2), it has a subsequence which converges uniformly on compact subsets of $\Delta$ to an analytic function $g$.

The function $g$ has its image in $\overline{D}$. However, unless $g$ is constant, this image must be an open set (by the Open Mapping Theorem – Theorem 4.5.8) and, hence, must be contained in $D$. In this case $f = h \circ g$ is an analytic function on $\Delta$ with values in $\mathbb{C} \setminus \{0, 1\}$ and is the uniform limit on compact subsets of $\Delta$ of a subsequence of $\{f_n\}$.

On the other hand, if $g$ is constant, then this constant must lie in the closure of $V_1$, since each $g_n(z_0)$ lies in $V_1$. This means the constant is either in $D$ or is one of the six points of $\overline{V}_1$ which lie on the unit circle. This means that $f = h \circ g$ has constant value 0, 1, or $\infty$.

Thus, each sequence $\{f_n\}$ has a subsequence which converges uniformly on compact subsets of $\Delta$ to $\infty$ or to an analytic function. Since there is no subsequence converging to $\infty$, it has a subsequence converging uniformly on compact subsets to an analytic function. This completes the proof of the theorem in the case where $U$ is a disc. The general case follows from Exercise 7.4.11. □

**Theorem 7.4.6** (Big Picard Theorem). *Let $f$ be a function which is analytic in a neighborhood $U$ of $z_0$ except at $z_0$ itself, where it has an essential singularity. Then $f$ takes on every value but one infinitely often in every neighborhood of $z_0$.*

**Proof.** We may as well assume that $z_0 = 0$. If $f$ does not have the property stated in the conclusion of the theorem, then there is a disc $D_r(0)$ on which $f$ fails to take on at least two complex values. We may assume these values are 0 and 1, since otherwise we may compose $f$ with a linear fractional transformation which takes the two values missed by $f$ to 0 and 1.

Thus, we may assume $f$ is an analytic function from $D_r(0) \setminus \{0\}$ to $\mathbb{C} \setminus \{0, 1\}$ with an essential singularity at 0. We define a sequence of functions with the same properties by setting

$$f_n(z) = f(z/n) \quad \text{for} \quad n = 1, 2, \cdots .$$

By the previous theorem, this sequence converges uniformly on compact subsets of $D_r(0)$ to $\infty$ or it has a subsequence converging uniformly on compact subsets of $D_r(0)$ to an analytic function. In the first case, $1/f$ is bounded and, hence, has a removable singularity at 0. This means it extends to an analytic function with a zero of some finite order at 0. This is impossible, since it implies that $f$ has a pole at 0 rather than an essential singularity. The second case implies that $f$ itself is bounded and, hence, has a removable singularity at 0. This also violates the hypothesis that $f$ has an essential singularity at 0. Thus, our assumption that $f$ misses two values in some disc centered at 0 has led to a contradiction. This completes the proof. □

## Exercise Set 7.4

1. Prove that if two circles intersect in two points and their tangents are perpendicular at one point of intersection, then the tangents are also perpendicular at the other point of intersection.
2. Prove that, given two distinct points on a circle $C_1$, there is a unique circle $C_2$ which meets $C_1$ at right angles at these two points.
3. Prove that if $C_r$ and $C_R$ are circles of radius $r$ and $R$, respectively, which are tangent at a point, then the reflection of $C_R$ through $C_r$ is a circle of radius
$$\frac{rR}{2R+r}$$
if neither circle is inside the other and is
$$\frac{rR}{2R-r}$$
if $C_r$ is inside $C_R$.
4. The set $V_n$ in the construction preceding Theorem 7.4.3 has boundary consisting of a set of circular arcs of different sizes with endpoints on the boundary of the unit disc. Prove by induction that the largest of these circular arcs has radius $(2n+1)^{-1}R_0$, where $R_0$ is the radius of each of the three circular arcs comprising the boundary of $V_0$. Hint: Use the calculation of the preceding exercise.
5. Prove that the union of the open sets $V_n$, described in the paragraph preceding Theorem 7.4.3, is the unit disc $D$. Hint: Use the result of the preceding exercise.
6. Prove that if $f$ and $1/f$ are both entire functions, then the image of $f$ is exactly $\mathbb{C} \setminus \{0\}$.
7. Prove that if $f$ is a non-constant entire function and $b^2 \neq 4ac$, then the function
$$g(z) = af^2(z) + bf(z) + c$$
must have a zero.
8. Is there a non-constant analytic function from the plane with one point removed to the plane with two points removed? Justify your answer.
9. If $U$ is a simply connected open set, show that the set of all analytic functions on $U$ with values in $\mathbb{C} \setminus \{0, 1\}$ is a normal family (see Section 6.4).
10. Suppose $r < 1$, $R < 1$ and $h$ is a conformal equivalence from the annulus $A_R = \{z : R < |z| < 1\}$ to the annulus $A_r = \{z : r < |z| < 1\}$. Suppose also that $h$ extends to a continuous map from $\overline{A}_R$ to $\overline{A}_r$ which takes the unit circle to itself and the disc of radius $R$ to the disc of radius $r$. Prove that $h$ can be continued to a conformal equivalence of the unit disc onto the unit disc that takes 0 to 0. Conclude from this that $R = r$. Hint: Do the analytic continuation through a series of steps involving reflection through a circle, as in Exercise 7.1.11; the first of these steps involves reflecting the annulus $A_R$ through the circle of radius $R$.
11. Prove that if $\mathcal{F}$ is a family of analytic functions on a connected open set $U$ and if for each disc $\Delta$ in $U$ the family of restrictions of elements of $\mathcal{F}$ to $\Delta$ is a normal family, then $\mathcal{F}$ itself is a normal family.

12. Prove that an entire function which is not a polynomial takes on every complex value but one infinitely often.

Chapter 8

# Infinite Products

As is amply demonstrated by power series expansions, a highly useful technique in complex analysis is to express an analytic function as an infinite sum of much simpler functions. Likewise, it can be useful to express an analytic function as an infinte product of simpler functions. This is especially true in the study of the zeroes of analytic functions. For example, a polynomial $p$ of degree $n$ can be written as a product

$$p(z) = a \prod_{j=1}^{n} (z - z_j),$$

where $\{z_1, z_2, \cdots, z_n\}$ are the zeroes of $p$. It turns out that product expansions of a similar type (but with infinitely many factors) are possible for other analytic functions.

Since the exponential function converts sums to products, we can expect that the theory of infinite products will be closely related to the theory of infinite sums.

## 8.1. Convergence of Infinite Products

In the following discussion, log will be the principal branch of the log function.

**Definition 8.1.1.** If $\{u_k\}$ is a sequence of complex numbers and

$$p_n = \prod_{k=1}^{n} u_k, \tag{8.1.1}$$

then we will say that the infinite product

$$\prod_{k=1}^{\infty} u_k \tag{8.1.2}$$

converges to the complex number $p$ if $\lim_{n\to\infty} p_n = p$.

**Theorem 8.1.2.** *If $\{u_k\}$ is a sequence of complex numbers, then the infinite product (8.1.2) converges to a non-zero number $p$ if and only if the infinite sum*

$$\sum_{k=1}^{\infty} \log u_k \tag{8.1.3}$$

*converges to a number $\lambda$. In this case, $p = e^{\lambda}$. Furthermore, if the infinite series converges absolutely, then the infinite product is unchanged by a rearrangement of the factors.*

**Proof.** We have to be careful here, because the log function converts products to sums only up to $\pm 2\pi i$. However, it is true that $\log uv = \log u + \log v$ if $u$ and $v$ have positive real parts since, in this case, $\log u$ and $\log v$ have imaginary parts in $(-\pi/2, \pi/2)$ and $\log uv$ has imaginary part in $(-\pi, \pi)$. Thus, $\log uv$ and $\log u + \log v$ cannot differ by a non-zero multiple of $2\pi i$.

We define the partial products $p_n$ as in (8.1.1). If $p = \lim_{n\to\infty} p_n$ exists and is non-zero, then

$$\lim_{n\to\infty} \log(p_n/p) = 0,$$

since log is continuous at 1. In particular, there is an $N$ such that

$$-\pi/4 < \operatorname{Im}(\log(p_n/p)) < \pi/4 \quad \text{whenever} \quad n \geq N.$$

It follows that $p_n/p_m = (p_n/p)(p_m/p)^{-1}$ is in the right half-plane for $n, m \geq N$. In particular, $u_{n+1} = p_{n+1}/p_n$ is in the right half-plane for $n \geq N$. Thus,

$$\log(p_{n+1}/p_N) = \log((p_n/p_N)u_{n+1}) = \log(p_n/p_N) + \log u_{n+1}$$

whenever $n \geq N$. This equation and an induction argument beginning with $n = N$ show that

$$\log(p_n/p_N) = \sum_{k=N+1}^{n} \log u_k$$

for all $n > N$. Since the left side of this equality converges as $n \to \infty$, so does the right side. This implies the convergence of the series (8.1.3).

Conversely, if this series converges and we let

$$\lambda_n = \sum_{k=1}^{n} \log u_n$$

be its $n$th partial sum, then the sequence $\{\lambda_n\}$ converges to a number $\lambda$. Since

$$p_n = e^{\lambda_n},$$

and the exponential function is continuous, the sequence $\{p_n\}$ converges to $e^{\lambda}$.

If the series (8.1.3) converges absolutely, then each of its rearrangements converges to the same number. It follows that each rearrangement of the infinite product (8.1.2) also converges to the same number. □

**Uniform Convergence of Products.** We will be primarily interested in infinite products of analytic functions. In this situation, whether or not the product converges uniformly is of critical importance. We say that the infinite product of a sequence of functions $\{u_k\}$ converges uniformly on a set $S$ if the sequence $p_n$ of partial products converges uniformly on $S$.

**Theorem 8.1.3.** *Let $u_k$ be a sequence of complex-valued functions defined and bounded on a set $S$. If the series*

$$\sum_{k=1}^{\infty} \log u_k(z)$$

*converges uniformly to $\lambda(z)$ on $S$, then the infinite product*

$$\prod_{k=1}^{\infty} u_k(z)$$

*converges uniformly to $e^{\lambda(z)}$ on $S$.*

**Proof.** Let $\lambda_n(z)$ be the $n$th partial sum of the infinite sum and $p_n(z)$ the $n$th partial product of the infinite product. Then the uniform convergence of the series on $S$ implies that $\lambda_n(z) - \lambda(z)$ converges uniformly to 0 on $S$. Since the exponential function is continuous at 0, this implies that

$$\frac{p_n(z)}{p(z)} = e^{\lambda_n(z) - \lambda(z)}$$

converges uniformly to 1.

The fact that each $p_n$ is bounded on $S$ implies that $\mathrm{Re}(\lambda_n)$ is bounded above on $S$. This and the uniform convergence imply that $\mathrm{Re}(\lambda)$ is bounded above on $S$, which implies that $p(z) = e^{\lambda(z)}$ is bounded on $S$. It follows that $p_n = (p_n/p)p$ converges uniformly to $p$ on $S$. □

**Theorem 8.1.4.** *Let $\{a_k(z)\}$ be a sequence of complex-valued functions defined on a set $S$. If the series*

$$\sum_{k=1}^{\infty} |a_k(z)| \tag{8.1.4}$$

*converges uniformly on $S$, then the infinite product*

$$\prod_{k=1}^{\infty} (1 + a_k(z)) \tag{8.1.5}$$

*converges uniformly on $S$. Each rearrangement of the infinite product converges to the same function. If the infinite product converges to $p(z)$, then each zero of $p(z)$ is a zero, with the same order, of some finite product of the factors $1 + a_k(z)$.*

**Proof.** If $|w| < 1/2$, then (Exercise 8.1.1)

$$\frac{2}{3}|w| \le |\log(1+w)| \le 2|w|. \tag{8.1.6}$$

If the series (8.1.4) converges uniformly on $S$, then there is a $K$ such that $|a_k(z)| \leq 1/2$ for $k \geq K$ and for all $z \in S$. If we use (8.1.6) with $w = a_k$, it follows that one of the two series

$$\sum_{k=K}^{\infty} |\log(1 + a_k(z))| \quad \text{and} \quad \sum_{k=K}^{\infty} |a_k(z)|$$

converges uniformly on $S$ if and only if the other one does also. Hence, if (8.1.4) converges uniformly, then

$$\sum_{k=K}^{\infty} \log(1 + a_k(z))$$

converges uniformly and absolutely. By the previous two theorems, this implies the uniform convergence of

$$\prod_{k=K}^{\infty} (1 + a_k(z))$$

to a function on $S$ with no zeroes. It follows that (8.1.5) converges uniformly on $S$, the limit is unaffected by rearrangements of the factors, and each of its zeroes is a zero, with the same order, of the product of the factors $1 + a_k(z)$ for $k < K$. □

**Example 8.1.5.** Prove that the infinite product

$$\prod_{k=1}^{\infty} (1 - z^2/k^2) \tag{8.1.7}$$

converges uniformly on each bounded subset of $\mathbb{C}$.

**Solution:** We have $|-z^2/k^2| \leq R^2/k^2$ for all $z$ in the disc $D_R(0)$. Since the positive termed series

$$\sum_{k=1}^{\infty} \frac{R^2}{k^2}$$

converges, it follows that the series

$$\sum_{k=1}^{\infty} \frac{|z^2|}{k^2}$$

converges uniformly on $D_R(0)$. Hence, by the previous theorem, the infinite product (8.1.7) also converges uniformly on $D_R(0)$ for each $R$ and, hence, on each bounded subset of $\mathbb{C}$.

**Logarithmic Derivative of a Product.** If an analytic function $f$ on an open set $U$ has an analytic logarithm $g$ on $U$ – that is, if $f = e^g$ on $U$ with $g$ analytic – then $g' = f'/f$. The expression $f'/f$ is independent of which logarithm is chosen for $f$. Furthermore, as long as $f$ is not identically zero on any component of $U$, $f'/f$ exists (as a meromorphic function on $U$) even if $f$ does not have an analytic logarithm on $U$. Note that $f$ cannot have an analytic or even a meromorphic logarithm in any neighborhood of a point where it has the value zero (Exercise 8.1.4).

**Definition 8.1.6.** Let $f$ be an analytic function on an open set $U$ and suppose that $f$ is not identically 0 on any component of $U$. Then the meromorphic function $f'/f$ is called the *logarithmic derivative* of $f$ on $U$.

Logarithmic derivative is quite a well-behaved notion. The logarithmic derivative of the product of two functions is the sum of their logarithmic derivatives (Exercise 8.1.5). Furthermore, the following theorem states that logarithmic derivative is preserved by uniform limits. The proof is left to the exercises (Exercise 8.1.6).

**Theorem 8.1.7.** *Let $\{f_n\}$ be a sequence of analytic functions on a connected open set $U$. If this sequence converges uniformly to $f$ on $U$, then the sequence $\{f_n'/f_n\}$ converges uniformly to $f'/f$ on compact subsets of $U \setminus S$, where $S$ is the set of zeroes of $f$.*

When applied to infinite products, this immediately implies the following corollary.

**Corollary 8.1.8.** *Let $\{u_k\}$ be a sequence of analytic functions on a connected open set $U$. If the product*

$$f(z) = \prod_{k=1}^{\infty} u_k(z)$$

*converges uniformly on compact subsets of $U$ to a function $f$ which is not identically 0, then the infinite sum*

$$\sum_{k=1}^{\infty} \frac{u_k'(z)}{u_k(z)}$$

*converges uniformly to $f'/f$ on compact subsets of $U \setminus S$, where $S$ is the set of zeroes of $f$.*

**Example 8.1.9.** Show that the function

$$f(z) = \pi z \prod_{k=1}^{\infty} \left(1 - \frac{z^2}{k^2}\right)$$

has a logarithmic derivative which can be written as

$$\frac{f'(z)}{f(z)} = \frac{1}{z} + \sum_{k=1}^{\infty} \frac{2z}{z^2 - k^2} \tag{8.1.8}$$

or as

$$\frac{f'(z)}{f(z)} = \lim_{n \to \infty} \sum_{k=-n}^{n} \frac{1}{z - k}. \tag{8.1.9}$$

**Solution:** Note that the infinite product in the expression for $f$ converges uniformly on each compact disc in the plane by Example 8.1.5.

By the previous theorem,

$$\frac{f'(z)}{f(z)} = \frac{1}{z} + \sum_{k=1}^{\infty} \frac{-2z/k^2}{1 - z^2/k^2} = \frac{1}{z} + \sum_{k=1}^{\infty} \frac{2z}{z^2 - k^2}.$$

This proves (8.1.8). Since

$$\frac{2z}{z^2 - k^2} = \frac{1}{z - k} + \frac{1}{z + k},$$

the $n$th partial sum of the series (8.1.8) can be rewritten as the sum that appears in (8.1.9).

The logarithmic derivative of $f$, as computed in the above example, will be used in the exercise set to prove that $f(z) = \sin \pi z$. That is,

$$\sin(\pi z) = \pi z \prod_{k=1}^{\infty} \left(1 - \frac{z^2}{k^2}\right). \tag{8.1.10}$$

## Exercise Set 8.1

1. Prove that if $w$ is a complex number with $|w| \le 1/2$, then
$$\frac{2}{3}|w| \le |\log(1+w)| \le 2|w|.$$
2. Does the infinite product
$$\prod_{k=1}^{\infty}\left(1+\frac{1}{k}\right)$$
converge? How about the product
$$\prod_{k=1}^{\infty}\left(1+\frac{1}{k^{3/2}}\right)?$$
3. Show that the infinite product
$$\prod_{k=1}^{\infty}\left(1-\frac{z}{k}\right)e^{z/k}$$
converges uniformly on compact subsets of the plane.
4. Prove that if $f$ is analytic on $U$ and has a zero at $z_0 \in U$, then there is no meromorphic function $g$ defined in a neighborhood $V$ of $z_0$ such that $f = e^g$ on $V$.
5. Prove that the logarithmic derivative of the product $fg$ of two analytic functions is the sum of the logarithmic derivative of $f$ and the logarithmic derivative of $g$. Also prove the analogous statement for the quotient $f/g$.
6. Prove Theorem 8.1.7. Hint: First prove that it is true on any disc in $U$ on which $f$ has no zeroes.
7. Prove that the logarithmic derivative of a meromorphic function $f$ on $\mathbb{C}$ is also a meromorphic function on $\mathbb{C}$ and is odd (even) if $f$ is odd (even).
8. If $f$ is the function defined in Example 8.1.9, prove that the logarithmic derivative of $f$ is an odd meromorphic function which is periodic of period 1. Observe that the logarithmic derivative of $\sin \pi z$ has the same properties.
9. Prove that if $f$ is the function of the previous exercise, and we set
$$g(z) = \frac{\sin(\pi z)}{f(z)},$$
then $g$ is an entire function with no zeroes and, hence, has a logarithm $h$ which is entire. Then, $\sin(\pi z) = f(z)\,e^{h(z)}$.

10. Prove that if $f$, $g$ and $h$ are the functions of the previous exercise, then the logarithmic derivative of $g$ is $h'(z) = \pi \cot \pi z - f'(z)/f(z)$.
11. With $h$ as above, prove that $h'$ is bounded on the strip $0 \leq \text{Re}(z) \leq 1$ (use (8.1.8)). Show that this implies it is bounded on the entire plane and, hence, is constant.
12. With $h$ as above, prove that $h'(0) = 0$ and, hence, that $h'$ is identically 0 and $h$ is a constant. Then use the fact that $\lim_{z\to 0} z^{-1} \sin z = 1$ to show that this constant is 0. Conclude that

$$\sin(\pi z) = \pi z \prod_{k=1}^{\infty} \left(1 - \frac{z^2}{k^2}\right).$$

## 8.2. Weierstrass Products

In this section we will show that, given any sequence of points of an open set $U \subset \mathbb{C}$, with no limit point in $U$, there is an analytic function on $U$ with exactly the points of this sequence as its zeroes, with each zero having order equal to the number of times it appears in the sequence. The analytic function will be constructed as an infinite product of certain simple functions, each of which has exactly one zero. These simple functions are constructed as follows.

For $p = 0, 1, 2, \cdots$ we define entire functions $E_p(z)$ by $E_0(z) = 1 - z$ and

$$E_p(z) = (1 - z)\, e^{z + z^2/2 + \cdots + z^p/p} \quad \text{for} \quad p > 0.$$

Note that $z + z^2/2 + \cdots + z^p/p$ is the $p$th partial sum for the power series expansion of $-\log(1 - z)$ about $z = 0$ and so, although $E_p(1) = 0$, the sequence $E_p(z)$ will converge uniformly to $(1 - z)(1 - z)^{-1} = 1$ on each disc of radius less than 1 centered at 0. More precisely:

**Theorem 8.2.1.** *Each $E_p(z)$ is an entire function with the following properties:*

(a) *the only zero of $E_p(z)$ occurs at $z = 1$;*
(b) *if $|z| \leq 1$, then $|E_p(z) - 1| < |z|^{p+1}$.*

**Proof.** Part (a) is obvious. To prove Part (b), we note that the derivative of $1 - E_p(z)$ is (Exercise 8.2.1)

$$(1 - E_p(z))' = -E_p'(z) = z^p\, e^{z + z^2/2 + \cdots + z^p/p}. \tag{8.2.1}$$

Since this has a zero of order $p$ at $z = 0$, the function $1 - E_p(z)$ has a zero of order $p + 1$ at $z = 0$.

The function (8.2.1) has a power series expansion about 0 with all of its coefficients non-negative real numbers, since this is true of the exponential function and the function $z + z^2/2 + \cdots + z^p/p$. It follows that the function

$$h(z) = \frac{1 - E_p(z)}{z^{p+1}}$$

also has non-negative real numbers as coefficients for its power series expansion about 0. This implies that the maximum value achieved by $|h(z)|$ for $|z| \leq 1$ is $h(1) = 1$. Part (b) follows from this. $\square$

If $f$ is an analytic function on $U$, then we will say that a sequence $\{z_k\} \subset U$ is a list of the zeroes of $f$ *counting multiplicity* if each $z_k$ is a zero of $f$, and if each zero $w$ of $f$ occurs $m(w)$ times in this sequence, where $m(w)$ is the order of the zero $w$.

Let $\{z_k\}$ be a sequence of non-zero complex numbers converging to $\infty$. The next theorem shows how to use scaled versions of the functions $E_p$ to construct an entire function with this sequence as a list of its zeroes counting multiplicity. The resulting product is called a *Weierstrass product*.

**Theorem 8.2.2.** *Let $A$ be a subset of $\mathbb{C}$. If $\{z_k\}$ is a sequence of non-zero complex numbers and $\{p_k\}$ is a sequence of integers such that*

$$\sum_{k=1}^{\infty} \left| \frac{r}{z_k} \right|^{p_k+1} < \infty \quad \textit{for all} \quad r > 0, \tag{8.2.2}$$

*then the Weierstrass product*

$$f(z) = \prod_{k=1}^{\infty} E_{p_k}(z/z_k) \tag{8.2.3}$$

*converges uniformly on compact subsets of $\mathbb{C}$ to an entire function which has $\{z_k\}$ as a list of its zeroes counting multiplicity.*

**Proof.** By part (b) of the previous theorem, we have

$$|E_{p_k}(z/z_k) - 1| \le \left| \frac{z}{z_k} \right|^{p_k+1} \quad \text{if} \quad |z| \le |z_k|.$$

The condition $|z| \le |z_k|$ must be satisfied for all sufficiently large $k$ if the series (8.2.2) converges. The theorem follows by applying Theorem 8.1.4 with $a_k(z) = E_{p_k}(z/z_k) - 1$. □

### The Weierstrass Theorem.

**Theorem 8.2.3.** *If $\{z_k\}$ is any sequence of complex numbers converging to infinity, then there is an entire function with $\{z_k\}$ as a list of its zeroes counting multiplicity.*

**Proof.** Suppose $m$ of the $z_k$ are equal to 0 ($m$ might be 0). We may as well assume these are the first $m$ terms of the sequence. Then $\{z_k\}_{k=m+1}^{\infty}$ is a sequence of non-zero complex numbers.

If $R > 0$, then, since $z_k \to \infty$, there is a $K > m$ such that $|z_k| > 2R$ for all $k \ge K$. Then the series

$$\sum_{k=m+1} \left| \frac{z}{z_k} \right|^k$$

converges uniformly on $|z| \le R$ by comparison with the geometric series with ratio 1/2. Thus, the hypotheses of the previous theorem are satisfied if we choose $p_k = k - 1$ for each $k$. The resulting Weierstrass product

$$\prod_{k=m+1}^{\infty} E_p(z/z_k)$$

converges uniformly on compact subsets of $\mathbb{C}$ to an entire function which has $\{z_k\}_{k=m+1}^{\infty}$ as a list of its zeroes counting multiplicity. Then

$$f(z) = z^m \prod_{k=m+1}^{\infty} E_p(z/z_k)$$

is an entire function with $\{z_k\}_{k=1}^{\infty}$ as a list of its zeroes counting multiplicity. □

**Example 8.2.4.** Find an entire function which has a zero of order $k$ at each positive integer $k$.

**Solution:** We construct a sequence

$$1, 2, 2, 3, 3, 3, 4, 4, 4, 4, \cdots$$

in which each positive integer $n$ appears $n$ times and the terms are arranged in increasing order. Then, for this sequence $\{z_k\}$ and a given positive integer $p$, we have

$$\sum_{k=1}^{\infty} \frac{1}{|z_k|^{p+1}} = \sum_{n=1}^{\infty} n \frac{1}{n^{p+1}} = \sum_{n=1}^{\infty} \frac{1}{n^p}.$$

If we choose $p = 2$, then the right side is the convergent series

$$\sum_{n=1}^{\infty} \frac{1}{n^2}.$$

The Weierstrass product for the sequences $\{z_k\}$ and $\{p_k = 2\}$ is

$$\prod_{k=1}^{\infty} \left((1 - z/k)\, e^{z/k + z^2/(2k^2)}\right)^k.$$

By Theorem 8.2.2 this infinite product converges to an entire function with the required zeroes.

**Weierstrass Factorization.** The Weierstrass Theorem for the plane leads immediately to the Weierstrass Factorization Theorem for entire functions:

**Theorem 8.2.5.** *Let $f$ be an entire function which is not identically zero. Let $m$ be the order of the zero of $f$ at 0, and let $\{z_k\}$ be a list of the non-zero zeroes of $f$ counting multiplicity. Then there exist non-negative integers $p_1, p_2, \cdots$ and an entire function $h$ such that*

$$f(z) = e^{h(z)}\, z^m \prod_{k=1}^{\infty} E_{p_k}(z/z_k).$$

*The sequence $\{p_k\}$ may be chosen in any way which satisfies (8.2.2).*

**Proof.** The product

$$g(z) = z^m \prod_{k=1}^{\infty} E_{p_k}(z/z_k)$$

converges uniformly on compact sets if $\{p_k\}$ is chosen such that (8.2.2) holds ($p_k = k - 1$ is one choice which always works, but there may be better choices for a given $f$). Furthermore, the resulting function $g$ has the same zeroes as $f$ with the same multiplicities. Thus, $fg^{-1}$ is an entire function with no zeroes (after removable singularities are removed). It follows that

$$fg^{-1} = e^h$$

for some entire function $h$. The theorem follows from this. □

In many cases, the sequence $\{p_k\}$ can be chosen to be constant.

**Example 8.2.6.** Find a Weierstrass factorization for $\sin(\pi z)$.

**Solution:** This function has a zero of order 1 at each integer and has no other zeroes. Since

$$\sum_{k=1}^{\infty} \frac{1}{k^2} < \infty,$$

the condition (8.2.2) holds if we choose $p_k = 1$ for every $k$. Then the above theorem tells us that

$$\sin(\pi z) = e^{h(z)} \, z \prod_{k \neq 0} E_1(z/k) = e^{h(z)} \, z \prod_{k \neq 0} (1 - z/k) \, e^{z/k},$$

where the product is over all non-zero integers $k$. Note that if the factors for $k$ and $-k$ in this product are paired, the result is

$$(1 - z/k) \, e^{z/k} (1 + z/k) \, e^{-z/k} = 1 - z^2/k^2.$$

We conclude from Exercise 8.1.12 that $e^{h(z)} = \pi$, that

$$\sin(\pi z) = \pi z \prod_{k \neq 0} (1 - z/k) \, e^{z/k}$$

is a Weierstrass factorization of $\sin(\pi z)$, and that this factorization is equivalent to the factorization

$$\sin(\pi z) = \pi z \prod_{k=1}^{\infty} (1 - z^2/k^2).$$

**The General Weierstrass Theorem.** If $\mathbb{C}$ is replaced by an arbitrary non-empty, proper open subset of $S^2$, the analogue of Theorem 8.2.3 holds with only a slightly more complicated proof.

**Theorem 8.2.7.** *Let $U$ be a non-empty, proper open subset of $S^2$. If $\{z_k\}$ is any sequence of points of $U$ with no limit points in $U$, then there is an analytic function $f$ on $U$ with $\{z_k\}$ as a list of its zeroes counting mulltiplicity.*

**Proof.** Either $U$ or its image under some linear fractional transformation will contain $\infty$. Thus, we may as well assume $\infty \in U$. Then the complement of $U$ in $S^2$ is a compact subset $K$ of the plane.

Since $\{z_k\}$ has no limit point in $U$, the distance between $z_k$ and $K$ must approach 0 as $k \to \infty$. It follows that we may choose a sequence $\{w_k\}$ of points of $K$ such that $\lim |z_k - w_k| = 0$.

We set

$$f(z) = \prod_{n=1}^{\infty} E_k \left( \frac{z_k - w_k}{z - w_k} \right).$$

The product converges uniformly on compact subsets of $U$, since $\lim |z_k - w_k| = 0$ implies the uniform convergence on compact subsets of $U$ of

$$\sum_{k=1}^{\infty} \left| \frac{z_k - w_k}{z - w_k} \right|^{k+1}.$$

The function $f$ is analytic in $U$ and has $\{z_k\}$ as a list of its zeroes counting multiplicity. □

**Meromorphic Functions.** On a connected open set $U$, the set of analytic functions forms an integral domain – that is, it is a commutative ring with the property that the product of two elements is zero if and only if one of them is zero. The set of meromorphic functions of $U$ forms a field – that is, a commutative ring in which every non-zero element has an inverse. The next theorem shows that the field of meromorphic function is actually the quotient field of the ring of analytic functions. That is, every meromorphic function is the quotient $f/g$ of two analytic functions.

**Theorem 8.2.8.** *If $U$ is a connected open subset of $\mathbb{C}$, then each meromorphic function on $U$ has the form $f/g$, where $f$ and $g$ are analytic on $U$ and $g$ is not identically zero.*

**Proof.** Let $h$ be a meromorphic function on $U$ and let $\{z_k\}$ be a sequence consisting of the poles of $h$, with each $z_k$ listed as many times as the order of the pole at $z_k$. By Theorem 8.2.7 there is an analytic function $g$ on $U$ with $\{z_k\}$ as a list of its zeroes counting multiplicity. Then, after removing removable singularities, $f = gh$ is an analytic function on $U$. Thus, $h = f/g$ with $f$ and $g$ analytic on $U$. □

**The Mittag-Leffler Theorem.** The Weierstrass Theorem (Theorem 8.2.7) gives the existence of an analytic function with a specified list of zeroes counting multiplicity. The Mittag-Leffler Theorem is a companion theorem. It gives the existence of a meromorphic function with a specified list of poles and principal parts. We prove it only for discs, although it is true for general open sets.

**Theorem 8.2.9.** *Let $R$ be a positive number or $\infty$. Let $S$ be a discrete set of points of $D_R(0)$ and $\{h_w : w \in S\}$ a set of polynomials with no constant terms. Then there exists a meromorphic function $f$ with a pole at $w$ with principal part $h_k((z-w)^{-1})$ for each $w \in S$ and with no other poles.*

**Proof.** We choose an increasing sequence of radii $\{r_n\}$ with $r_n \to R$, we let $S_1$ be the subset of $S$ which lies in $\overline{D}_{r_1}(0)$ and, for $n > 1$, we let

$$S_n = \{w \in S : r_{n-1} < |w| \le r_n\}.$$

Then, for each $n$,

$$g_n(z) = \sum_{w \in S_n} h_k((z-w)^{-1})$$

is a meromorphic function on the plane with a pole at $w$ with the required principal part for each $w \in S_n$ and with no other poles.

We might hope to construct the function we are after by simply taking the infinite sum of the functions $g_n$. Unfortunately, there is no reason to think this sequence should converge on $D_R(0)$. However, we can modify each $g_n$, without changing its poles and principal parts, in such a way as to end up with an infinite series which does converge.

For each $n > 1$, the function $g_n$ is analytic on an open set containing the closed disc $\overline{D}_{r_{n-1}}(0)$. Hence, it is the uniform limit on this closed disc of its power series at 0. It follows that there is a polynomial $p_n$ such that

$$|g_n(z) - p_n(z)| < 2^{-n} \quad \text{for} \quad |z| \leq r_{n-1}.$$

If we set $f_1 = g_1$ and $f_n = g_n - p_n$ for $n > 1$, then, for each $m > 1$, the series

$$\sum_{n=m+1}^{\infty} f_n(z)$$

converges uniformly to an analytic function on $D_{r_m}(0)$. This means that

$$f(z) = \sum_{n=1}^{\infty} f_n(z)$$

is defined as a meromorphic function on $D_{r_m}(0)$ and has the required poles and principal parts at those points of $S$ which lie in this disc. Since this is true for each $m$, and $\lim r_m = R$, $f$ is meromorphic on all of $D_R(0)$ and has the required poles and principal parts. □

## Exercise Set 8.2

1. Show that the derivative of $E_p(z)$ is $z^p\, e^{z+z^2/2+\cdots+z^p/p}$.
2. Compute the logarithmic derivative of $E_p(z)$.
3. Find an entire function (given by a Weierstrass product) that has a zero of order 1 at $\sqrt{n}$ for $n = 1, 2, 3, \cdots$ and no other zeroes.
4. Find an entire function (given by a Weierstrass product) that has a zero of order 2 at $\sqrt{n}$ for $n = 1, 2, 3, \cdots$ and has no other zeroes.
5. Find an entire function (given by a Weierstrass product) that has a zero of order $n$ at $n^2$ for $n = 1, 2, 3, \cdots$ and has no other zeroes.
6. If $f$ is an entire function, show that $f = g^n$ for some entire function $g$ if and only if the order of each zero of $f$ is divisible by $n$.
7. Suppose $f$ is an entire function such that $\{z_k\}$ is a list of its non-zero zeroes counting multiplicity and suppose that

$$\sum_{k=1}^{\infty} \frac{1}{|z_k|} < \infty.$$

   Describe the simplest Weierstrass factorization of $f$.
8. Suppose $f$ is an odd entire function, the order of the zero at 0 is $m$, and $\{z_k\}$ is a list of the other zeroes of $f$ counting multiplicity. Show that $m$ is positive and odd. If

$$\sum_{k=1}^{\infty} \frac{1}{|z_k|^2} < \infty,$$

prove that $f$ has a factorization of the form

$$f(z) = z^m \, e^{h(z)} \prod_{k=1}^{\infty} \left(1 - \frac{z^2}{z_k^2}\right),$$

where $h$ is an entire function.

9. Show that, if $U$ is any non-empty open subset of the plane, then there is an analytic function on $U$ which cannot be extended to be analytic on any larger open set. Hint: Use the general Weierstrass Theorem to construct an analytic function on $U$ with a lot of zeroes.
10. Prove that, given a sequence $\{z_k\}$ of complex numbers converging to infinity and a sequence $\{n_k\}$ of integers, there is an entire function $f$ with given values for $f$ and its derivatives up to order $n_k$ at $z_k$ for each $k$. Hint: Use the Mittag-Leffler and Weierstrass Theorems together.
11. Prove that if $f_1$ and $f_2$ are two entire functions with no common zeroes, then there exist entire functions $g_1$ and $g_2$ such that
$$g_1 f_1 + g_2 f_2 = 1.$$
Hint: Use the Mittag-Leffler Theorem to show that an entire function $g_2$ can be chosen so that, at each zero of $f_1$, the function $1 - g_2 f_2$ has a zero of order at least as large.
12. Let $f_1, f_2, \cdots, f_n$ be entire functions. Show that there are entire functions $h_1, h_2, \cdots, h_n$, and $u$ such that $f_j = u h_j$ for $j = 1, \cdots, n$ and the functions $h_1, h_2, \cdots, h_n$ have no common zeroes. Hint: Use the Weierstrass Theorem.
13. Let $f_1, f_2, \cdots, f_n$ be entire functions with no common zero. Use induction and the preceding two exercises to show that there are entire functions $g_1, g_2, \cdots, g_n$ such that
$$g_1 f_1 + g_2 f_2 + \cdots + g_n f_n = 1.$$
14. Those who are familiar with commutative ring theory may want to do this exercise. Let $\mathcal{E}$ be the ring of entire functions. Show that the following ring-theoretic properties of $\mathcal{E}$ are consequences of the preceding two exercises:
    (a) every finitely generated ideal of $\mathcal{E}$ is a principal ideal;
    (b) every finitely generated maximal ideal of $\mathcal{E}$ is of the form
$$M_w = \{f \in \mathcal{E} : f(w) = 0\} \quad \text{for some} \quad w \in \mathbb{C}.$$
15. The conclusions of the last four exercises actually hold for the ring of analytic functions on any open subset of the plane. However, to prove them all in this generality would require a stronger form of the Mittag-Leffler Theorem than the one proved here. Prove these results for the largest class of open sets that you can using the machinery developed in this text.

## 8.3. Entire Functions of Finite Order

**Definition 8.3.1.** An entire function $f$ is said to be of *finite order* if there is a number $t$ such that

$$|f(z)| \le e^{|z|^t}$$

for all $z$ with $|z|$ sufficiently large. The infimum of all such numbers $t$ is called the *order* of $f$.

For each non-negative integer $p$, the function $e^{z^p}$ is an entire function of finite order $p$. More generally:

**Example 8.3.2.** Show that $e^{h(z)}$ is an entire function of finite order $p$ if $h$ is a polynomial of degree $p$.

**Solution:** If $t > p$, then $\lim_{z\to\infty} |z|^{-t}|h(z)| = 0$. This implies that there is an $R > 0$ such that

$$|h(z)| < |z|^t \quad \text{for} \quad |z| > R.$$

Then

$$|\,e^{h(z)}\,| \le e^{|z|^t} \quad \text{for} \quad |z| > R. \tag{8.3.1}$$

Since such a statement is true for all $t > p$, by definition $e^{h(z)}$ has finite order at most $p$.

On the other hand, if $t < p$, then $\lim_{z\to\infty} |z|^{-t}|h(z)| = +\infty$. Hence, there is no $R$ for which (8.3.1) holds. We conclude that the order of $f$ is at least $p$ and, hence, is equal to $p$.

**Non-Vanishing Entire Functions of Finite Order.** It turns out that the functions $e^{h(z)}$ of the preceding example are the only entire functions of finite order which are non-vanishing.

To prove this, we will need the following theorem of Borel-Carathéodory relating the growth of the real part of an analytic function to the growth of the the absolute value of the function.

**Theorem 8.3.3.** *Suppose $0 < r < R$ and let $g$ be a function analytic on an open set containing $D_R(0)$. Then*

$$|g(z)| \le \frac{2r}{R-r}\sup\{\mathrm{Re}(g(w)) : |w| = R\} + \frac{R+r}{R-r}|g(0)| \quad \textit{if} \quad |z| \le r.$$

**Proof.** We suppose first that $g(0) = 0$. We set

$$m = \sup\{\mathrm{Re}(g(w)) : |w| = R\}.$$

Note that the Mean Value Theorem for harmonic functions implies that $m \ge 0$.

If $|w| = R$ and $u = \mathrm{Re}(g(w))$, then $u \le m$ and

$$u - 2m \le u \le 2m - u.$$

Thus, $|u| \le |2m - u|$, from which it follows that

$$|g(w)| \le |2m - g(w)|,$$

since the numbers $g(w)$ and $2m - g(w)$ have the same imaginary parts of the same magnitude and have real parts $u$ and $2m - u$, respectively.

We conclude from the above that the function

$$h(z) = \frac{g(z)}{z(2m - g(z))}$$

satisfies the inequality

$$|h(w)| \le \frac{1}{R} \quad \text{for} \quad |w| = R.$$

Since the analytic function $h$ has a removable singularity at 0, this inequality holds throughout the disc $D_R(0)$ by the Maximum Modulus Theorem. Thus,

$$\frac{|g(z)|}{r|2m - g(z)|} \le \frac{1}{R} \quad \text{whenever} \quad |z| = r,$$

which implies

$$|g(z)| \le \frac{r}{R}(2m + |g(z)|).$$

If we collect terms involving $|g(z)|$ on the left and divide by $1 - r/R$, the result is

$$|g(z)| \le \frac{2r}{R - r} m \quad \text{for} \quad |z| \le r.$$

This concludes the proof in the case where $g(0) = 0$. This general case follows from applying this result to the function $g_0(z) = g(z) - g(0)$. The details are left to the exercises. □

**Theorem 8.3.4.** *An entire function $f$ with no zeroes has finite order $p$ if and only if $p$ is a non-negative integer and $f$ has the form*

$$f(z) = \mathrm{e}^{h(z)},$$

*where $h$ is a polynomial of degree $p$.*

**Proof.** In view of Example 8.3.2, we need only show that every non-vanishing entire function $f$ of finite order $p$ has the above form.

Since $f$ has no zeroes and the plane is simply connected, there is an entire function $h$ such that

$$f(z) = \mathrm{e}^{h(z)} \quad \text{for all} \quad z \in \mathbb{C}.$$

Since $f$ has finite order $p$, for each $t > p$ there is an $M > 0$ such that

$$\mathrm{e}^{\mathrm{Re}(h(z))} = |f(z)| \le \mathrm{e}^{|z|^t} \quad \text{for} \quad |z| \ge M.$$

This implies

$$\mathrm{Re}(h(z)) \le |z|^t \quad \text{for} \quad |z| \ge M.$$

We apply the previous theorem with $r > M$ and $R = 2r$ to conclude

$$|h(z)| \le 2|z|^t + 3|h(0)| \quad \text{if} \quad |z| = r.$$

Since this is true for all $r > M$, Exercise 3.3.9 implies that $h$ must be a polynomial of degree at most $t$. Since $t$ was an arbitrary number greater than $p$, we conclude that $h$ is a polynomial of degree at most $p$. If it were a polynomial of degree less than $p$, then $f$ would have order less than $p$. Hence, the degree of the polynomial $h$ is exactly $p$. This, of course, implies that $p$ is a non-negative integer. □

**Canonical Products.** Given a sequence $\{z_k\}$, we let $\mu$ be the inf of the numbers $t$ such that

$$\sum_{k=1}^{\infty} \frac{1}{|z_k|^t} < \infty. \tag{8.3.2}$$

If there is no such $t$, then we set $\mu = \infty$. The number $\mu$ is called the *exponent of convergence* for the sequence $\{z_k\}$.

If $\{z_k\}$ has finite exponent of convergence $\mu$, then we can write down a convergent Weierstrass product (8.2.3), using $\{z_k\}$, in which the sequence $\{p_k\}$ is a constant $p$. We choose $p$ to be the smallest integer such that $\mu < p+1$. Then the condition

$$\sum_{k=1}^{\infty} \frac{1}{|z_k|^{p+1}} < \infty \tag{8.3.3}$$

is satisfied. Hence, by Theorem 8.2.2, the Weierstrass product

$$f(z) = \prod_{k=1}^{\infty} E_p(z/z_k) \tag{8.3.4}$$

converges. This is called the *canonical product* for the sequence $\{z_k\}$.

The significance of the choice of $p$ made for the canonical product is that, with this choice, the resulting product is an entire function with order $\lambda$ equal to the exponent of convergence $\mu$ of the sequence $\{z_k\}$. The next theorem yields part of what is needed to prove this. The remainder of the proof will come in the next section.

**Theorem 8.3.5.** *The canonical product for a sequence $\{z_k\}$, with finite exponent of convergence $\mu$, is an entire function of finite order $\lambda \le \mu$.*

**Proof.** We choose $p$ to be the smallest integer such that $\mu < p+1$, and let $t$ be any number in the range $\mu < t < p+1$.

We claim that there is a positive constant $A$ such that

$$|E_p(z)| \le e^{A|z|^t} \tag{8.3.5}$$

for all $z$.

If $|z| \le 1/2$, this follows from (8.1.6) with $w = E_p(z) - 1$ and Theorem 8.2.1. These combine to show that

$$|\log E_p(z)| \le 2|z|^{p+1} \le 2|z|^t,$$

and this implies (8.3.5) holds with $A = 2$.

If $|z| > 1/2$, then $|z|^k \le 2^{t-k}|z|^t$, and so

$$\log|E_p(z)| = \log|1-z| + \sum_{k=1}^{p} \frac{\operatorname{Re}(z^k)}{k}$$

$$\le |z| + \sum_{k=1}^{p} |z|^k \le (p+1)2^t|z|^t.$$

Thus, (8.3.5) holds with $A = (p+1)2^t$ in this case.

To prove the theorem, we note that, if $f$ is given by the canonical product (8.3.4), then by (8.3.5),

$$|f(z)| \leq \prod_{k=1}^{\infty} e^{A|z/z_k|^t} = e^{B|z|^t},$$

where

$$B = A \sum_{k=1}^{\infty} 1/|z_k|^t.$$

The series in this expression converges because $t$ is larger than the exponent of convergence $\mu$.

Since for any $s > t$, we have $B|z|^t \leq |z|^s$ for $|z|$ sufficiently large, it follows that $f$ has finite order at most $t$. Since $t$ was an arbitrary number strictly between $\mu$ and $p+1$, we conclude that $f$ has order at most $\mu$. □

One might guess, based on Theorem 8.3.4, that the order of an entire function of finite order must be a non-negative integer. This is not the case, as is shown by the following example.

**Example 8.3.6.** Find an entire function with finite order 1/2.

**Solution:** The function

$$\frac{\sin \pi z}{\pi z} = \prod_{k=1}^{\infty} \left(1 - \frac{z^2}{k^2}\right)$$

has order 1 (Exercise 8.3.3). It seems reasonable that if we replace $z^2$ by $z$ in this product, then the result would be an entire function of order 1/2. In fact, the resulting function has a zero of order 1 at $k^2$ for each positive integer $k$ and

$$\sum_{k=1}^{\infty} \frac{1}{(k^2)^t} < \infty$$

for every $t > 1/2$ and for no smaller values of $t$. Hence, the sequence $\{1/k^2\}$ has exponent of convergence 1/2. Since 0 is the smallest integer $p$ such that $1/2 < p+1$, the preceding theorem implies that the canonical product

$$f(z) = \prod_{k=1}^{\infty} \left(1 - \frac{z}{k^2}\right)$$

is an entire function of finite order at most 1/2.

In fact, it is easy to directly compute the order of $f$ if we note that

$$f(z) = \frac{\sin \pi\sqrt{z}}{\pi\sqrt{z}}.$$

The expression on the right is entire and is independent of the choice of the square root function because the function $(\pi z)^{-1} \sin \pi z$ is an even function. It is easy to see from this that, since $(\pi z)^{-1} \sin \pi z$ has order 1, $f$ has order 1/2 (Exercise 8.3.5).

## Exercise Set 8.3

1. Finish the proof of Theorem 8.3.3 by showing that, if it is true in the case where $g(0) = 0$, then it is true in general.
2. Show that a polynomial has finite order 0.
3. Show that $\sin z$, $z^{-1}\sin z$, and $\cos z$ all have finite order 1.
4. If $f$ is an entire function of order $\lambda(f)$, $k$ is a non-negative integer, and $g(z) = f(z^k)$, then prove that $\lambda(g) = k\lambda(f)$, where $\lambda(g)$ is the order of $g$.
5. Prove that if $g(z)$ is an even entire function of finite order $\lambda$ and $f(z) = g(\sqrt{z})$, then $f$ is an entire function of finite order $\lambda/2$. In particular, show that $\cos\sqrt{z}$ has order $1/2$.
6. Prove that the order of the sum or product of two entire functions is less than or equal to the maximum of the orders of the two functions.
7. What is the order of the entire function $e^{\sin z}$?
8. Suppose $f$ is an entire function which satisfies the inequality $|f(z)| \le |z|^{|z|}$ for $|z|$ sufficiently large. Prove that $f$ has finite order at most 1.
9. Find the exponent of convergence of the following sequences: $\{2^k\}$, $\{k^r\}$ $(r > 0)$, $\{\log k\}$.
10. Given an arbitrary non-negative real number $\mu$, show that there is a sequence of complex numbers $\{z_k\}$ with exponent of convergence $\mu$.
11. Does the order of an entire function necessarily have to be the same as the exponent of convergence of its sequence of zeroes? Justify your answer.

## 8.4. Hadamard's Factorization Theorem

Our goal in this section is to complete the characterization of entire functions of finite order $\lambda$. We will prove a theorem of Hadamard which asserts that every such function factors as a power of $z$ times a canonical product of order at most $\lambda$ times the exponential of a polynomial of degree at most $\lambda$. The key ingredient in the proof is Jensen's Formula relating the density of the zeroes of an entire function to the rate of growth at infinity of the function.

### Jensen's Formula.

**Theorem 8.4.1.** *If $f$ is analytic in an open set containing the disc $\overline{D}_r(0)$, $f$ has no zeroes on the boundary of this disc, $f(0) \neq 0$, and $z_1, z_2, \cdots, z_n$ are the zeroes, counting multiplicity, of $f$ in $D_r(0)$, then*

$$\log\left(\frac{|f(0)|r^n}{|z_1|\cdot|z_2|\cdots|z_n|}\right) = \frac{1}{2\pi}\int_0^{2\pi} \log(|f(r\,e^{i}\,\theta)|)\,d\theta.$$

**Proof.** We first prove this in the case where $r = 1$. We divide $f$ by a product of linear fractional transformations which preserve the unit circle and have zeroes at the points $z_i$. This yields a function

$$g(z) = f(z)\frac{1-\overline{z}_1 z}{z-z_1}\frac{1-\overline{z}_2 z}{z-z_2}\cdots\frac{1-\overline{z}_n z}{z-z_n}.$$

This function is analytic and non-vanishing in an open set containing the closed unit disc $\overline{D}$, and has the same modulus on the unit circle as does $f$. Thus, $g$ has an analytic logarithm in an open set containing $\overline{D}$. Then $\log|g(x)|$ is the real part of an analytic function in this set and, hence, is harmonic. The Mean Value Theorem for harmonic functions implies that

$$\log\left(\frac{|f(0)|}{|z_1|\cdot|z_2|\cdots|z_n|}\right) = \log|g(0)| = \frac{1}{2\pi}\int_0^{2\pi}\log|f(e^{i\theta})|\,d\theta. \tag{8.4.1}$$

To prove the theorem for general $r$, it suffices to apply (8.4.1) with $f$ replaced by the function $f(rz)$. If $f$ has zeroes at $z_1, z_2, \cdots, z_n$ in the disc $D_r(0)$, then $f(rz)$ has zeroes $z_1/r, z_2/r, \cdots, z_n/r$ in the unit disc $D$. Thus, the equation of the theorem follows directly from (8.4.1) applied to $f(rz)$. □

This leads to the following estimate on the number of zeroes of an entire function inside a disc $D_r(0)$.

**Theorem 8.4.2.** *If $f$ is an entire function with $|f(0)| = 1$, $n(r)$ is the number of zeroes of $f$ inside a disc $D_r(0)$, and $M(2r)$ is the supremum of $|f(z)|$ on the boundary of $D_{2r}(0)$, then*

$$n(r) \le \frac{\log M(2r)}{\log 2}.$$

**Proof.** Let $n = n(r)$ and $m = n(2r)$, and let $z_1, z_2, \cdots, z_m$ be the zeroes of $f$ inside the disc $D_{2r}(0)$ ordered so that $|z_j| \le |z_k|$ for $j \le k$. Then Jensen's Theorem with $r$ replaced by $2r$ implies that

$$\log\left|f(0)\frac{2r}{z_1}\frac{2r}{z_2}\cdots\frac{2r}{z_n}\cdots\frac{2r}{z_m}\right| \le \log M(2r).$$

Since

$$2 < \frac{2r}{|z_j|} \quad \text{if} \quad j \le n \quad \text{and} \quad 1 < \frac{2r}{|z_j|} \quad \text{if} \quad n < j \le m,$$

this implies that

$$\log|f(0)2^n| \le \log M(2r), \quad \text{or} \quad \log|f(0)| + n\log 2 \le \log M(2r).$$

Since $\log|f(0)| = 0$, it follows that

$$n \le \frac{\log M(2r)}{\log 2},$$

which completes the proof. □

**Zeroes of Functions of Finite Order.** The preceding theorem has the following consequence for entire functions of finite order.

**Theorem 8.4.3.** *Let $f$ be an entire function of finite order $\lambda$ and with $f(0) \ne 0$. Let $\{z_k\}$ be a list of the zeroes of $f$, counted according to multiplicity and indexed in the order of increasing modulus, and let $\mu$ be the exponent of convergence of $\{z_k\}$; then $\mu \le \lambda$.*

**Proof.** We claim that, for each $t > \lambda$, there are constants $N$, $C > 0$ and $q > 1$ such that

$$|z_k|^t \geq Ck^q \quad \text{for all} \quad k \geq N. \tag{8.4.2}$$

Assuming this, we conclude that the series

$$\sum_{k=1}^{\infty} \frac{1}{|z_k|^t}$$

converges for all $t > \lambda$, by comparison with the series

$$\sum_{k=1}^{\infty} \frac{1}{k^q},$$

which converges for $q > 1$. This, in turn, implies the exponent of convergence $\mu$ is at most $\lambda$.

To complete the proof, we must verify the claim concerning (8.4.2). In doing this, we may as well assume that $|f(0)| = 1$, since, if this is not so, we may make it so by replacing $f$ by $f$ divided by a constant times a power of $z$. Such a replacement will have no effect on whether the above claim is true.

Let $r_k = |z_k|$. Since the zeroes are indexed in such a way that the modulus is a non-decreasing function of $k$, there are at least $k$ zeroes of $f$ with modulus less than or equal to $r_k$. By Theorem 8.4.2,

$$k \leq \frac{\log M(2r_k)}{\log 2},$$

where $M(2r_k)$ is the sup of $|f(z)|$ on the circle $|z| = 2r_k$.

We choose $s$ with $\lambda < s < t$. Since $f$ has order $\lambda$, there is an $R$ such that $r_k \geq R$ implies

$$M(2r_k) \leq e^{(2r_k)^s}.$$

Hence, for $r_k \geq R$,

$$k \leq \frac{(2r_k)^s}{\log 2}.$$

This implies

$$r_k^t \geq \frac{(\log 2)^{t/s}}{2^t} k^{t/s} = Ck^q,$$

where

$$C = \frac{(\log 2^{t/s})}{2^t} \quad \text{and} \quad q = \frac{t}{s} > 1.$$

This is true provided $r_k = |z_k| > R$. However, since $\lim z_k = \infty$, there is an $N$ such that $k > N$ implies $|z_k| > R$. This completes the proof. $\square$

The above theorem, when combined with Theorem 8.3.5, yields the following corollary.

**Corollary 8.4.4.** *The canonical product for a sequence $\{z_k\}$ with exponent of convergence $\mu$ has finite order $\lambda = \mu$.*

**Hadamard's Theorem.** In the proof of the next theorem, we will need the following estimates on the size of the inverse $E_p^{-1}(z)$ of the function $E_p(z)$.

**Lemma 8.4.5.** *If $p$ is a non-negative integer, $p \le t \le p+1$, and $z \in \mathbb{C}$, then there is a constant $A$ such that*

$$\frac{1}{|E_p(z)|} \le e^{A|z|^t} \tag{8.4.3}$$

*if $|z| \ge 2$ or $|z| \le 1/2$.*

**Proof.** If $|z| \ge 2$, then $|1-z| \ge 1$ and $|z^k/k| \le |z|^t$ for $k \le p$. Hence,

$$|E_p^{-1}(z)| = |1-z|^{-1}|\, e^{-z-z^2/2-\cdots-z^p/p}\,| \le e^{p|z|^t}$$

and so (8.4.3) holds with $A = p$ in this case.

On the other hand, the definition of $E_p(z)$ and the Taylor series expansion of $\log(1-z)$ lead to

$$\log E_p(z) = \log(1-z) + \sum_{k=1}^{\infty} z^k/k = -\sum_{k=p+1}^{\infty} z^k/k,$$

and so if $|z| \le 1/2$, then

$$\log|E_p^{-1}(z)| = -\operatorname{Re}(\log E_p(z)) \le |z|^{p+1}\sum_{j=0}^{\infty}|z|^k \le 2|z|^{p+1} \le 2|z|^t.$$

Thus, (8.4.3) holds with $A = 2$ in this case. If we choose $A = \max\{2, p\}$, then (8.4.3) holds in both cases. □

We are now in a position to prove Hadamard's Theorem characterizing entire functions of finite order. This will be used in the proof of the Prime Number Theorem in the next chapter.

**Theorem 8.4.6.** *If $f$ is an entire function of order $\lambda$, and $p$ is the smallest integer such that $p+1 > \lambda$, then $f$ factors as*

$$f(z) = z^m\, e^{h(z)} \prod_{k=1}^{\infty} E_p(z/z_k), \tag{8.4.4}$$

*where $m$ is the order of the zero of $f$ at 0, $\{z_k\}$ is a list of the other zeroes of $f$ counting multiplicity, and $h(z)$ is a polynomial of degree at most $p$.*

**Proof.** According to the Weierstrass Factorization Theorem (Theorem 8.2.5) $f$ has a factorization of the form (8.4.4), where $h$ is an entire function. Thus, the only thing to be proved is that $h$ is a polynomial of degree at most $p$. This will follow from Theorem 8.3.4 if we can show that the function

$$g(z) = e^{h(z)} = \frac{f(z)}{z^m \prod_{j=1}^{\infty} E_p(z/z_k)}$$

has finite order at most $\lambda$.

Let $t$ be any number with $\lambda < t \le p+1$ and let $r \ge 1$ be any radius which is not one of the numbers $|z_k|$. We factor $g(z)$ as $g(z) = g_1(z)g_2(z)$, where

$$g_1(z) = f(z)z^{-m} \prod_{|z_k| \le 2r} E_p^{-1}(z/z_k) \tag{8.4.5}$$

and

$$g_2(z) = \prod_{|z_k| > 2r} E_p^{-1}(z/z_k). \tag{8.4.6}$$

Suppose $|z| = 4r = R$. Then $|z/z_k| \ge 2$ for all $k$ with $|z_k| \le 2r$. By the previous lemma, there is a positive constant $A_1$ such that

$$|g_1(z)| \le |f(z)| \prod_{|z_k| \le 2r} e^{A_1|z/z_k|^t}.$$

Since $f$ has finite order $\lambda$, for sufficiently large $r$ we have

$$|f(z)| \le e^{|z|^t}$$

and, hence,

$$|g_1(z)| \le e^{B_1 r^t}, \tag{8.4.7}$$

where

$$B_1 = 4^t \left(1 + A_1 \sum_{k=1}^{\infty} \frac{1}{|z_k|^t}\right).$$

The infinite series in this expression converges by Theorem 8.4.3. Since $g_1(z)$ is an entire function (once the removable singularities at the $z_k$ with $|z_k| < 2r$ are removed), if the inequality (8.4.7) holds for $|z| = 4r = R$, it must hold for all $z$ in the disc $|z| \le R$, by the Maximum Modulus Principle. In particular, this inequality holds for all $z$ with $|z| = r$.

Also if $|z| = r$, then $|z/z_k| < 1/2$ if $|z_k| > 2r$, and the previous lemma implies that there is a constant $A_2$ such that

$$|g_2(z)| \le \prod_{|z_k| > 2r} e^{A_2|z/z_k|^t} \le e^{B_2 r^t}, \tag{8.4.8}$$

where

$$B_2 = A_2 \sum_{k=1}^{\infty} \frac{1}{|z_k|}.$$

If we set $B = B_1 + B_2$ and combine (8.4.7) and (8.4.8), we obtain

$$|g(z)| \le e^{B|z|^t}.$$

Since $t$ is an arbitrary number larger than $\lambda$ and less than or equal to $p+1$, $g$ has order at most $\lambda$. This completes the proof. □

## Exercise Set 8.4

1. What does Hadamard's Factorization Theorem say about an entire function of order $\lambda < 1$?
2. If a non-constant entire function of finite order $\lambda$ has zeroes at the points $i\sqrt{n}$, what are the possible values for $\lambda$?
3. Show that an even entire function of order 1 has the form
$$Cz^m \prod_k \left(1 - \frac{z^2}{z_k^2}\right),$$
where $m$ is even, $C$ is a non-zero constant, and the sequence
$$\{z_1, -z_1, z_2, -z_2, \cdots, z_k, -z_k, \cdots\}$$
is a list of the zeroes of $f$ counting multiplicity.
4. What is the exponent of convergence for the sequence of zeroes in the preceding exercise?
5. State and prove the analogues of the previous two exercises for odd entire functions of order 1.
6. Prove that if $f$ is an entire function of order $\lambda$ and $\lambda$ is not an integer, then $f$ has infinitely many zeroes.
7. Under the hypotheses of the preceding exercise, prove that $f$ takes on every complex value infinitely many times.
8. Prove that if $f$ and $g$ are entire functions of finite order $\lambda$ and if $f(z_k) = g(z_k)$ on a sequence which satisfies
$$\sum_{k=1} \frac{1}{|z_k|^t} = \infty$$
for some $t > \lambda$, then $f(z) = g(z)$ identically.
9. Use the previous exercise to prove that if two functions of finite order agree at the points of the sequence $\{\log n\}_{n=1}^{\infty}$, then they agree identically. Thus, $e^z$ is the only entire function of finite order which has the value $n$ at the point $\log n$ for $n = 1, 2, \cdots$.
10. Find an entire function which has a zero at each point of the sequence $\{\log n\}$ for $n = 1, \cdots, \infty$. Does it have finite order?
11. Suppose $f$ is an entire function of finite order $\lambda$ and $\mu$ is the exponent of convergence of the list of zeroes of $f$. Prove that if $\mu < \lambda$, then $\lambda$ is an integer.
12. Is there an entire function of order $3/2$ which has the integers as its list of zeroes, counting multiplicity?

Chapter 9

# The Gamma and Zeta Functions

This chapter is devoted to developing some of the properties of two special functions of a complex variable – the gamma function and the zeta function. These functions are of great importance in modern mathematics. The developement of their properties provides a very instructive practical application of many of the techniques developed in the preceding chapters.

The zeta function is the subject of one of the most famous unsolved problems in mathematics – the Riemann Hypothesis. This conjecture arose from Riemann's attempt to settle an old conjecture concerning the rate of growth of the number $\pi(x)$ of primes less than or equal to $x$ as the positive number $x$ increases. In the process, Riemann developed (but did not completely prove) a formula for $\pi(x)$. This formula involves the zeroes of the zeta function in the strip $0 < \mathrm{Re}(z) < 1$, and its study led Riemann to conjecture that all these zeroes lie on the line $\mathrm{Re}(z) = 1/2$. If true, this would have been helpful in both the proof of Riemann's Formula and its use in analyzing the growth of $\pi(x)$.

The methods introduced by Riemann eventually led to proofs by others of the result on the growth of $\pi(x)$ that he was seeking. This result is now known as the Prime Number Theorem. These proofs use information about the location of the zeroes of the zeta function, but not, of course, the information proposed in the Riemann Hypothesis, since it has never been proved.

One of the reasons this chapter is included in the text is so that we may describe the Riemann Hypothesis and its connection to the Prime Number Theorem. For completeness, we conclude the chapter with a proof of the Prime Number Theorem. This proof makes strong use of the results on infinite products presented in the previous chapter.

We begin the chapter with a discussion of the gamma function.

## 9.1. Euler's Gamma Function

We define Euler's gamma function for $\mathrm{Re}(z) > 0$ by the integral formula

$$\Gamma(z) = \int_0^\infty e^{-t}\, t^{z-1}\, dt. \tag{9.1.1}$$

Of course, $\Gamma$ is defined by an improper integral and so we must show that this integral actually converges if $\mathrm{Re}(z) > 0$. In fact, it not only converges, but the resulting function of $z$ is analytic.

**Theorem 9.1.1.** *The integral* (9.1.1) *converges and defines an analytic function* $\Gamma(z)$ *for* $\mathrm{Re}(z) > 0$.

**Proof.** For $0 < r < s$ we define a function $\Gamma_{r,s}$ on the right half-plane by

$$\Gamma_{r,s}(z) = \int_r^s e^{-t}\, t^{z-1}\, dt.$$

The function $e^{-t}\, t^{z-1} = e^{-t+(z-1)\log t}$ is continuous as a function of $(t, z)$ in $[r, s] \times \mathbb{C}$ and is analytic in $z$ for each fixed value of $t$. By Exercise 3.2.16, $\Gamma_{r,s}$ is analytic on the entire plane. We will show that as $s \to \infty$ and $r \to 0$ the functions $\Gamma_{r,s}$ converge uniformly on each strip of the form

$$S = \{z : a \leq \mathrm{Re}(z) \leq b\} \quad \text{with} \quad 0 < a < b.$$

The limit function is then necessarily analytic on the right half-plane and is, by definition, Euler's function $\Gamma$.

If $x = \mathrm{Re}(z)$, then

$$|e^{-t}\, t^{z-1}| = e^{-t}\, t^{x-1}.$$

Thus, if $z$ is in the strip $S$, then

$$|e^{-t}\, t^{z-1}| \leq t^{a-1} \quad \text{for} \quad t \leq 1. \tag{9.1.2}$$

If $t \geq 1$, then

$$|e^{-t}\, t^{z-1}| \leq e^{-t}\, t^{b-1} \quad \text{on} \quad S.$$

The function $e^{-t}\, t^{b+1}$ is continuous and has limit 0 at infinity. It is, therefore, bounded on $[1, \infty)$, by a positive number $K$. Thus,

$$|e^{-t}\, t^{z-1}| \leq Kt^{-2} \quad \text{for} \quad t \geq 1. \tag{9.1.3}$$

Since $t^{a-1}$ is integrable on $(0, 1]$ and $Kt^{-2}$ is integrable on $[1, \infty)$, inequalities (9.1.2) and (9.1.3) imply that the improper integrals of $e^{-t}\, t^{z-1}$ on $(0, 1]$ and on $[1, \infty)$ both exist (see Theorem 5.2.2). Hence, the improper integral defining $\Gamma$ exists for each $z \in S$.

To show that $\Gamma(z)$ is analytic, we will show that $\Gamma_{r,s}$ converges uniformly to $\Gamma$ on each strip of the form $S$ as $r \to 0$ and $s \to \infty$. In fact, from (9.1.2) and (9.1.3) we conclude

$$|\Gamma(z) - \Gamma_{r,s}(z)| \leq \int_0^r t^{a-1}\, dt + \int_s^\infty Kt^{-2}\, dt \leq r^a/a + K/s.$$

Given $\epsilon > 0$ the right side of this inequality is less than $\epsilon$ whenever $r < (a\epsilon/2)^{1/a}$ and $s > 2K/\epsilon$. It follows from this that $\Gamma_{r,s}(z)$ converges uniformly to $\Gamma(z)$ for $z \in S$ as $r \to 0$ and $s \to \infty$. This completes the proof. $\square$

**Analytic Continuation of Gamma.** We will continue $\Gamma$ to a meromorphic function defined on the entire plane. The key to doing this is the fact that $\Gamma$ satisfies a functional equation, as specified in the following theorem.

**Theorem 9.1.2.** *The gamma function satisfies the functional equation*

$$\Gamma(z+1) = z\Gamma(z)$$

*for all $z$ in the right half-plane.*

**Proof.** We have

$$\Gamma(z+1) = \int_0^\infty e^{-t} t^z \, dt.$$

Integrating by parts with $u = t^z$ and $dv = e^{-t} \, dt$ yields

$$\Gamma(z+1) = -t^z e^{-t} \Big|_0^\infty + \int_0^\infty e^{-t} z t^{z-1} \, dt = z\Gamma(z).$$ □

**Corollary 9.1.3.** *If $n$ is a positive integer, then $\Gamma(n) = (n-1)!$.*

We leave the proof of this corollary as an exercise (Exercise 9.1.2).

**Theorem 9.1.4.** *The gamma function has a meromorphic continuation to the complex plane which has simple poles at the points $\{0, -1, -2, \cdots\}$.*

**Proof.** We prove by induction that $\Gamma$ has a meromorphic continuation with the indicated poles and satisfying the functional equation $\Gamma(z+1) = z\Gamma(z)$ on the set $\{z : \text{Re}(z) > -n\}$ for $n = 0, 1, 2, \cdots$. This is trivially true for $n = 0$. If it is true for $n$, then

$$\Gamma(z) = \frac{\Gamma(z+1)}{z}$$

defines $\Gamma$ on $\{z : \text{Re}(z) > -n-1\}$ in a fashion which is consistent with its definition on the smaller set $\{z : \text{Re}(z) > -n\}$, because of the functional equation. Clearly the poles of this continuation are as required and the functional equation continues to hold. □

**Zeroes of Gamma.** It turns out that $\Gamma$ has no zeroes. To prove this requires deriving another functional equation. The derivation involves a pair of computational lemmas.

We define Euler's beta function, $B(z, w)$, by

$$B(z, w) = \int_0^1 (1-s)^{z-1} s^{w-1} \, ds,$$

for $z$ and $w$ with positive real part.

**Lemma 9.1.5.** *If $z$ and $w$ have positive real parts, then*

$$\frac{\Gamma(z)\Gamma(w)}{\Gamma(z+w)} = B(z, w).$$

**Proof.** For $z$ with $\mathrm{Re}(z) > 0$, the substitution $t = u^2$ leads to

$$\Gamma(z) = \int_0^\infty t^{z-1} e^{-t} dt = 2\int_0^\infty u^{2z-1} e^{-u^2} du.$$

Then for two points $z, w$ with positive real parts we have

$$\Gamma(z)\Gamma(w) = 4\int_0^\infty \int_0^\infty e^{-(u^2+v^2)} u^{2z-1} v^{2w-1}\, du\, dv.$$

We pass to polar coordinates, setting $u = r\cos(\theta)$ and $v = r\sin(\theta)$. Then

$$\begin{aligned}\Gamma(z)\Gamma(w) &= 4\int_0^\infty \int_0^{\pi/2} e^{-r^2} r^{2(z+w)-2} \cos^{2z-1}(\theta)\sin^{2w-1}(\theta)\; r\, dr\, d\theta \\ &= 2\int_0^\infty e^{-r^2} r^{2(z+w)-1}\, dr \cdot 2\int_0^{\pi/2} \cos^{2z-1}(\theta)\sin^{2w-1}(\theta)\, d\theta \\ &= \Gamma(z+w)\cdot 2\int_0^{\pi/2} \cos^{2z-1}(\theta)\sin^{2w-1}(\theta)\, d\theta.\end{aligned}$$

The substitution $s = \sin^2(\theta)$ leads to

$$2\int_0^{\pi/2} \cos^{2z-1}(\theta)\sin^{2w-1}(\theta)\, d\theta = \int_0^1 (1-s)^{z-1} s^{w-1}\, ds.$$

The latter expression is Euler's beta function $B(z, w)$. Thus,

$$\frac{\Gamma(z)\Gamma(w)}{\Gamma(z+w)} = B(z, w). \qquad \square$$

We will use the above identity in the case $w = 1 - z$ to derive a functional equation for $\Gamma$. To do this, we will need to evaluate $B(z, 1 - z)$. This involves the following integral evaluation.

**Lemma 9.1.6.** *If $x \in (0, 1)$, then* $\displaystyle\int_0^\infty \frac{t^{-x}}{1+t}\, dt = \frac{\pi}{\sin \pi x}$.

**Proof.** We essentially proved this back in Chapter 5 where we discussed the Mellin transform. In fact, the integral in the theorem is just the Mellin transform of $\frac{1}{1+t}$ evaluated at $1 - x$. In Example 5.4.3, as a special case of Theorem 5.4.2, we proved that the Mellin transform of $\frac{1}{1+x}$ is

$$\int_0^\infty \frac{x^{t-1}}{1+x}\, dx = \frac{\pi}{\sin \pi t}.$$

If the roles of $t$ and $x$ are reversed, this becomes

$$\int_0^\infty \frac{t^{x-1}}{1+t}\, dt = \frac{\pi}{\sin \pi x}.$$

The identity of the theorem is then obtained by replacing $x$ by $1 - x$ and using the identity $\sin(\pi - \pi x) = \sin \pi x$. $\square$

**Theorem 9.1.7.** *The gamma function satisfies the functional equation*

$$\Gamma(z)\Gamma(1-z) = \frac{\pi}{\sin \pi z}.$$

**Proof.** If we set $z = x$ and $w = 1 - x$ for $x \in (0,1)$, then since $\Gamma(1) = 1$, Lemma 9.1.5 implies

$$\Gamma(x)\Gamma(1-x) = \frac{\Gamma(x)\Gamma(1-x)}{\Gamma(1)} = B(x, 1-x) = \int_0^1 (1-s)^{x-1} s^{-x}\, ds.$$

Then the substitution $s = \dfrac{t}{t+1}$ leads to

$$\Gamma(x)\Gamma(1-x) = \int_0^\infty \left(1 - \frac{t}{t+1}\right)^{x-1} \frac{t^{-x}}{(t+1)^{-x}} \frac{dt}{(t+1)^2} = \int_0^\infty \frac{t^{-x}}{1+t}\, dt.$$

We use the previous lemma to evaluate the last integral and conclude that

$$\Gamma(x)\Gamma(1-x) = \frac{\pi}{\sin \pi x}.$$

Since this identity holds for $x \in (0,1)$, the Identity Theorem implies that it continues to hold when $x$ is replaced by any complex number $z$ for which the functions involved are defined. □

This theorem has the following corollary, the proof of which is left as an exercise (Exercise 9.1.4).

**Corollary 9.1.8.** *The gamma function has no zeroes.*

**Product Formula for $\Gamma$.** The fact that $e^{-t} = \lim_{n \to \infty}(1 - t/n)^n$ can be exploited to express $\Gamma$ as an infinite product of the sort studied in the previous chapter. The first step in deriving this formula is to show the following:

**Theorem 9.1.9.** *The identity*

$$\Gamma(x) = \lim_{n \to \infty} \int_0^n (1 - t/n)^n t^{x-1}\, dt$$

*holds for all $x > 0$.*

**Proof.** The function $e^{-s} - 1 + s$ is 0 at $s = 0$ and has a positive derivative $1 - e^{-s}$ for $s > 0$. It is, therefore positive for $s > 0$. Thus, $1 - s \le e^{-s}$ for $s > 0$. With $s = t/n$ this implies $1 - t/n < e^{-t/n}$ for $t > 0$ and, on taking $n$th powers,

$$(1 - t/n)^n \le e^{-t}.$$

Furthermore, an elementary calculus argument (Exercise 9.1.8) shows that

$$e^{-t} - (1 - t/n)^n \le \frac{1}{n\,e}, \tag{9.1.4}$$

for $t \ge 0$.

If we fix $a > 0$, then

$$\Gamma(x) - \int_0^n (1 - t/n)^n t^{x-1}\, dt \le \int_0^a (e^{-t} - (1 - t/n)^n) t^{x-1}\, dt + \int_a^\infty e^{-t} t^{x-1}\, dt$$

for $n > a$. The first term on the right converges to 0 as $n \to \infty$ by (9.1.4) and the second term can be made less than any given $\epsilon$ by choosing $a$ large enough, because the improper integral defining $\Gamma$ converges. □

**Theorem 9.1.10.** *The entire function* $1/\Gamma$ *can be represented as the infinite product*

$$\frac{1}{\Gamma(z)} = z\prod_{k=1}^{\infty}\frac{1+z/k}{(1+1/k)^z},$$

*where this product converges uniformly on each disc of finite radius.*

**Proof.** The integral in the previous theorem may be evaluated using a repeated application of integration by parts (Exercise 9.1.9). The result is

$$\Gamma(x) = \lim_{n\to\infty}\frac{n^x n!}{x(x+1)\cdots(x+n)}.$$

for $x > 0$.

If we invert this, divide both numerater and denomenator by $n!$, and note that $n^x = \prod_{k=1}^{n-1}(1+1/k)^x$, we obtain

$$\frac{1}{\Gamma(x)} = x\prod_{k=1}^{\infty}\frac{1+x/k}{(1+1/k)^x}$$

for $x > 0$.

If we can show that this infinite product converges, not just for $x > 0$, but uniformly on each disc of finite radius in the complex plane, then the result will be an entire function which agrees with the entire function $1/\Gamma(z)$ on the positive real axis. This implies the two entire functions agree on all of $\mathbb{C}$. Thus, the proof will be complete if we can show that

$$z\prod_{k=1}^{\infty}\frac{1+z/k}{(1+1/k)^z} \tag{9.1.5}$$

converges uniformly on each compact disc. This product is very nearly a Weierstrass product, as studied in the previous chapter. This fact can be used to prove the uniform convergence on compact discs. The details are left to Exercise 9.1.10.

□

## Exercise Set 9.1

1. Show that, for $z$ real and positive, $\Gamma(z)$ is the Mellin transform of a certain function. What function? (see Section 5.4).
2. Prove that $\Gamma(n) = (n-1)!$ if $n$ is a positive integer.
3. Prove that $z(z+1)(z+2)\cdots(z+n) = \dfrac{\Gamma(z+n+1)}{\Gamma(z)}$.
4. Prove Corollary 9.1.8.
5. Prove that the residue of $\Gamma(z)$ at $-n$ is $\dfrac{(-1)^n}{n!}$ for $n = 0, 1, 2, \cdots$.
6. Prove that, for $r > 0$ and $\mathrm{Re}(z) > 0$, $\int_0^\infty e^{-rt}t^{z-1}\,dt = r^{-z}\Gamma(z)$.
7. Prove that $\Gamma(z)\Gamma(-z) = \dfrac{-\pi}{z\sin\pi z}$.

8. Prove that $e^{-t} - (1 - t/n)^n \leq \dfrac{1}{n\,e}$ for all $t \in [0, n]$. Hint: Show that the maximum of the function $h(t) = e^{-t} - (1 - t/n)^n$ on $[0, n]$ occurs at a point $t_0$ where $h(t_0) = e^{-t_0} t_0/n$. Then show that this number is less than or equal to $\dfrac{1}{ne}$.
9. Using integration by parts, prove that if $x > 0$, then
$$\int_0^n \left(1 - \frac{t}{n}\right)^n t^{x-1}\, dt = \frac{n^x n!}{x(x+1)\cdots(x+n)}.$$
10. Prove that the infiinite product
$$\frac{1}{\Gamma(z)} = z \prod_{k=1}^{\infty} \frac{1 + z/k}{(1 + 1/k)^z}$$
converges uniformly on each compact disc. Hint: Show that
$$\frac{1 + z/k}{(1 + 1/k)^z} = (1 + z/k)\, e^{-z/k}\, e^{a_k z},$$
where $a_k = 1/k - \log(1 + 1/k)$. Then show that $\prod_k (1 + z/k)\, e^{-z/k}$ is a convergent Weierstrass product and $\sum_k |a_k|$ converges.

## 9.2. The Riemann Zeta Function

The Riemann zeta function is defined on the set $\{z : \mathrm{Re}(z) > 1\}$ by the infinite series

$$\zeta(z) = \sum_{n=1}^{\infty} n^{-z}. \tag{9.2.1}$$

If $z = x + iy$, then $|n^{-z}| = n^{-x}$, and so this series converges uniformly absolutely on each set of the form $\{z : \mathrm{Re}(z) \geq r\}$ for $r > 1$. It follows that $\zeta(z)$ is defined and analytic on the set $\{z : \mathrm{Re}(z) > 1\}$.

**A Product Formula for the Zeta Function.** Let $\{p_1, p_2, p_3, \cdots\}$ be the set of prime numbers written in increasing order. Then we have

**Theorem 9.2.1.** *For* $\mathrm{Re}(z) > 1$, $\zeta(z) = \prod_{n=1}^{\infty} (1 - p_n^{-z})^{-1}$.

**Proof.** The fact that the infinite product converges for $\mathrm{Re}(z) > 1$ follows from Theorem 8.1.4 and the fact that $\sum_{n=1}^{\infty} p_n^{\mathrm{Re}(z)}$ converges if $\mathrm{Re}(z) > 1$. Then

$$\zeta(z)(1 - 2^{-z}) = \sum_1^{\infty} n^{-z} - \sum_1^{\infty} (2n)^{-z} = \sum_{n \in S_1} n^{-z},$$

where $S_1$ is the set of odd natural numbers. An induction argument using the same technique then shows that for each natural number $k$

$$\zeta(z) \prod_{n=1}^{k} (1 - p_n^{-z}) = \sum_{n \in S_k} n^{-z},$$

where $S_k$ is the set of natural numbers not divisible by any of the first $k$ primes. Since the right side of this equation has limit 1 as $k \to \infty$, the theorem follows. $\square$

This theorem has the following two corollaries. We leave the proofs to the exercises (Exercises 9.2.2 and 9.2.3).

**Corollary 9.2.2.** *There are infinitely many primes.*

**Corollary 9.2.3.** *The zeta function has no zeroes in the region* $\mathrm{Re}(z) > 1$.

**The Function $\xi$.** Our next goal is to extend the zeta function to be a meromorphic function on the entire plane. We will do this by expressing the zeta function in terms of the gamma function and a certain entire function $\xi$.

We begin the development of $\xi$ by making the substitution $t = n^2 s^2 \pi$ in the formula (9.1.1) defining $\Gamma$. The result is

$$\Gamma(z) = 2n^{2z}\pi^z \int_0^\infty e^{-n^2 s^2 \pi} s^{2z} \frac{ds}{s}.$$

If we divide by $n^{2z}\pi^z$ and sum over $n = 1, 2, 3, \cdots$, we obtain

$$\zeta(2z)\Gamma(z)\pi^{-z} = 2\sum_{n=1}^\infty \int_0^\infty e^{-n^2 s^2 \pi} s^{2z} \frac{ds}{s} \quad \text{if} \quad \mathrm{Re}(z) > 1. \tag{9.2.2}$$

We will use the result of Exercise 9.2.7 to prove that it is legitimate to move the summation inside the integral in the expression on the right. We estimate the size of each integrand in this series as follows:

$$\left| e^{-n^2 s^2 \pi} s^{2z-1} \right| \le e^{-ns^2} s^{2\,\mathrm{Re}(z)-1}.$$

The functions on the right are positive and their sum is

$$\sum_{n=1}^\infty e^{-ns^2} s^{2\,\mathrm{Re}(z)-1} = \frac{s^{2\,\mathrm{Re}(z)-1}}{e^{s^2} - 1}. \tag{9.2.3}$$

For each $z$, this series converges uniformly on each closed subinterval of $(0, \infty)$. Furthermore, since $e^{s^2} - 1 \ge s^2$, if $\mathrm{Re}(z) > 2$, the function on the right in (9.2.3) is less than or equal to $s^{2\,\mathrm{Re}(z)-3}$ and, hence, has finite integral over $[0, 1]$ if $\mathrm{Re}(z) > 2$. Since $e^{s^2} - 1 \ge e^{s^2}/2$ if $s \ge 1$, the function on the right in (9.2.3) is less than or equal to $2\,e^{-s^2} s^{\mathrm{Re}(z)-1}$ on $[1, \infty)$ and, hence, has finite integral on $[1, \infty)$. It follows that this function has finite integral on $[0, \infty)$. Thus, by the result of Exercise 9.2.7, the sum can be taken inside the integral in (9.2.2). Doing so yields

$$\zeta(2z)\Gamma(z)\pi^{-z} = 2\int_0^\infty \sum_{n=1}^\infty e^{-n^2 s^2 \pi} s^{2z} \frac{ds}{s} \quad \text{for} \quad \mathrm{Re}(z) > 2. \tag{9.2.4}$$

If we replace $z$ by $z/2$ and set

$$H(s) = \sum_{n=1}^\infty e^{-n^2 s^2 \pi}, \tag{9.2.5}$$

then (9.2.4) may be rewritten as

$$\zeta(z)\Gamma(z/2)\pi^{-z/2} = 2\int_0^\infty H(s) s^z \frac{ds}{s} \quad \text{for} \quad \mathrm{Re}(z) > 2. \tag{9.2.6}$$

The function $\xi$ is obtained by multiplying this expression by $z(z-1)/2$. Thus, for $\mathrm{Re}(z) > 2$,

$$\xi(z) = \frac{z(z-1)}{2}\zeta(z)\Gamma(z/2)\pi^{-z/2} = z(z-1)\int_0^\infty H(s)s^z \frac{ds}{s}. \tag{9.2.7}$$

**The Poisson Summation Formula.** We pause to prove a technical result about Fourier transforms (see Section 5.3). It will be used in the upcoming proof that $\xi$ extends to an entire function.

**Theorem 9.2.4.** *If $f$ is a continuous function on $\mathbb{R}$ with the property that the series $\sum_{n=-\infty}^{\infty} f(x+2\pi n)$ converges absolutely and uniformly for $x \in [-\pi,\pi]$ and the series $\sum_{n=-\infty}^{\infty} f\hat{}(n)$ converges absolutely, then*

$$\sum_{n=1}^{\infty} f(2\pi n) = \frac{1}{\sqrt{2\pi}}\sum_{n=1}^{\infty} f\hat{}(n),$$

*where $f\hat{}$ is the Fourier transform of $f$.*

**Proof.** We set $g(\theta) = \sum_{n=1}^{\infty} f(\theta + 2\pi n)$ whenever $\theta \in [-\pi,\pi]$, and then integrate this function against the Poisson kernel (see Section 6.5)

$$P_r(\theta) = \frac{1-r^2}{1-2r\cos\theta + r^2} = \sum_{-\infty}^{\infty} r^{|n|}\, e^{in\theta}$$

over the interval $[-\pi,\pi]$. The result is

$$\begin{aligned}
\int_{-\pi}^{\pi} g(\theta)P_r(\theta)\,d\theta &= \int_{-\pi}^{\pi}\sum_{-\infty}^{\infty} f(\theta+2\pi n)P_r(\theta)\,d\theta \\
&= \sum_{-\infty}^{\infty}\int_{-\pi}^{\pi} f(\theta+2\pi n)P_r(\theta)\,d\theta \\
&= \sum_{-\infty}^{\infty}\int_{(2n-1)\pi}^{(2n+1)\pi} f(\theta)P_r(\theta)\,d\theta \\
&= \int_{-\infty}^{\infty} f(\theta)P_r(\theta)\,d\theta \\
&= \sqrt{2\pi}\sum_{-\infty}^{\infty} r^{|n|} f\hat{}(n).
\end{aligned} \tag{9.2.8}$$

Note that the third step in this calculation uses the hypothesis that the series defining $g$ converges uniformly absolutely .

As $r \to 1$ the integral on the left in (9.2.8) converges to

$$2\pi g(0) = 2\pi\sum_{n=1}^{\infty} f(2\pi n),$$

by Lemma 6.5.5, while the sum on the right converges to

$$\sqrt{2\pi}\sum_{-\infty}^{\infty} f\hat{}(n)$$

since, by hypothesis, the series $\sum_{-\infty}^{\infty} f\hat{}(n)$ converges absolutely. We conclude that

$$\sum_{n=-\infty}^{\infty} f(2\pi n) = \frac{1}{\sqrt{2\pi}} \sum_{n=-\infty}^{\infty} f\hat{}(n),$$

as required. □

**The Symmetry of $\xi$.** We can now prove that $\xi$ extends to be an entire function and that it satisfies the symmetry relation $\xi(z) = \xi(1-z)$. We first derive a symmetry relation for $H$.

**Lemma 9.2.5.** *The function $H$ satisfies the relation*

$$H(s^{-1}) = sH(s) + (s-1)/2. \tag{9.2.9}$$

**Proof.** Note that $H(s) = (G(s) - 1)/2$, where

$$G(s) = \sum_{-\infty}^{\infty} \mathrm{e}^{-n^2 s^2 \pi} = 1 + 2\sum_{n=1}^{\infty} \mathrm{e}^{-n^2 s^2 \pi}.$$

If $g(x) = \frac{1}{\sqrt{2\pi}} \mathrm{e}^{-x^2/2}$ is the normal distribution function from Example 5.3.6, then

$$G(s) = \sqrt{2\pi} \sum_{-\infty}^{\infty} g(ns\sqrt{2\pi}).$$

Note that the function $g$ is its own Fourier transform, by Example 5.3.6. This implies that the function $f$, defined by $f(x) = g(xs/\sqrt{2\pi})$, satisfies the hypotheses of the previous theorem. If we apply that theorem to $f$, we conclude that

$$G(s) = \sum_{-\infty}^{\infty} f(2\pi n) = \frac{1}{\sqrt{2\pi}} \sum_{-\infty}^{\infty} f\hat{}(n).$$

A change of variables in the integral defining the Fourier transform shows that $f\hat{}(n) = s^{-1}\sqrt{2\pi} g\hat{}(ns^{-1}\sqrt{2\pi})$. Thus,

$$G(s) = \sum_{-\infty}^{\infty} s^{-1}\sqrt{2\pi} g(ns^{-1}\sqrt{2\pi}) = s^{-1}G(s^{-1}).$$

The identity (9.2.9) follows from this. □

The consequence for the function $\xi$ is the following:

**Theorem 9.2.6.** *The function $\xi$ extends to an entire function which is symmetric about the point $z = 1/2$; that is, $\xi(1-z) = \xi(z)$.*

**Proof.** If we break the integral on the right side of (9.2.6) into an integral over $[1, \infty)$ and an integral over $[0, 1]$ and make the substitution $s \to s^{-1}$ in the latter integral, the result is

$$\int_1^{\infty} H(s)s^z \frac{ds}{s} + \int_1^{\infty} H(s^{-1})s^{-z} \frac{ds}{s}.$$

Using (9.2.9), this becomes

$$\int_1^\infty H(s)s^z \frac{ds}{s} + \int_1^\infty (sH(s) + (s-1)/2)s^{-z}\frac{ds}{s}$$
$$= \int_1^\infty H(s)s^z \frac{ds}{s} + \int_1^\infty H(s)s^{(1-z)}\frac{ds}{s} + \frac{1}{2}\int_1^\infty (s^{(1-z)} - s^{-z})\frac{ds}{s}.$$

Since

$$\frac{1}{2}\int_1^\infty (s^{(1-z)} - s^{-z})\frac{ds}{s} = \frac{1/2}{z(z-1)} \quad \text{if} \quad \operatorname{Re}(z) > 1,$$

we conclude that

$$\xi(z) = 1/2 - z(1-z)\int_1^\infty H(s)(s^z + s^{1-z})\frac{ds}{s} \quad \text{for} \quad \operatorname{Re}(z) > 2. \tag{9.2.10}$$

However, the right side of this equation is defined and analytic on the entire plane, since $H(s)$ times any power of $s$ is absolutely integrable on $[1, \infty)$. It is also obviously symmetric about $z = 1/2$. Thus, $\xi$ has an extension to the whole plane with the required properties. □

**Meromorphic Extension of $\zeta$.** The next theorem is an immediate consequence of the preceding theorem and (9.2.7).

**Theorem 9.2.7.** *The function $\zeta$ has a meromorphic extension to the plane given by the formula*

$$\zeta(z) = \frac{2\pi^{z/2}\xi(z)}{z(z-1)\Gamma(z/2)}. \tag{9.2.11}$$

It is useful to note that the above formula can be put in a slightly different form by using Theorem 9.1.2. This theorem, with $z$ replaced by $z/2$, implies that

$$(z/2)\Gamma(z/2) = \Gamma(z/2+1).$$

Then (9.2.11) becomes

$$\zeta(z) = \frac{\pi^{z/2}\xi(z)}{(z-1)\Gamma(z/2+1)}. \tag{9.2.12}$$

## Exercise Set 9.2

1. Show that $\lim_{x\to\infty} \zeta(x+iy) = 1$ and the convergence is uniform in $y$.
2. Use Theorem 9.2.1 to prove Corollary 9.2.2.
3. Use Theorem 9.2.1 to prove Corollary 9.2.3.
4. If $z = s + it$ with $s > 1$, prove that

$$\left|\frac{1}{\zeta(z)}\right| < \zeta(s).$$

Hint: Use Theorem 9.2.1.

5. Use the result of the previous exercise to prove that

$$\left|\frac{1}{\zeta(z)}\right| < \zeta(2)$$

if $\mathrm{Re}(z) \geq 2$.

6. Let $u(t) = \sum_{n=1}^{\infty} u_n(t)$ be the sum of a series of positive continuous functions on $(0, \infty)$ and suppose this series converges uniformly on closed bounded intervals of $(0, \infty)$. Prove that

$$\sum_{n=1}^{\infty} \int_0^{\infty} u_n(t)\, dt = \int_0^{\infty} u(t)\, dt.$$

Hint: Either both sides are infinite or one of them is finite.

7. Let $h(t) = \sum_{n=1}^{\infty} h_n(t)$ be the sum of an infinite series of continuous functions on $(0, \infty)$ and suppose

$$|h_n(t)| \leq u_n(t) \quad \text{for all} \quad n, t,$$

where $\sum_{n=1}^{\infty} u_n$ is a positive termed series which satisfies the conditions of the previous exercise. Prove that the improper integral of $h$ on $\mathbb{R}$ converges and

$$\int_0^{\infty} h(t)\, dt = \sum_{n=1}^{\infty} \int_0^{\infty} h_n(t)\, dt.$$

8. Show that the Poisson Summation Formula can, under appropriate hypotheses on $f$ and $f\hat{}$, be reformulated as

$$\sum_{n=-\infty}^{\infty} f(n) = \sqrt{2\pi} \sum_{n=-\infty}^{\infty} f\hat{}(2\pi n).$$

9. Use the form of the Poisson Summation Formula derived in the previous exercise to show that

$$\sum_{-\infty}^{\infty} \frac{1}{1+n^2} = \pi \frac{e^{2\pi}+1}{e^{2\pi}-1}.$$

Hint: The Fourier transform of $\dfrac{1}{1+x^2}$ is calculated in Example 5.3.3.

The next five exercises outline an alternative to the approach used in Theorem 9.2.7 to prove that $\zeta$ extends to be meromorphic in the plane.

10. Formula (9.2.4) was developed using the substitution $t = n^2 s^\pi$ in the integral defining $\Gamma$. Use the substitution $t = ns$ in a similar way to derive the formula

$$\zeta(z)\Gamma(z) = \int_0^{\infty} \frac{1}{e^s - 1} s^{z-1}\, ds \quad \text{for} \quad \mathrm{Re}(z) > 1.$$

11. For complex numbers $w$ and $z$, define $(-w)^{z-1}$ to be $e^{(z-1)\log(-w)}$, where log is the principal branch of the log function. Show that this function is analytic except for a cut on the positive real line, that its limit as $w$ approaches the positive real number $s$ from above is $e^{(z-1)(\log s - \pi i)}$, and that its limit as $w$ appproaches $s$ from below is $e^{(z-1)(\log s + \pi i)}$.

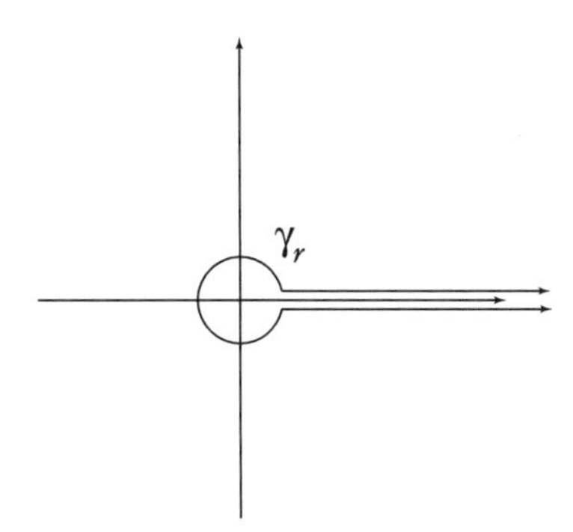


**Figure 9.2.1.** Contour for Exercise 9.2.12.

12. Using this definition for $(-w)^{z-1}$, consider the contour integral

$$\eta(z) = \int_{\gamma_r} \frac{1}{e^w - 1}(-w)^{z-1}\, dw, \tag{9.2.13}$$

where $\gamma_r$ is the contour indicated in Figure 9.2.1, $r < 2\pi$ is the radius of the indicated circle, and the two horizontal lines are a distance $\epsilon \le r$ above and below the positive real axis. Prove that this integral exists for all $z$, is independent of $r$ and $\epsilon$, and defines an entire function $\eta(z)$.
13. By passing to the limit as $\epsilon \to 0$ and $r \to 0$ in the integral 9.2.13, prove that $\eta(z) = -2\pi i \sin(\pi z)\Gamma(z)\zeta(z)$ if $\mathrm{Re}(z) > 1$.
14. Use the previous exercise and an identify involving $\Gamma$ to prove that

$$\zeta(z) = -\frac{1}{2\pi i}\Gamma(1-z)\eta(z) \quad \text{for} \quad \mathrm{Re}(z) > 1.$$

Conclude that $\zeta$ has a meromorphic extension to the plane with a single simple pole at $z = 1$.

## 9.3. Properties of $\zeta$

The expressions (9.2.11) and (9.2.12) for $\zeta$ and the properties of $\Gamma$ and $\xi$ lead to a wealth of information about $\zeta$. Ultimately, this information will lead to a proof of the Prime Number Theorem in the last section of this chapter.

### Zeroes and Poles.

**Theorem 9.3.1.** *The zeta function has a simple pole at $z = 1$, with residue 1, and no other poles.*

**Proof.** The function $\Gamma(z/2 + 1)$ has no zeroes, and the function $\xi$ is entire. Thus, (9.2.12) implies that the only pole of $\zeta$ is at $z = 1$ and it is a simple pole. The fact that the residue is 1 follows from the fact, proved in the exercises, that $\Gamma(1/2) = \sqrt{\pi}$, and from (9.2.10), which implies $\xi(1) = 1/2$. ☐

**Theorem 9.3.2.** *The zeta function has a zero of order* 1 *at each negative even integer.*

This is Exercise 9.3.1.

**Corollary 9.3.3.** *The zeta function has no zeroes outside the strip*

$$0 \leq \mathrm{Re}(z) \leq 1$$

*except the ones that occur at negative even integers.*

**Proof.** Except for the zeroes of $\zeta$ that occur at negative even integers (due to the poles of $\Gamma$), the functions $\zeta$ and $\xi$ have the same zeroes. Since $\zeta$ has no zeroes in the region $\mathrm{Re}(z) > 1$ by Corollary 9.2.3, and since $\xi$ is symmetric about $1/2$, it follows that $\zeta$ has no zeroes outside the strip $0 \leq \mathrm{Re}(z) \leq 1$ except the negative even integers. □

This result can be strengthened to exclude the existence of zeroes of $\zeta$ on the lines $\mathrm{Re}(z) = 0, 1$. The proof uses the following lemma:

**Lemma 9.3.4.** *For* $\mathrm{Re}(z) > 1$ *there is an analytic logarithm for* $\zeta(z)$*, defined by*

$$\log \zeta(z) = \sum_p \sum_{m=1}^{\infty} \frac{p^{-mz}}{m}. \tag{9.3.1}$$

*The derivative of this function is*

$$\frac{\zeta'(z)}{\zeta(z)} = -\sum_p \sum_{m=1}^{\infty} p^{-mz} \log p. \tag{9.3.2}$$

*Here, in both equations, the summation on the left is over all primes* $p$.

**Proof.** If $\mathrm{Re}(z) > 1$, then $|p^{-z}| < 1/2$ for all primes $p$ and so $\log(1 - p^{-z})$ is defined and analytic if log is the principal branch of the log function. Furthermore, by Theorem 9.2.1 we have for $\mathrm{Re}(z) > 1$

$$\exp\left(-\sum_p \log(1 - p^{-z})\right) = \prod (1 - p^{-z})^{-1} = \zeta(z). \tag{9.3.3}$$

Hence, $-\sum_p \log(1 - p^{-z})$ is an analytic logarithm for $\zeta(z)$ on $\mathrm{Re}(z) > 1$. If we expand this function in a power series in $p^{-z}$, the result is (9.3.1). On differentiating (9.3.1), we obtain (9.3.2). □

**Theorem 9.3.5.** *The zeta function has no zeroes outside the strip*

$$0 < \mathrm{Re}(z) < 1$$

*except those which occur at negative even integers.*

**Proof.** We first note that the inequality

$$0 \leq 3 + 4\cos\theta + \cos 2\theta \tag{9.3.4}$$

holds for all real numbers $\theta$ due to the identity

$$3 + 4\cos\theta + \cos 2\theta = 2(1+\cos\theta)^2,$$

which follows from $\cos 2\theta = 2\cos^2\theta - 1$.

If $z = x + iy$, then $\mathrm{Re}(p^{-mz}) = p^{-mx}\cos(-my\log p)$. It follows from (9.3.4) that, for $x > 1$,

$$0 \le 3p^{-mx} + 4\,\mathrm{Re}(p^{-mx-myi}) + \mathrm{Re}(p^{-mx-2myi}),$$

which, when combined with (9.3.1), implies that, for $x > 1$,

$$3\,\mathrm{Re}(\log\zeta(x)) + 4\,\mathrm{Re}(\log\zeta(x+iy)) + \mathrm{Re}(\log\zeta(x+i2y)) \ge 0,$$

or, on exponentiating,

$$|\zeta(x)|^3|\zeta(x+iy)|^4|\zeta(x+i2y)| \ge 1. \tag{9.3.5}$$

We divide both sides of this inequality by $x-1$ and write the result in the form

$$|(x-1)\zeta(x)|^3\left|\frac{\zeta(x+iy)}{x-1}\right|^4|\zeta(x+i2y)| \ge \frac{1}{x-1}. \tag{9.3.6}$$

Since $\zeta(z)$ has a simple pole at $z = 1$, $(1-z)\zeta(z)$ has a removable singularity at $z = 1$. This implies that the first factor on the left in (9.3.6) is bounded as $x \to 1$. Since there are no other poles of $\zeta$, the factor on the right is also bounded as $x \to 1$, provided $y \neq 0$. If $\zeta$ has a zero at $1 + iy$, then the middle factor is also bounded as $x \to 1$. Since the right side of (9.3.6) is not bounded as $x \to 1$, we conclude that there can be no zero of $\zeta$ at $z = 1 + iy$.

Now that we know that there are no zeroes of $\zeta$ on the line $\mathrm{Re}(z) = 1$, we use the fact that the set of zeroes of $\zeta$ in the strip $0 \le \mathrm{Re}(z) \le 1$ is symmetric about the line $\mathrm{Re}(z) = 1/2$ (since these are also the zeroes of $\xi$) to conclude that there are no zeroes on the line $\mathrm{Re}(z) = 0$. In view of Corollary 9.3.3, this completes the proof. □

**Estimate on the growth of $\xi$.** Integration by parts in the integral appearing in (9.2.10) leads to

$$\begin{aligned}\xi(z) &= 1/2 + H(1) + \int_1^\infty H'(s)((1-z)s^z + zs^{1-z})\,ds\\ &= 1/2 + H(1) + \int_1^\infty s^2H'(s)((1-z)s^{(z-1)} + zs^{-z})\,\frac{ds}{s}.\end{aligned}$$

Another integration by parts leads to

$$\xi(z) = 1/2 + H(1) + 2H'(1) + \int_1^\infty (s^2H'(s))'(s^{z-1} + s^{-z})\,ds.$$

If we differentiate (9.2.9) and set $s = 1$, the result is

$$\frac{1}{2} + H(1) + 2H'(1) = 0.$$

Thus,

$$\xi(z) = \int_1^\infty (s^2H'(s))'(s^{z-1} + s^{-z})\,ds. \tag{9.3.7}$$

Since

$$(s^{z-1}+s^{-z}) = s^{-1/2}(s^{z-1/2}+(s^{-z+1/2})$$
$$= 2s^{-1/2}\cosh((z-1/2)\log(s)),$$

(9.3.7) can be rewrittten as

$$\xi(z) = \int_1^\infty (s^2H'(s))'s^{-1/2}\cosh((z-1/2)\log(s))\,ds. \tag{9.3.8}$$

**Theorem 9.3.6.** *If $\xi$ is expanded as a power series in $z-1/2$, the coefficients of the power series are all real and non-negative.*

**Proof.** A direct calculation using (9.2.5) shows that

$$(s^2H'(s))'s^{-1/2} = \sum_{n=1}^{\infty}(4n^4\pi^2s^4 - 6n^2\pi s^2)s^{-1/2}\,e^{-n^2s^2\pi}.$$

The terms of this series are clearly positive for $s \geq 1$ and so the function itself is positive. Also, the power series coefficients in the expansion of $\cosh w$ about 0 are real and non-negative. Since $\log(s) \geq 0$ for $s \geq 1$, the theorem follows (see Exercise 9.37). □

**Theorem 9.3.7.** *There is a constant $R$ such that $|\xi(1/2+z)| \leq r^r$ for all $z \in \mathbb{C}$ with $|z| = r > R$.*

**Proof.** Since $\xi(z+1/2)$ has a power series expansion in $z$ with real non-negative coefficients, its maximum absolute value on any disc $D_r(0)$ is achieved at $z = r$. However, if $n$ is an integer such that $1/2 + r \leq 2n \leq 5/2 + r$, then by (9.2.7) and the fact that $\xi$ is increasing on the positive real line (Exercise 9.3.2),

$$\xi(1/2+r) \leq \xi(2n) = n(2n-1)\zeta(2n)\Gamma(n)\pi^{-n}.$$

Now $\zeta$ is decreasing on $(1,\infty)$. Thus,

$$\zeta(2n) \leq \zeta(2) \quad \text{if} \quad n \geq 1,$$

and

$$\Gamma(n) = (n-1)!$$

if $n$ is a positive integer. Thus,

$$\xi(1/2+r) \leq 2n\,n!\,\zeta(2) \leq 2\zeta(2)\,n^{n+1} \leq r^r$$

if $r$ is sufficiently large (since $n \leq 5/4 + r/2$) (Exercise 9.3.9). □

The following corollary is a direct consequence of the above theorem and the results of sections 8.3 and 8.4. We leave the details to the exercises.

**Corollary 9.3.8.** *The function $\xi$ is an entire function of finite order at most 1. Consequently, the zeroes of $\xi$ (and, hence, the zeroes of $\zeta$ that lie in the strip $0 < \mathrm{Im}(z) < 1$) form a sequence with exponent of convergence at most 1.*

**A Product Expansion for $\xi$.** In what follows, it will be convenient to make the change of variables $w = z - 1/2$. Then

$$\xi(z) = \xi(1/2 + w).$$

Since $\xi$ is symmetric about 1/2, the function $\xi(1/2 + w)$ is symmetric about 0.

By Corollary 9.3.8, $\xi(1/2 + w)$ is an entire function of finite order at most 1. Hence, the Hadamard Factorization Theorem applies with $\lambda = 1$. It tells us that $\xi$ has a factorization of the form

$$\xi(1/2 + w) = \mathrm{e}^{q(w)} \prod_{\sigma} (1 - w/\sigma)\, \mathrm{e}^{w/\sigma},$$

where $q$ is a polynomial of degree at most 1, and the product is over all zeroes $\sigma$ of $\xi(1/2+w)$. However, the zeroes of this function are symmetric about 0 and so, if a zero $\sigma$ appears in this product, then so does its negative. The exponential factors $\mathrm{e}^{w/\sigma}$ and $\mathrm{e}^{-w/\sigma}$ cancel and so

$$\xi(1/2 + w) = \mathrm{e}^{q(w)} \prod_{\sigma} (1 - w/\sigma)$$

as long as it is understood that the factors involving a given $\sigma$ and its negative $-\sigma$ are to be grouped together. If this is done, then the product expansion becomes

$$\xi(1/2 + w) = \mathrm{e}^{q(w)} \prod_{\mathrm{Im}(\sigma)>0} (1 - w^2/\sigma^2).$$

It is this product that actually converges (see the discussion of the product expansion of $\sin(\pi z)$ in Example 8.2.6).

Now $\xi(1/2 + w)$ is an even function of $w$ (symmetric about 0) and so is the above product. It follows that the polynomial $q$ must also be an even function. Since the only even polynomials of degree at most 1 are constants, we conclude that

$$\xi(1/2 + w) = c \prod_{\sigma} (1 - w/\sigma) \tag{9.3.9}$$

for some constant $c$.

If we recall that $w = z - 1/2$ and $\sigma = \rho - 1/2$ and we use the identity

$$1 - \frac{(z - 1/2)}{(\rho - 1/2)} = \left(1 - \frac{z}{\rho}\right)\left(1 + \frac{1/2}{\rho - 1/2}\right) \tag{9.3.10}$$

(Exercise 9.3.10), then (9.3.9) becomes

$$\xi(z) = \prod_{\rho} \left(1 - \frac{z}{\rho}\right)\left(1 + \frac{1/2}{\rho - 1/2}\right).$$

Since the product

$$c_1 = c \prod_{\rho} \left(1 + \frac{1/2}{\rho - 1/2}\right) \tag{9.3.11}$$

converges as long as the factors involving $\rho - 1/2$ and its negative are grouped together (Exercise 9.3.11), we obtain an infinite product expansion

$$\xi(z) = c_1 \prod_{\rho} \left(1 - \frac{z}{\rho}\right).$$

By evaluating at $z = 0$, we see that $c_1 = \xi(0) = 1/2$. This proves the following theorem.

**Theorem 9.3.9.** *The function $\xi$ has the following infinite product expansion which converges uniformly on each disc of finite radius:*

$$\xi(z) = \frac{1}{2} \prod_{\rho} \left(1 - \frac{z}{\rho}\right),$$

*where the numbers $\rho$ are the zeroes of $\xi(z)$ and the factors are arranged so that $\rho$ and $1 - \rho$ are grouped together. The product converges uniformly on each disc of finite radius.*

This, in turn, implies that the zeta function has the following infinite product expansion:

**Theorem 9.3.10.** *The zeta function satisfies*

$$\zeta(z) = \frac{1}{z(z-1)} \frac{\pi^{z/2}}{\Gamma(z/2)} \prod_{\rho} \left(1 - \frac{z}{\rho}\right),$$

*where the product is over the zeroes $\rho$ of $\zeta$ in the strip $0 < \mathrm{Re}(z) < 1$, and the factors involving $\rho$ and $1 - \rho$ are grouped together in the infinite product.*

## Exercise Set 9.3

1. Prove that $\zeta$ has a zero of order 1 at each negative even integer.
2. Prove that $\xi$ is increasing on the positive real line.
3. Show that $\xi(0) = 1/2$ even though (9.2.7) suggests it should be 0. Why is there no contradiction here?
4. Calculate $\zeta(0)$ using (9.2.12).
5. Calculate $\zeta(2)$ directly from the definition of $\zeta$, using results from Section 5.5.
6. Prove Corollary 9.3.8 (see Section 8.3 and Exercise 8.3.8).
7. Suppose $f$ is an entire function and the coefficients of the power series expansion of $f$ about 0 are all non-negative. Show that if $g(t)$ and $h(t)$ are positive continuous functions on $[1, \infty)$, then

$$\int_1^\infty g(t) f(zh(t))\, dt$$

is also an entire function of $z$ with non-negative coefficients in its power series expansion about 0, provided this integral exists for all positive real values of $z$.

8. Show that
$$\int_1^\infty e^{-t}\, t^z \frac{dt}{t}$$
is an entire function of $z$ with all non-negative coefficients in its power series expansion about 0.
9. Verify the claim made in the last sentence of the proof of Theorem 9.3.7 – that is, show that $2\zeta(2)n^{n+1} \leq r^r$ if $r$ is sufficiently large and $n \leq 5/4 + r/2$.
10. Prove the identity (9.3.10).
11. Show why the product in (9.3.11) converges if the terms are grouped as indicated.

## 9.4. The Riemann Hypothesis and Prime Numbers

The Riemann Hypothesis is the conjecture that all zeroes of the zeta function in the strip $0 < \mathrm{Re}(z) < 1$ lie on the line $\mathrm{Re}(z) = 1/2$. The significance of this conjecture lies in its connection with the problem of estimating the density of the prime numbers in the set of natural numbers. In this section we will discuss some of the history of the two problems and attempt to illustrate the connection between them without going into too much computational detail. For a more comprehensive and detailed account of the subject see the book by H. M. Edwards [**4**].

We let $\pi(x)$ denote the number of primes less than or equal to the positive real number $x$. It was recognized by Riemann that there is a connection between the rate of growth of $\pi(x)$ as $x$ increases and the zeroes of the zeta function. That there is some connection between the growth of $\pi(x)$ and the zeta function is seen in the proof that there are infinitely many primes (Exercise 9.2.2).

Based on experimental evidence, Gauss and Legendre conjectured that

$$\lim_{x\to\infty} \frac{\pi(x)}{x} \log x = 1. \tag{9.4.1}$$

This means that the fraction of the natural numbers up to $x$ that are prime, $\dfrac{\pi(x)}{x}$, is asymptotic to $\dfrac{1}{\log x}$ in the sense that their ratio has limit 1.

In his famous 1859 paper Riemann introduced a formula for $\pi(x)$. For a certain constant $c$, the function Li is defined to be

$$\mathrm{Li}(x) = \begin{cases} 0, & x \leq 2; \\ \displaystyle\int_2^x \frac{dt}{\log t} + c, & x > 2. \end{cases}$$

The formula of Riemann for $\pi(x)$ is then

$$\pi(x) = \sum_{n=1}^{\infty} \mathrm{Li}(x^{1/n}) + \sum_\rho \sum_{n=1}^{\infty} \mathrm{Li}(x^{\rho/n}) + \text{other}, \tag{9.4.2}$$

where the "other" terms are of lower order in $x$, and $\rho$ ranges over all zeroes of $\zeta$ in the strip $0 < \mathrm{Re}(z) < 1$. Note that the sums over $n$ are actually finite sums for each fixed $x$ and $\rho$, since $x^{1/n}$ and $x^{\rho/n}$ are less than 2 if $n$ is large enough.

An integration by parts argument shows that the first (and presumably dominant) term in the expansion (9.4.2) may be rewritten as

$$\mathrm{Li}(x) = \frac{x}{\log x} - \int_2^x \frac{dt}{(\log t)^2} + c.$$

The second term in this expression, when divided by $\dfrac{x}{\log x}$, has limit 0 at infinity (Exercise 9.4.1). Thus, if the remaining terms in Riemann's Formula for $\pi(x)$, when divided by $\dfrac{x}{\log x}$, also have limit 0 at infinity, and if it is legitimate to take the limit inside the sum in the first term, then (9.4.1) follows.

In dealing with the terms involving the zeroes $\rho$ of $\zeta$ in Riemann's Formula, it would be useful if the zeroes of $\zeta$ in the strip $0 < \mathrm{Re}(z) < 1$ all satisfied $\mathrm{Re}(z) < r$ for some $r < 1$. Riemann suspected that this was true and, in fact, he conjectured that all such zeroes actually lie on the line $\mathrm{Re}(z) = 1/2$. This is the Riemann Hypothesis.

Actually, Riemann did not give a complete proof of his formula (9.4.2) for $\pi(x)$ and, in fact, he did not even prove that the infinite series in this formula converges. Both facts were eventually proved, but the difficulties involved in these proofs and in determining the contribution of the terms involving the zeroes of $\zeta$ to the asymptotic behavior of $\pi(x)$ led to the introduction of another function $\psi(x)$ which also measures the density of primes and which satisfies a simpler and more natural formula analogous to (9.4.2).

Eventually, Hadamard and de la Vallée-Poussin in 1896 proved (9.4.1). It is now known as the Prime Number Theorem. The proofs of Hadamard and de la Vallée-Poussin as well as other classical proofs of this result are based on the fact that there are no zeroes of the zeta function on the line $\mathrm{Re}(z) = 1$. We will present one such proof in the next section.

**The Function $\psi$.** As mentioned above, the function $\pi(x)$ is closely related to another function $\psi(x)$ which also measures the density of primes and which has a more straightforward connection to the zeta function. The following equation gives two ways of expressing this function:

$$\psi(x) = \sum_{p^m \le x} \log p = \sum_{p \le x} m_p \log p, \tag{9.4.3}$$

where the first sum is over all prime powers $p^m \le x$. For a given prime $p$, the term $\log p$ appears in this sum as many times as there are positive powers of $p$ less than or equal to $x$. This number is the $m_p$ which appears in the second sum. It can also be described as the largest number $m$ such that $p^m \le x$.

We will show that if the function $\psi$ satisfies $\lim_{x\to\infty} \psi(x)/x = 1$, then the Prime Number Theorem follows.

**Lemma 9.4.1.** *Let $x$ and $y$ be real numbers greater than 1. Then*

$$\frac{\psi(x)}{\log x} \le \pi(x) \le y + \frac{\psi(x)}{\log y}.$$

**Proof.** We have

$$\pi(x) = \pi(y) + \sum_{y<p\le x} 1 \le y + \sum_{y<p\le x} \frac{\log p}{\log y} \quad \text{if} \quad y < x, \tag{9.4.4}$$
$$\pi(x) \le y \quad \text{if} \quad y \ge x,$$

where the sums are over primes $p$ in the indicated range.

By definition,

$$\psi(x) = \sum_{p\le x} m_p \log p = \sum_{p\le x} \log p^{m_p}. \tag{9.4.5}$$

The sum on the right satisfies the inequalities

$$\sum_{y<p\le x} \log p \le \sum_{p\le x} \log p^{m_p} \le \pi(x)\log x,$$

and so, by (9.4.5),

$$\sum_{y<p\le x} \log p \le \psi(x) \le \pi(x)\log x.$$

The left side of this inequality, combined with (9.4.4), yields

$$\pi(x) \le y + \frac{\psi(x)}{\log y}.$$

The right side, when divided by $\log x$, yields

$$\frac{\psi(x)}{\log x} \le \pi(x).$$

This completes the proof of the lemma. □

**Theorem 9.4.2.** *If* $\lim_{x\to\infty} \frac{\psi(x)}{x} = 1$, *then* $\lim_{x\to\infty} \frac{\pi(x)}{x}\log x = 1$.

**Proof.** Given $x > 1$, we use the previous lemma with

$$y = \frac{x}{(\log x)^2}.$$

According to that lemma,

$$\frac{\psi(x)}{\log x} \le \pi(x) \le \frac{x}{(\log x)^2} + \frac{\psi(x)}{\log x - 2\log\log x},$$

and so

$$1 \le \pi(x)\frac{\log x}{\psi(x)} \le \frac{1}{\log x}\frac{x}{\psi(x)} + \frac{\log x}{\log x - 2\log\log x}.$$

The first term on the right has limit 0 as $x \to \infty$, while the second term has limit 1. This is because

$$\lim_{x\to\infty} \frac{\psi(x)}{x} = 1,$$

by hypothesis, while

$$\lim_{x\to\infty} \frac{\log x}{\log x - 2\log\log x} = 1.$$

The latter statement is left as an exercise (Exercise 9.4.4). It follows that

$$\lim_{x\to\infty} \pi(x)\frac{\log x}{\psi(x)} = 1,$$

and, from this, that

$$\lim_{x\to\infty} \frac{\pi(x)}{x} \log x = 1.$$

This completes the proof. □

The above result says that the Prime Number Theorem will follow if we can prove that $\lim_{x\to\infty} \psi(x)/x = 1$. There are direct proofs of this; however, they involve serious difficulties with improper integrals and conditionally convergent series. It turns out that these difficulties can be made to disappear if we follow a similar approach, but use the integral of $\psi$ rather than $\psi$ itself as the main focus of attention. Thus, we set

$$\phi(x) = \int_1^x \psi(u)\, du. \tag{9.4.6}$$

This is another function which measures the density of primes. Furthermore, the Prime Number Theorem follows from an appropriate estimate on its asymptotic behavior.

**Properties of $\phi$.** It turns out that, $\lim\limits_{x\to\infty} \dfrac{\psi(x)}{x} = 1$, from which the Prime Number Theorem follows, provided

$$\lim_{x\to\infty} \frac{\phi(x)}{x^2} = \frac{1}{2}. \tag{9.4.7}$$

This is proved using a kind of reverse L'Hôpital's Rule. Specifically:

**Lemma 9.4.3.** *Let $f$ be a positive, increasing function on $[1,\infty)$ and suppose $r > 0$. Then*

$$\lim_{x\to\infty} \frac{f(x)}{x^r} = \lim_{x\to\infty} \frac{r+1}{x^{r+1}} \int_1^x f(u)\, du,$$

*provided the limit on the right exists and is finite.*

**Proof.** If we were to assume that the limit on the left exists, then the equality would follow from applying L'Hôpital's rule to the limit on the right. However, we are assuming only that the limit on the right exists, and so we must proceed differently (see Exercise 9.4.6).

Using the fact that $f$ is increasing and positive, we conclude that, for $\alpha < 1$ and $\beta > 1$,

$$\frac{1}{(1-\alpha)x} \int_{\alpha x}^x f(u)\, du \le f(x) \le \frac{1}{(\beta-1)x} \int_x^{\beta x} f(u)\, du.$$

On dividing by $x^r$, this becomes

$$\frac{1}{(1-\alpha)x^{r+1}} \int_{\alpha x}^x f(u)\, du \le \frac{f(x)}{x^r} \le \frac{1}{(\beta-1)x^{r+1}} \int_x^{\beta x} f(u)\, du.$$

If we set $F(x) = \int_1^x f(u)\, du$, then this can be rewritten as

$$\frac{F(x) - F(\alpha x)}{(1-\alpha)x^{r+1}} \le \frac{f(x)}{x^r} \le \frac{F(\beta x) - F(x)}{(\beta-1)x^{r+1}},$$

or as

$$\frac{1}{1-\alpha}\left(\frac{F(x)}{x^{r+1}} - \alpha^{r+1}\frac{F(\alpha x)}{(\alpha x)^{r+1}}\right) \le \frac{f(x)}{x^r} \le \frac{1}{\beta-1}\left(\beta^{r+1}\frac{F(\beta x)}{(\beta x)^{r+1}} - \frac{F(x)}{x^{r+1}}\right).$$

If we set

$$L = \lim_{x\to\infty}\frac{F(x)}{x^{r+1}} = \lim_{x\to\infty}\frac{F(\alpha x)}{(\alpha x)^{r+1}} = \lim_{x\to\infty}\frac{F(\beta x)}{(\beta x)^{r+1}},$$

then the above inequality implies that

$$\frac{1-\alpha^{r+1}}{1-\alpha}L \le \liminf_{x\to\infty}\frac{f(x)}{x^r} \le \limsup_{x\to\infty}\frac{f(x)}{x^r} \le \frac{\beta^{r+1}-1}{\beta-1}L.$$

The lemma then follows from this on taking the limit as $\alpha$ and $\beta$ approach 1, since both $\dfrac{1-\alpha^{r+1}}{1-\alpha}$ and $\dfrac{\beta^{r+1}-1}{\beta-1}$ have limit $r+1$. □

This leads directly to the following theorem. The details are left to the exercises.

**Theorem 9.4.4.** *If* $\lim_{x\to\infty}\dfrac{\phi(x)}{x^2} = 1/2$, *then* $\lim_{x\to\infty}\dfrac{\pi(x)}{x}\log x = 1$.

### Exercise Set 9.4

1. Prove that $\displaystyle\lim_{x\to\infty}\frac{\log x}{x}\int_2^x \frac{dt}{(\log t)^2} = 0.$
2. Prove Part (a) of Lemma 9.5.1.
3. Prove Part (b) of Lemma 9.5.1.
4. Prove that $\displaystyle\lim_{x\to\infty}\frac{\log x}{\log x - 2\log\log x} = 1.$
5. Use Lemma 9.4.3 to prove Theorem 9.4.4.
6. Find functions $f$ and $g$ on $\mathbb{R}$ such that

$$\lim_{x\to\infty}\frac{f(x)}{g(x)} \quad \text{exists, but} \quad \lim_{x\to\infty}\frac{f'(x)}{g'(x)} \quad \text{does not.}$$

## 9.5. A Proof of the Prime Number Theorem

In view of Theorems 9.4.2 and 9.4.4, to prove the Prime Number Theorem, it suffices to show that $\lim_{x\to\infty}\phi(x)/x^2 = 1/2$. The strategy for doing this involves expressing $\phi(x)$ as an integral involving $\zeta'/\zeta$. This is where the zeroes of the zeta function come in.

The integral formula relating $\zeta'/\zeta$ and $\phi$ is derived from the series expansion (9.3.2) and the following integral formula.

**Lemma 9.5.1.** *Suppose $p(z)$ is a non-constant polynomial and $b$ a real number such that no zero of $p$ lies on the line* $\mathrm{Re}(z) = b$. *If $y > 1$, let $A = \{z_1, z_2, \cdots, z_n\}$ be the set of zeroes of $p(z)$ that lie to the left of the line* $\mathrm{Re}(z) = b$. *Then*

$$\frac{1}{2\pi i}\int_{b-i\infty}^{b+i\infty}\frac{y^z}{p(z)}\,dz = \sum_{k=1}^{n}\mathrm{Res}(y^z/p(z), z_k). \tag{9.5.1}$$

*If the set $A$ is empty, then the integral is zero.*

*If $y < 1$, a similar formula holds, the only differences being: $A$ is replaced by the set of zeroes to the right of* $\mathrm{Re}(z) = b$ *and the expression on the right is multiplied by* $-1$.

**Proof.** In Chapter 5 we showed how to use residue theory to calculate the Fourier transforms of certain functions (Theorem 5.3.2). The integral that appears in (9.5.1) is actually the Fourier transform of a function to which Theorem 5.3.2 applies. To see this, we write

$$\frac{y^z}{p(z)} = \frac{e^{z \log y}}{p(z)} = \frac{y^b}{p(b+it)} e^{it \log y}$$

for $z = b + it$. Then

$$\frac{1}{2\pi i}\int_{b-i\infty}^{b+i\infty} \frac{y^z}{p(z)}\, dz = \frac{1}{2\pi}\int_{-\infty}^{\infty} f(t)\, e^{it\log y}\, dt = \frac{1}{\sqrt{2\pi}} \hat{f}(-\log y),$$

where $f$ is the restriction to the real line of the meromorphic function

$$f(z) = \frac{y^b}{p(b+iz)}.$$

The function $f$ has limit 0 at infinity since $p$ is a non-constant polynomial. Thus, by Theorem 5.3.2, if $y > 1$, then

$$\frac{1}{2\pi i}\int_{b-i\infty}^{b+i\infty} \frac{y^z}{p(z)}\, dz = i \sum_{w\in B} \mathrm{Res}(f(z)\, e^{iz\log y}, w),$$

where $B$ is the set of poles of $f$ in the upper half-plane. Since

$$f(z)\, e^{z\log y} = \frac{y^{b+iz}}{p(b+iz)},$$

a calculation of the effect on a residue of the change of variables $z \to b+iz$ (Exercise 9.5.8) shows that

$$\frac{1}{2\pi i}\int_{b-i\infty}^{b+i\infty} \frac{y^z}{p(z)}\, dz = \sum_{\lambda\in A} \mathrm{Res}(f(z)\, e^{iz\log y}, \lambda),$$

where $A = \{\lambda = b + iw : w \in B\}$ – that is, $A$ is the set of zeroes of $p$ that lie to the left of the line $\mathrm{Re}(z) = b$. This completes the proof in the case $y > 1$. The proof in the case $y < 1$ proceeds in the same way. □

**Example 9.5.2.** Prove that if $b > 0$ and $y > 0$, then

$$\frac{1}{2\pi i}\int_{b-i\infty}^{b+i\infty} \frac{y^z}{z(z+1)}\, dz = \begin{cases} 1 - 1/y, & \text{if } y > 1; \\ 0, & \text{if } y < 1. \end{cases}$$

**Solution:** Since $b > 0$, by the previous lemma, if $y > 1$, the integral is the sum of the residues of $\dfrac{y^z}{z(z+1)}$ at 0 and $-1$, which is $1 - 1/y$. If $y < 1$, the lemma implies that the integral is 0, since $z(z+1)$ has no zeroes to the right of $\mathrm{Re}(z) = 0$.

**Theorem 9.5.3.** *If $x > 0$ is not a power of a prime and $b > 1$, then*

$$\phi(x) = -\frac{1}{2\pi i}\int_{b-i\infty}^{b+i\infty} \frac{x^{z+1}}{z(z+1)}\frac{\zeta'(z)}{\zeta(z)}\,dz.$$

**Proof.** We multiply equation (9.3.2) by $\dfrac{x^{z+1}}{z(z+1)}$ and integrate. The result is

$$\begin{aligned}(9.5.2)\qquad &-\frac{1}{2\pi i}\int_{b-i\infty}^{b+i\infty} \frac{x^{z+1}}{z(z+1)}\frac{\zeta'(z)}{\zeta(z)}\,dz\\ &= \sum_{p^m}\left(\frac{1}{2\pi i}\int_{b-i\infty}^{b+i\infty}\left(\frac{x}{p^m}\right)^z \frac{x\log p}{z(z+1)}\,dz\right),\end{aligned}$$

provided the integrals exist and the integral can be moved inside the summation on the right. Assuming these things for the moment, we have, by the previous example,

$$-\frac{1}{2\pi i}\int_{b-i\infty}^{b+i\infty} \frac{x^{z+1}}{z(z+1)}\frac{\zeta'(z)}{\zeta(z)}\,dz = \sum_{p^m\le x}(x-p^m)\log p.$$

The expression on the right is $\phi(x) = \int_1^x \psi(u)\,du$ (Exercise 9.5.1), and so the proof will be complete if we can verify that the integrals in (9.5.2) exist and the integral can be brought inside the summation on the right.

The integrand corresponding to $n = p^m$ on the right in (9.5.2) is less than or equal in modulus to

$$\frac{\log n}{n^b}\frac{x^{b+1}}{b^2+t^2}$$

on the vertical line $z = b + it$. This has the form $c_n f(t)$, where $f$ is a positive integrable function of $t$ on $(-\infty,\infty)$ and $\sum_1^n c_n$ is a convergent series of positive numbers (see Exercise 9.5.2). By Exercise 9.5.4 this implies that the series of integrals on the right in (9.5.2) converges and it converges to the integral on the left. □

**A Series Expansion of $\phi$.** The Prime Number Theorem will follow directly from the following infinite series expansion of $\phi$.

**Theorem 9.5.4.** *There are constants $A$ and $B$ such that*

$$\phi(x) = \frac{x^2}{2} - \sum_{k=1}^{\infty}\frac{x^{1-2k}}{2k(2k-1)} - \sum_{\rho}\frac{x^{\rho+1}}{\rho(\rho+1)} - Ax + B,$$

*where $\rho$ ranges over the zeroes of $\zeta$ in the strip $0 < \mathrm{Re}(z) < 1$.*

**Proof.** The integral that appears in Theorem 9.5.3 can also be evaluated by using the infinite product expansion of Theorem 9.3.10. This theorem implies that the logarithmic derivative of $\zeta$ can be written as

$$\frac{\zeta'(z)}{\zeta(z)} = \frac{1}{1-z} - \frac{1}{z} + \frac{\log\pi}{2} - \frac{\Gamma'(z/2)}{\Gamma(z/2)} + \sum_{\rho}\frac{1}{z-\rho},$$

where, in the last sum, terms involving $\rho$ and $1-\rho$ must be grouped together for the series to converge. The product formula for $1/\Gamma$ given in Theorem 9.1.10 leads to

$$-\frac{\Gamma'(z/2)}{\Gamma(z/2)} = \frac{1}{z} + \sum_{k=1}^{\infty}\left(\frac{1}{2k+z} - \frac{1}{2}\log(1+1/k)\right).$$

Thus,

$$\frac{\zeta'(z)}{\zeta(z)} = -\frac{1}{z-1} + \frac{\log \pi}{2} + \sum_{k=1}^{\infty}\left(\frac{1}{z+2k} + \frac{1}{2}\log(1+1/k)\right) + \sum_{\rho}\frac{1}{z-\rho}.$$

This simplifies significantly if we subtract $\zeta'(0)/\zeta(0)$:

$$\frac{\zeta'(z)}{\zeta(z)} - \frac{\zeta'(0)}{\zeta(0)} = -1 - \frac{1}{z-1} + \sum_{k=1}^{\infty}\left(\frac{1}{z+2k} - \frac{1}{2k}\right) + \sum_{\rho}\left(\frac{1}{z-\rho} + \frac{1}{\rho}\right),$$

or

$$\frac{\zeta'(z)}{\zeta(z)} = -\frac{z}{z-1} - \sum_{k=1}^{\infty}\frac{z}{2k(z+2k)} + \sum_{\rho}\frac{z}{\rho(z-\rho)} + \frac{\zeta'(0)}{\zeta(0)}. \tag{9.5.3}$$

We next multiply equation (9.5.3) by $\dfrac{1}{2\pi i}\dfrac{x^{z+1}}{z(z+1)}$ and integrate along the line $\mathrm{Re}(z) = b > 1$, obtaining

(9.5.4)

$$\begin{aligned}
\phi(x) &= -\frac{1}{2\pi i}\int_{b-i\infty}^{b+i\infty}\frac{x^{z+1}}{z(z+1)}\frac{\zeta'(z)}{\zeta(z)}\,dz \\
&= \frac{1}{2\pi i}\int_{b-i\infty}^{b+i\infty}\frac{x^{z+1}}{(z-1)(z+1)}\,dz + \frac{1}{2\pi i}\int_{b-i\infty}^{b+i\infty}\sum_{k=1}^{\infty}\frac{x^{z+1}}{2k(z+2k)(z+1)}\,dz \\
&\quad - \frac{1}{2\pi i}\int_{b-i\infty}^{b+i\infty}\sum_{\rho}\frac{x^{z+1}}{\rho(z-\rho)(z+1)}\,dz - \frac{1}{2\pi i}\int_{b-i\infty}^{b+i\infty}\frac{\zeta'(0)}{\zeta(0)}\frac{x^{z+1}}{z(z+1)}\,dz.
\end{aligned}$$

Assuming for the moment that the integral can be taken inside each of the infinite sums, the result is

$$\begin{aligned}
\phi(x) &= -\frac{1}{2\pi i}\int_{b-i\infty}^{b+i\infty}\frac{x^{z+1}}{z(z+1)}\frac{\zeta'(z)}{\zeta(z)}\,dz \\
&= \frac{1}{2\pi i}\int_{b-i\infty}^{b+i\infty}\frac{x^{z+1}}{(z-1)(z+1)}\,dz + \sum_{k=1}^{\infty}\frac{1}{2\pi i}\int_{b-i\infty}^{b+i\infty}\frac{x^{z+1}}{2k(z+2k)(z+1)}\,dz \\
&\quad - \sum_{\rho}\frac{1}{2\pi i}\int_{b-i\infty}^{b+i\infty}\frac{x^{z+1}}{\rho(z-\rho)(z+1)}\,dz - \frac{1}{2\pi i}\int_{b-i\infty}^{b+i\infty}\frac{\zeta'(0)}{\zeta(0)}\frac{x^{z+1}}{z(z+1)}\,dz.
\end{aligned}$$

Each of these integrals can be evaluated using Lemma 9.5.1. This leads to

$$\phi(x) = \frac{x^2}{2} - \sum_{k=1}^{\infty}\frac{x^{1-2k}-1}{2k(2k-1)} - \sum_{\rho}\frac{x^{\rho+1}-1}{\rho(\rho+1)} - \frac{\zeta'(0)}{\zeta(0)}x,$$

or

$$\phi(x) = \frac{x^2}{2} - \sum_{k=1}^{\infty} \frac{x^{1-2k}}{2k(2k-1)} - \sum_{\rho} \frac{x^{\rho+1}}{\rho(\rho+1)} - Ax + B,$$

where $A = \dfrac{\zeta'(0)}{\zeta(0)}$ and $B = \displaystyle\sum_{k=1}^{\infty} \frac{1}{2k(2k-1)} + \sum_{\rho} \frac{1}{\rho(\rho+1)}$.

It remains to prove that the integral can be taken inside the infinite sums in (9.5.4). The $k$th term of the first sum is

$$\frac{x^{z+1}}{2k(z+2k)(z+1)}. \tag{9.5.5}$$

The numerator of this fraction is bounded on the vertical line $\mathrm{Re}(z) = b$. With $z = b + it$, we estimate the middle factor of the denominator as follows:

$$\begin{aligned} |z+2k|^2 &= t^2 + (b+2k)^2 = t^2 + b^2 + 4bk + 4k^2 \\ &\geq t^2 + b^2 + 4k^2 = |z|^2 + (2k)^2 \geq 4|z|k. \end{aligned}$$

Thus,

$$|z+2k| \geq 4|z|^{1/2}k^{1/2}.$$

The right factor of the denominator satisfies $|z+1| \geq |z|$ since $z = b+it$ has positive real part. Hence, the fraction(9.5.5) has modulus less than or equal to a constant times $|z|^{-3/2}k^{-3/2}$. It follows that, in the first infinite sum the integral of each term over $\mathrm{Re}(z) = b$ exists and the integral of the sum is the sum of the integrals and the latter sum is absolutely convergent (see Exercise 9.5.4).

The term involving $\rho$ of the second infinite sum in (9.5.4) is

$$\frac{x^{z+1}}{\rho(z-\rho)(z+1)}. \tag{9.5.6}$$

The numerator is bounded by $x^{b+1}$ on $\mathrm{Re}(z) = b$. With $z = b + iy$, $\rho = \beta + i\gamma$, and $c = b - 1$, we estimate the denominator as follows: Since $|z - \rho| \geq (|y - \gamma| + c)/2$ and $|z+1| \geq (|y| + b + 1)/2 \geq (|y| + c)/2$, we have

$$|\rho(z-\rho)(z+1)| \geq |\gamma|(|y-\gamma| + c)(|y| + c)/4.$$

Thus, the modulus of (9.5.6) is less than or equal to

$$g(y) = 4x^{b+1}|\gamma|^{-1}h(y) \quad \text{where} \quad h(y) = \frac{1}{(|y-\gamma| + c)(|y| + c)}.$$

If we divide the real line into three subintervals by cutting at $y = 0$ and $y = \gamma$, then, on each of these subintervals, the absolute values in $h$ can be eliminated, and the integral of $h$ with respect to $y$ can be evaluated using the method of partial fractions. The result (Exercise 9.5.5) is that the integral of $h$ over each of the unbounded subintervals is

$$|\gamma|^{-1} \log(|\gamma|/c + 1),$$

while the integral of $h$ over the bounded subinterval is

$$2(|\gamma| + 2c)^{-1} \log(|\gamma|/c + 1) \leq 2|\gamma|^{-1} \log(|\gamma|/c + 1).$$

Thus, the integral of $g$ over $(-\infty, \infty)$ is less than or equal to

$$16x^{b+1}|\gamma|^{-2} \log(|\gamma|/c + 1)$$

and, since $\log(|\gamma|/c) \leq (|\gamma|/c)^{1/2}$, this integral is less than or equal to

$$16c^{1/2}x^{b+1}|\gamma|^{-3/2}. \tag{9.5.7}$$

Since, by Corollary 9.3.8, the sequence of roots $\rho = \beta + i\gamma$ has exponent of convergence at most 1, the same is true of the sequence of imaginary parts $\gamma$. It follows that the terms (9.5.7) are the terms of a convergent series. Thus, Exercise 9.2.7, modified to cover integrals over $(-\infty, \infty)$, implies that the integral of the sum is the sum of the integrals for the series of terms given by (9.5.6). □

### The Prime Number Theorem.

**Theorem 9.5.5.** *If $\pi(x)$ is the number of primes less than or equal to $x$, then*

$$\lim_{x \to \infty} \pi(x) \frac{\log x}{x} = 1.$$

**Proof.** By Theorems 9.4.4 and 9.4.2, it suffices to prove that $\lim_{x\to\infty} \phi(x)/x^2 = 1/2$.

By Theorem 9.5.4,

$$\frac{\phi(x)}{x^2} = \frac{1}{2} - \sum_{k=1}^{\infty} \frac{x^{-1-2k}}{2k(2k-1)} - \sum_{\rho} \frac{x^{\rho-1}}{\rho(\rho+1)} - \frac{A}{x} + \frac{B}{x^2}. \tag{9.5.8}$$

Each of the infinite sums in this expression involves only negative powers of $x$ and each of them is absolutely convergent at $x = 1$. It follows that both infinite series converge uniformly in $x$ on $[1, \infty)$. Thus, in taking the limit of $\phi(x)/x^2$ as $x \to \infty$, we may take the limit inside the infinite sums. Since each term on the right side of (9.5.8) has limit 0 except the term 1/2, the theorem is proved. □

## Exercise Set 9.5

1. Verify that $\int_1^x \psi(u)\, du = \sum_{p^m \leq x}(x - p^m) \log p$, where $p$ is prime and $m$ is a positive integer.
2. Prove that if $p$ and $r$ are arbitrary positive numbers, there is a constant $C$ such that $\log^p(t) \leq Ct^r$ for all $t > 1$.
3. Give a direct proof of Lemma 9.5.1, using residue theory, but without interpreting the integal as a Fourier transform and applying Theorem 5.3.2.
4. For each $n$, let $g_n$ be be a continuous function on $(-\infty, \infty)$ satisfying $|g_n(t)| \leq c_n f(t)$, where $f \geq 0$ is integrable and $\sum_1^\infty c_n$ is a convergent series of positive numbers. Use Exercise 9.2.7 to prove that
$$\sum_{n=1}^{\infty} \int_{-\infty}^{\infty} g_n(t)\, dt = \int_{-\infty}^{\infty} \sum_{n=1}^{\infty} g_n(t)\, dt.$$
5. Verify the claims made near the end of the proof of Theorem 9.5.4 regarding the integral of
$$h(y) = \frac{1}{(|y-\gamma| + c)(|y| + c)}$$
over each of the subintervals of $(-\infty, \infty)$ created by cutting at $y = 0$ and $y = \gamma$.

6. Prove that $\log(s+1) \le s^{1/2}$ for $s \in (0, \infty)$. This was used near the end of the proof of Theorem 9.5.4.
7. Verify the statement about taking the limit inside the integral in the proof of Theorem 9.5.5. That is, prove that if a series $\sum_{n=1}^{\infty} u_n(x)$ of functions on $[1, \infty)$ converges uniformly absolutely on $[1, \infty)$, then
$$\lim_{x\to\infty} \sum_{n=1}^{\infty} u_n(x) = \sum_{n=1}^{\infty} \lim_{x\to\infty} u_n(x),$$
provided each limit on the right converges.
8. Prove that if $f$ is a function analytic in an open set containing $b + iw$ and $g(z) = f(b + iz)$, then $g$ is analytic in an open set containing $w$ and
$$\operatorname{Res}(g, w) = -i \operatorname{Res}(f, b + iw).$$

# Bibliography


[1] Ahlfors, L., *Complex Analysis*, 2nd ed., McGraw-Hill, 1966.

[2] Ash, R., *Complex Variables*, Academic Press, 1971.

[3] Brown, J. W. and Churchill, R. V., *Complex Variables and Applications*, McGraw-Hill, 2009.

[4] Edwards, H. M., *Riemann's Zeta Function*, Dover, 1974.

[5] Jeffrey, A., *Complex Analysis and Applications*, CRC Press, 1992.

[6] Lang, S., *Complex Analysis*, Addison-Wesley, 1977.

[7] Marsden, M. H. and Hoffman, M. J., *Basic Complex Analysis*, 3rd ed., W. H. Freeman, 1999.

[8] Rudin, W., *Real and Complex Analysis*, McGraw-Hill, 1987.

[9] Saff, E. B. and Snider, A. D., *Fundamentals of Complex Analysis for Mathematics, Science, and Engineering*, 3rd ed., Prentice Hall, 2003.

# Index

# Titles in This Series

**Volume**

16 **Joseph L. Taylor**
Complex Variables

15 **Mark A. Pinsky**
Partial Differential Equations and Boundary-Value Problems with Applications, Third Edition

14 **Michael E. Taylor**
Introduction to Differential Equations

13 **Randall Pruim**
Foundations and Applications of Statistics: An Introduction Using R

12 **John P. D'Angelo**
An Introduction to Complex Analysis and Geometry

11 **Mark R. Sepanski**
Algebra

10 **Sue E. Goodman**
Beginning Topology

9 **Ronald Solomon**
Abstract Algebra

8 **I. Martin Isaacs**
Geometry for College Students

7 **Victor Goodman and Joseph Stampfli**
The Mathematics of Finance: Modeling and Hedging

6 **Michael A. Bean**
Probability: The Science of Uncertainty with Applications to Investments, Insurance, and Engineering

5 **Patrick M. Fitzpatrick**
Advanced Calculus, Second Edition

4 **Gerald B. Folland**
Fourier Analysis and Its Applications

3 **Bettina Richmond and Thomas Richmond**
A Discrete Transition to Advanced Mathematics

2 **David Kincaid and Ward Cheney**
Numerical Analysis: Mathematics of Scientific Computing, Third Edition

1 **Edward D. Gaughan**
Introduction to Analysis, Fifth Edition